D0721932

AUTOIONIZATION
Recent Developments and Applications

PHYSICS OF ATOMS AND MOLECULES

ATOM–MOLECULE COLLISION THEORY: A Guide for the Experimentalist
Edited by Richard B. Bernstein

ATOMIC INNER-SHELL PHYSICS
Edited by Bernd Crasemann

ATOMS IN ASTROPHYSICS
Edited by P. G. Burke, W. B. Eissner, D. G. Hummer, and I. C. Percival

AUTOIONIZATION: Recent Developments and Applications
Edited by Aaron Temkin

COHERENCE AND CORRELATION IN ATOMIC COLLISIONS
Edited by H. Kleinpoppen and J. F. Williams

DENSITY MATRIX THEORY AND APPLICATIONS
Karl Blum

ELECTRON AND PHOTON INTERACTIONS WITH ATOMS
Edited by H. Kleinpoppen and M. R. C. McDowell

ELECTRON–ATOM AND ELECTRON–MOLECULE COLLISIONS
Edited by Juergen Hinze

ELECTRON-MOLECULE COLLISIONS
Edited by Isao Shimamura and Kazuo Takayanagi

INNER-SHELL AND X-RAY PHYSICS OF ATOMS AND SOLIDS
Edited by Derek J. Fabian, Hans Kleinpoppen, and Lewis M. Watson

INTRODUCTION TO THE THEORY OF LASER–ATOM INTERACTIONS
Marvin H. Mittleman

ISOTOPE SHIFTS IN ATOMIC SPECTRA
W. H. King

PROGRESS IN ATOMIC SPECTROSCOPY, Parts A, B, and C
Edited by W. Hanle, H. Kleinpoppen, and H. J. Beyer

VARIATIONAL METHODS IN ELECTRON–ATOM SCATTERING THEORY
R. K. Nesbet

AUTOIONIZATION

Recent Developments and Applications

Edited by

Aaron Temkin

Laboratory for Astronomy and Solar Physics
National Aeronautics and Space Administration
Goddard Space Flight Center
Greenbelt, Maryland

PLENUM PRESS • NEW YORK AND LONDON

Library of Congress Cataloging in Publication Data

Main entry under title:

Autoionization: recent developments and applications.

(Physics of atoms and molecules)
Includes bibliographies and index.
1. Auger effect. I. Temkin, Aaron. II. Series.
QC793.5.E627A96 1985 539.7′2112 85-6334
ISBN 0-306-41854-1

A Division of Plenum Publishing Corporation
233 Spring Street, New York, N.Y. 10013

Printed in the United States of America

CONTRIBUTORS

A. K. BHATIA
Atomic Physics Office
Laboratory for Astrophysical and Solar Physics
Goddard Space Flight Center
National Aeronautics and Space Administration
Greenbelt, MD

KWONG T. CHUNG
Department of Physics
North Carolina State University
Raleigh, NC

BRIAN F. DAVIS
Department of Physics
North Carolina State University
Raleigh, NC

GEORGE A. DOSCHEK
E. O. Hurlburt Center for Space Research
Naval Research Laboratory
Washington, D.C.

B. R. JUNKER
Office of Naval Research
Arlington, VA

C. WILLIAM MCCURDY
Department of Chemistry
Ohio State University
Columbus, OH

A. TEMKIN
Atomic Physics Office
Laboratory for Astronomy and Solar Physics
Goddard Space Flight Center
National Aeronautics and Space Administration
Greenbelt, MD

PREFACE

About five years ago, Professor P. G. Burke asked me to edit a sequel to an earlier book—*Autoionization: Theoretical, Astrophysical, and Laboratory Experimental Aspects*, edited by A. Temkin, Mono Book Corp., Baltimore, 1966. Because so much time had gone by and so much work had been done, the prospect of updating the 1966 volume seemed out of the question.

In 1965 the phenomenon of autoionization, although long known, was just starting to emerge from a comparatively intuitive stage of understanding. Three major developments characterized that development: In solar (astro-)physics, Alan Burgess (1960) had provided the resolution of the discrepancy of the temperature of the solar corona as observed versus that deduced from ionization balance calculations, by including the process of dielectronic recombination in the calculation; Madden and Codling (1963) had just performed their classic experiment revealing spectroscopically sharp lines in the midst of the photoionization continuum of the noble gases; and Feshbach (1962) had developed a theory with the explicit introduction of projection operators, which for the first time put the calculation of autoionization states on a firm theoretical footing. There were important additional contributions made at that time as well; nevertheless, without going into further detail, we were able to include in our 1966 volume, in spite of its modest size, a not too incomplete survey of the important developments at that time.

To do the equivalent now would be virtually impossible. In considering the alternatives, I felt that laboratory experimental developments in particular have far outstripped what can reasonably be included in the confines of a single book. Therefore, I have omitted them completely. The situation with regard to solar and astrophysical applications at first seemed also too vast for inclusion. However, the unlikely has become fact by virtue of a magnificent effort by Dr. George Doschek. His chapter, "Diagnostics of Solar and Astrophysical Plasmas Dependent on Autoionization Phenomena," is, in my opinion, a masterful exposition and summary of diagnostic analysis and applications of autoionization in almost the entire realm of space physics. It is necessarily a large part of this book. I hope the reader will find it enlightening and useful. It will surely have a vital place in the space physics literature.

The remaining chapters I have chosen to include are devoted to theory and calculation. Even here a severe limitation was required, but in the belief that good theory allows good calculations, and the value of calculations cannot exceed the quality of their theoretical underpinnings, we could be selective.

In the category of theory–calculation we could certainly have included an overview of methods and programs, developing mainly from the close-coupling formalism, that dealt directly with electron scattering including resonances. Fortunately, there have been a number of recent reviews, so we have not felt it mandatory to include such a review here. Too recent to be included here and in a somewhat different category are successful developments, primarily calculational in nature, including resonances in many-body diagrammatic and random phase approximation (RPA) techniques. In contrast, there has been very little written of a review nature on the calculation of electron–atom (atomic ion) resonances within the context of the Feshbach theory. Since that theory has long provided the theoretical basis for much of the work of the Goddard group, I believe the present volume is a very appropriate place to present such a review. In our first article, Dr. Bhatia and I have attempted to review, from a more pedagogical point of view rather than from one of completeness, our work on two-electron systems (one-electron targets) for which theory allows explicit and rigorous projection operators to be given. We have included, however, a more detailed exposition of a recent calculation of the line-shape parameter, because that requires a rather different approach to a part of the Feshbach theory known as the nonresonant continuum. I believe the idea of a more generally defined nonresonant continuum may be of value in other contexts as well.

In a second article, Dr. Bhatia and I have undertaken the process of implementing the Feshbach approach to more than two-electron systems. As a prerequisite for actually doing calculations, we have found it necessary to precisely define the projection operators (P and Q) in complete and explicit terms. We have chosen to include a part of that analysis here because it also serves the pedagogical aims we have also attempted to fulfill. In the second part of that chapter we have discussed approximations of these operators that we have called quasi-projection operators ($\hat{P}$ and $\hat{Q}$). Historically our introduction of these quasi-projection operators preceded our recent development of the projection operators themselves. Notwithstanding, quasi-projections allow for meaningful calculations to be done, and we have briefly reviewed some of them.

The fact that one can calculate with projectors without them being idempotent (the latter property usually being implicit in the definition of the name "projection operator") is not confined to the specific quasi-projection operators we have introduced. In the third chapter of this book Drs. K. T. Chung and Brian K. Davis describe a hole-projection formalism wherein electrons in inner orbitals are projected out of an otherwise general ansatz for the wave function of the total system by a projection-type operator, which can certainly be considered in the category of quasi projectors. Their theory relies on a mini–max theorem which (although rigorously proved only in a one-electron context) states that the physically meaningful state is realized when the energy is *minimized* with respect to the parameters of the complete wave function, but at the same time the energy is *maximized* with respect to the parameters describing the excluded orbitals (the holes). It is clear that the formalism should be particularly effective in calculating inner-shell vacancies of many-electron systems; however, even for 3-electron systems, to which the calculations have thus far been confined, as the article will show, the results are very impressive. In their summation the authors refer to a recent paper wherein they have combined their hole-projection method with elements of complex rotation to calculate widths. I expect that augmentation to become an important addition to the methodology.

Complex rotation is the subject of the last set of articles in this volume. The basic idea can be expressed in many ways, but for the purposes of this Preface one way is to notice that a stationary (i.e., bound)-state wave function has the time dependence $\exp(-iEt/h)$, where E is a real number. Therefore, if a state is decaying, it should be describable by a complex time dependence $W = E - i\Gamma/2$; then its imaginary part will automatically describe the decay width (the inverse of the decay time) of the resonance. From the calculational point of view this has the implication that certain states, which are not quadratically integrable on the real axis, do become integrable off the real axis. This (measure zero) set of discrete states are uncovered—according to a basic theorem of Balslev and Combes—if the electronic coordinates are rotated in the complex plane beyond a minimum amount which, not surprisingly, is related to the width of the resonance. From the calculational point of view, however, this is a most important fact, because it removes boundary conditions from the problem. Thus, it implies that one can, in principle, calculate a many-electron resonant state without knowing the wave function of the target system. This, in turn (and again, in principle), overcomes a major shortcoming of the projection-operator approach, wherein although the eigenfunctions of QHQ are discrete and exist on the real axis, the projection operator Q does depend on the eigenfunctions of the target system and therefore must be approximated for more than one-electron target systems. In addition, both shape and Feshbach resonances can emerge from the complex rotation approach.

Briefly stated, Dr. B. R. Junker concentrates on applications to atoms and ions, whereas Dr. C. M. McCurdy deals with molecular systems; both authors have made a concerted effort to coordinate their respective treatments. An important element of the approach of these two articles is the idea that it is preferable to retain the Hamiltonian in its real form and put the complex nature of the calculation completely in the ansatz for the wave function. How best to do this is not yet completely settled, but I believe these treatments go a long way in elucidating the technique. For simpler systems (e.g., He^-) the results are probably the most reliably accurate of any thus far obtained. We are pleased to have these two contributions from two expert practitioners whose interests are calculational as well as theoretical.

I would like to thank all the authors for their contributions, and Professors P. G. Burke and H. Kleinpoppen for their encouragement. I am as usual indebted to Dr. A. K. Bhatia, in this case for his additional help in preparing the index.

Silver Spring, Maryland AARON TEMKIN

CONTENTS

CHAPTER ONE
THEORY AND CALCULATION OF RESONANCES AND AUTOIONIZATION OF TWO-ELECTRON ATOMS AND IONS

A. TEMKIN AND A. K. BHATIA

CHAPTER TWO
PROJECTION AND QUASI-PROJECTION OPERATORS FOR ELECTRON IMPACT RESONANCES ON MANY-ELECTRON ATOMIC TARGETS

A. TEMKIN AND A. K. BHATIA

CHAPTER THREE
HOLE-PROJECTION METHOD FOR CALCULATING FESHBACH RESONANCES AND INNER-SHELL VACANCIES

Kwong T. Chung and Brian F. Davis

CHAPTER FOUR
COMPLEX STABILIZATION METHOD

B. R. Junker

CHAPTER FIVE
MOLECULAR RESONANCE CALCULATIONS: APPLICATIONS OF COMPLEX-COORDINATE AND COMPLEX BASIS FUNCTION TECHNIQUES

C. William McCurdy

CHAPTER SIX
DIAGNOSTICS OF SOLAR AND ASTROPHYSICAL PLASMAS DEPENDENT ON AUTOIONIZATION PHENOMENA

GEORGE A. DOSCHEK

CHAPTER ONE

THEORY AND CALCULATION OF RESONANCES AND AUTOIONIZATION OF TWO-ELECTRON ATOMS AND IONS

A. TEMKIN AND A. K. BHATIA

1. INTRODUCTION

In this chapter, we shall review and discuss the theory and calculation of autoionization states of two-electron systems from the point of view of projection operator formalism as used by the authors over many years. Autoionization states are resonances caused by scattering an electron impinging on a target ion or atom. They can also be produced by photoabsorption as well as heavy-particle collisions. It is only comparatively recently that such resonances have been studied in great detail as a result of important experimental and theoretical developments together with the advent of high-speed computers.

Here, we shall confine ourselves to theoretical and calculational developments emanating from the Feshbach formalism and its application to scattering electrons from one-electron (hydrogenlike) targets. Such resonances are by definition autoionization states of the composite—in this case, two-electron—system. One of the purposes of studying such systems is the hope of being able to achieve accuracy approaching what has been attained for ordinary two-electron bound states. To that extent we are testing quantitatively the continuum solutions of the Schrödinger equation to an accuracy far beyond what has previously been forthcoming from ordinary (i.e., nonresonant) scattering.

In Section 2, we briefly review the Feshbach formalism[1] with emphasis on how it applies to one-electron targets. Of particular interest is the fact that

A. TEMKIN and A. K. BHATIA ■ Atomic Physics Office, Laboratory for Astronomy and Solar Physics, Goddard Space Flight Center, National Aeronautics and Space Administration, Greenbelt, MD 20771.

the resonance parameters that emerge from the Feshbach approach are not identical to the corresponding Breit–Wigner quantities. Furthermore, we demonstrate how the two are related. Special attention will be devoted to the Fano line shape parameter[2] q (Section 3) and how it can be unambiguously formulated in projection-operator form.

These theoretical developments will be complemented by a review of our calculations for ^1S, ^3P, and ^1D states of H^- below the first excited threshold of the target H atom and by comparison with other accurate calculations and experiments. Similar calculations of $^{1,3}S$, $^{1,3}P$, and $^{1,3}D$ states for He, which is the second basic two-electron system we deal with, are made in Sections 4, 5, and 6.

2. THE FESHBACH FORMALISM

In this section, we describe briefly the Feshbach[1] theory for two-electron systems. Given a fixed hydrogen like target and a second electron, the Schrödinger equation is written in the usual way

$$(H - E)\Psi(\mathbf{r}_1, \mathbf{r}_2) = 0 \tag{2.1}$$

Here, we take $\Psi(\mathbf{r}_1, \mathbf{r}_2)$ to be an eigenfunction of definite angular momentum, spin, and parity. Spin is included by requiring the spatial Ψ to be antisymmetric with respect to the exchange of the $\mathbf{r}_1$ and $\mathbf{r}_2$ coordinates for triplet states, and symmetric for singlet states; H is the total Hamiltonian of the system

$$H = -\frac{\hbar^2}{2m}(\nabla_1^2 + \nabla_2^2) - \frac{Ze^2}{r_1} - \frac{Ze^2}{r_2} + \frac{e^2}{r_{12}} \tag{2.2}$$

Z is the total charge of the nucleus, and E in Eq. (2.1) is the total energy. (In general, we shall use rydberg units $\hbar = 1$, $m = 1/2$, $e^2 = 2$.)

The essence of the Feshbach formalism[1] is the introduction of projection operators P and Q such that

$$P + Q = 1 \quad \text{completeness} \tag{2.3a}$$

$$\left.\begin{aligned} P^2 &= P \quad (2.3\text{b}) \\ Q^2 &= Q \quad (2.3\text{c}) \end{aligned}\right\} \quad \text{idempotency}$$

$$PQ = 0 \qquad \text{orthogonality} \tag{2.3d}$$

Equations (2.3b) and (2.3c) are the conditions that operators P and Q are projection operators (i.e., idempotent). From Eq. (2.1), we obtain by premultiplication, using Eq. (2.3) where necessary,

$$P(H-E)(P+Q)\Psi=0 \tag{2.4}$$

$$Q(H-E)(P+Q)\Psi=0 \tag{2.5}$$

And from Eq. (2.5), we obtain by "division" the formal expression

$$Q\Psi=\frac{1}{Q(E-H)Q}QHP\Psi \tag{2.6}$$

Substituting Eq. (2.6) into Eq. (2.4) results in

$$P\left[H+HQ\frac{1}{Q(E-H)Q}QH-E\right]P\Psi=0 \tag{2.7}$$

The significance of the projection operators P and Q in the Feshbach theory comes from the condition that the $P\Psi$ have the same asymptotic form as Ψ

$$\lim_{r_1 \text{ or } r_2\to\infty} P\Psi=\Psi \tag{2.8}$$

Equation (2.8) implies that as far as the scattering is concerned, the same information is obtained from $P\Psi$ as Ψ. Equation (2.3a) leads to the asymptotic condition on $Q\Psi$

$$\lim_{r_1 \text{ or } r_2\to\infty} Q\Psi\to 0 \tag{2.9}$$

This is the same condition that is obtained for bound states. Thus, we have divided the total wave function Ψ into scatteringlike and bound-state-like components. The significance of this is, as we shall see, that Eq. (2.7) is essentially a one-body equation in which all the many-body effects have been condensed into a one-body optical potential

$$\mathscr{V}_{\text{op}}=PHQ\frac{1}{E-QHQ}QHP \tag{2.10a}$$

However, as useful as Eq. (2.7) and the optical potentials are, they are not a panacea; $\mathscr{V}_{\text{op}}$, for example, is nonlocal, and to construct it exactly would be as difficult as solving the original Schrödinger equation.

Let us first consider a specific form of projection operators (the preceding conditions do not specify P and Q uniquely) for systems containing two identical particles (electrons). It has been shown by Hahn, O'Malley, and Spruch[3] in a basic paper introducing the Feshbach formalism into the atomic-scattering problem (it was originally developed in the context of nuclear physics) that for two-electron systems, projection operators P and

Q satisfying the requirements in Eqs. (2.8) and (2.9) are given by

$$Q = Q_1 Q_2 \tag{2.11}$$

where, by definition,

$$Q_i \equiv 1 - P_i \tag{2.12a}$$

and

$$P_i = \phi_0(\mathbf{r}_i)\rangle\langle\phi_0(\mathbf{r}_i) \tag{2.12b}$$

Since from Eq. (2.12b) $P_i^2 = P_i$, we have from Eqs. (2.3a) and (2.12a) the explicit expressions for P and Q

$$Q = (1 - P_1)(1 - P_2) = 1 - P_1 - P_2 + P_1 P_2 \tag{2.13a}$$

$$P = 1 - Q = P_1 + P_2 - P_1 P_2 \tag{2.13b}$$

It may readily be verified that P and Q are idempotent and orthogonal ($P^2 = P$, $Q^2 = Q$; $PQ = 0$). It is useful to remember that P acting on the wave function Ψ projects *onto* the ground state of the target and Q projects *out of* the ground state. In Eq. (2.12b), ϕ_0 is the ground state of the hydrogen like target

$$\phi_0(\mathbf{r}) = \left(\frac{Z^3}{\pi}\right)^{1/2} e^{-Zr} \tag{2.14}$$

The major implication of the optical potential can be gleaned by expanding it into a complete set of eigenfunctions of the projected Hamiltonian QHQ, i.e.,

$$QHQ\Phi_n(\mathbf{r}_1, \mathbf{r}_2) = \mathscr{E}_n \Phi_n(\mathbf{r}_1, \mathbf{r}_2) \tag{2.15}$$

Since $Q^2 = Q$

$$Q^2 HQ\Phi_n(\mathbf{r}_1, \mathbf{r}_2) = QHQ\Phi_n(\mathbf{r}_1, \mathbf{r}_2) = \mathscr{E}_n Q\Phi_n(\mathbf{r}_1, \mathbf{r}_2)$$

This implies that the eigenfunctions Φ_n are also eigenfunctions of the operator Q, with the eigenvalues equal to 1 (we shall describe this by saying the Φ_n are in Q space). Inserting the completeness of Φ_n in Q space

$$Q = \sum_n \Phi_n\rangle\langle\Phi_n$$

into Eq. (2.10a), we obtain

$$\mathscr{V}_{\text{op}} = \sum_n \frac{PHQ\Phi_n\rangle\langle\Phi_n QHP}{E - \mathscr{E}_n} \tag{2.10b}$$

Throughout this chapter $\mathscr{V}_{\text{op}}$ as expanded in Eq. (2.10b) will play a central role. First and foremost for our purposes here is the fact that it can predict resonances. This is heuristically obvious from Eq. (2.10b) since $\mathscr{V}_{op}$ formally has singularities whenever $E = \mathscr{E}_n$ (later, we show that they are not real singularities); thus, we can anticipate that $\mathscr{V}_{op}$ will, in fact, undergo rapid variations (i.e., resonant behavior) in the vicinity of the 'singularities.' To this we should immediately add that Eq. (2.15) can have a discrete spectrum providing $\mathscr{E}_n < \varepsilon_1$, where $\mathscr{E}_1$ is the energy of the first excited state of the target. [That is because only in that region is P in Eq. (2.13a) such that $\lim_{r_i \to \infty} P\Psi = \lim_{r_i \to \infty} \Psi$ and $\lim_{r_i \to \infty} Q\Psi = 0$. The latter implies that if QHQ has eigenvalues $\mathscr{E}_n(<\varepsilon_1)$, they are discrete, and the corresponding eigenfunctions are quadratically integrable:

$$\Phi_n = Q\Phi_n \underset{r_i \to \infty}{\to} 0$$

The true spectrum of QHQ above ε_1 is continuous; but if we diagonalize QHQ with a quadratically integrable basis, we will obtain discrete eigenvalues even above ε_1, but the apparent singularities there do not correspond to resonances.

The second item concerning $\mathscr{V}_{op}$ we would like to add is that even for $\mathscr{E}_n < \varepsilon_1$, the apparent singularities $E = \mathscr{E}_n$ in Eq. (2.10b) are not real singularities. To see this , let $\mathscr{V}_{op}$ operate on the true solution Ψ of the Schrödinger equation

$$\mathscr{V}_{op}\Psi\rangle = \sum_n \frac{PHQ\Phi_n\rangle\langle\Phi_n QHP\Psi\rangle}{E - \mathscr{E}_n} \tag{2.16}$$

Using Eq. (2.6) in the form

$$\langle\Phi_n QHP\Psi\rangle = \langle\Phi_n Q(E-H)Q\Psi\rangle = (E - \mathscr{E}_n)\langle\Phi_n Q\Psi\rangle$$

on the right-hand side RHS of Eq. (2.16) gives

$$\begin{aligned}\mathscr{V}_{op}\Psi\rangle &= \sum_n \frac{PHQ\Phi_n\rangle(E - \mathscr{E}_n)\langle\Phi_n Q\Psi\rangle}{E - \mathscr{E}_n} \\ &= \sum_n PHQ\Phi_n\rangle\langle\Phi_n Q\Psi\rangle \end{aligned} \tag{2.17}$$

We see that the apparent singularity has canceled out! What is happening, in fact, is that near $\mathscr{E}_n$ the wave function varies rapidly but smoothly, so that the phase shift exhibits typical Breit–Wigner behavior.

Let us continue with this analysis of the optical potential by rewriting the optical-potential Eq. (2.7) using the spectral-decomposition Eq. (2.10b) of the optical potential. We write Eq. (2.7) in the form

$$(H' - E)P\Psi = -\frac{PHQ\Phi_s\rangle\langle\Phi_s QHP\Psi\rangle}{E - \mathscr{E}_s} \tag{2.18a}$$

where

$$H' = PHP + \sum_{n \neq s} \frac{PHQ\Phi_n\rangle\langle\Phi_n QHP}{E - \mathscr{E}_n} \tag{2.18b}$$

The contribution to the scattering from the resonant part comes from the RHS of Eq. (2.18a), and H' represents the nonresonant part of the scattering. In Eq. (2.18b), the summation is understood to include an integral over the continuous part of the spectrum.

A formal solution of Eq. (2.18), following Feshbach but for the case of electron–atom scattering, has been given in the classic paper by O'Malley and Geltman.[4] That treatment, however, is itself somewhat formal and, except for our own work, has really never been directly used except for variational solutions of the QHQ problem in Eq. (2.15). We, therefore, believe it will be more instructive in the context of the two-electron system to give a more incisive description of the procedure whereby the solution of Eq. (2.18) is effected and where at the same time formal definitions of the resonant quantities will also very explicitly emerge.

To effect a solution (and this is true of scattering equations in general), we premultiply Eq. (2.18a) by $\langle\phi_0 Y_{L0}$[meaning $\int d\mathbf{r}_2 d\Omega_1 \phi_0(r_2) Y_{L0}^*(\Omega_1)$] to get

$$\langle\phi_0 Y_{L0}[H_{PP} - E]P\Psi\rangle + \sum_{n(\neq s)} \frac{\langle\phi_0 Y_{L0} PHQ\Phi_n\rangle\langle\Phi_n QHP\Psi\rangle}{E - \mathscr{E}_n} = -\frac{\langle\phi_0 Y_{L0} PHQ\Phi_s\rangle\langle\Phi_s QHP\Psi\rangle}{E - \mathscr{E}_s} \tag{2.19a}$$

At this point, we explicitly deal with a partial wave solution Ψ_L whose P part is given by

$$P\Psi \to P\Psi_L = U(r_1) Y_{L0}(\Omega_1)\phi_0(r_2) \pm U(r_2) Y_{L0}(\Omega_2)\phi_0(r_1) \tag{2.20a}$$

where $\pm$ indicates singlet or triplet spin. The resultant integro-differential operator coming from the PHP part of H' is the well-known exchange approximation operator[5]

$$\mathscr{L}_{\text{ex}} U = \langle\phi_0 Y_{L0}|PHP - E|P\Psi\rangle \tag{2.21}$$

The integro-differential equation satisfied by U is, more explicitly,

$$\mathscr{L}_{\text{ex}} U + \sum_{n(\neq s)} \int \frac{\mathscr{V}_n(r, r') U(r') r'^2 dr'}{E - \mathscr{E}_n} = -\int \frac{\mathscr{V}_s(r, r') U(r') r'^2 dr'}{E - \mathscr{E}_s} \tag{2.19b}$$

where

$$\mathscr{V}_n = 2\langle\phi_0 Y_{L0} PHQ\Phi_n\rangle\langle\Phi_n QHP\phi_0 Y_{L0}\rangle \tag{2.19c}$$

The integral terms in Eq. (2.19b) are the explicit form of the optical-potential terms in Eq. (2.19b). They can be simplified even more by defining

$$v_{0n}^{\pm}(\mathbf{r}_1) \equiv \langle \phi_0(\mathbf{r}_2) Y_{L0}(\Omega_1) PHQ\Phi_n^{\pm}(1,2) \rangle \tag{2.22}$$

In Eq. (2.22), we have appended the ($\pm$) labels to distinguish, for the moment, the spatial parts of singlet and triplet wave functions for which

$$\Phi_n^{\pm}(1,2) = \pm\, \Phi_n^{\pm}(2,1) \tag{2.23}$$

It then becomes clear that using $P\Psi$ from Eq. (2.20a), the factor

$$\begin{aligned}\langle \Phi_n QHP\Psi \rangle &= \langle \Phi_n QH[U(r_1)Y_{L0}(\Omega_1)\phi_0(r_2) \pm U(r_2)Y_{L0}(\Omega_2)\phi_0(r_1)] \rangle \\ &= 2\langle v_{0n}^{\pm} U \rangle \end{aligned} \tag{2.24}$$

from which the optical-potential Eq. (2.19) can finally be written as

$$\mathscr{L}_{\text{ex}} U + \sum_{n(\neq s)} \frac{2v_{0n}(r)\langle v_{0n} U \rangle}{E - \mathscr{E}_n} = -\frac{2v_{0s}(r)\langle v_{0s} U \rangle}{E - \mathscr{E}_s} \tag{2.19d}$$

and concomitantly the kernel in Eq. (2.19b) becomes

$$\mathscr{V}_n(r, r') = 2v_{0n}(r)v_{0n}(r') \tag{2.25}$$

To solve Eq. (2.19), we use a Green's technique based on the left-hand side of Eq. (2.19) set equal to delta function

$$\mathscr{L}_{\text{ex}} G + \sum_{n(\neq s)} \int \frac{\mathscr{V}_n(r, r')[G(r, r'')]_{r=r'} r'^2 dr'}{E - \mathscr{E}_n} = -\frac{\delta(r - r'')}{rr''} \tag{2.26}$$

The G is constructed from a combination of regular ($g_<$) and irregular ($g_>$) solutions of the LHS of Eq. (2.26)

$$\mathscr{L}_{\text{ex}} g + \sum_{n(\neq s)} \int \frac{\mathscr{V}_n(r, r') g(r') r'^2 dr'}{E - \mathscr{E}_n} = 0 \tag{2.27a}$$

with boundary conditions

$$g_<(r) = \begin{cases} 0 & \mathbf{r} = 0 \\ \dfrac{\sin(kr - L\pi/2 + \eta_0)}{kr} & \mathbf{r} \to \infty \end{cases} \tag{2.27b}$$

$$\lim_{\mathbf{r} \to \infty} g_>(r) = -\frac{\cos(kr - L\pi/2 + \eta_0)}{kr} \tag{2.27c}$$

Note in particular that the nonresonant phase shift η_0 is determined from Eq. (2.27a) and $g_<(0) = 0$ and that η_0 so determined is used in starting the backward integration implicit in Eq. (2.27a). The complete Green's function solving Eq. (2.26) is then

$$k^{-1}G(r,r') = \begin{cases} g_<(r)g_>(r') & r<r' \\ g_<(r')g_>(r) & r>r' \end{cases} \tag{2.28}$$

And, finally, the complete solution of the (radial) optical-potential Eq. (2.19d) is

$$U(r) = U_0(r) + \int_0^\infty \frac{G(r,r')\mathscr{V}_s(r',r'')U(r'')r''^2dr''r'^2dr'}{E-\mathscr{E}_s} \tag{2.29}$$

The $U_0(r)$ in Eq. (2.29) is the solution of the nonresonant optical-potential equation that is identical to $g_<(r)$ as previously noted

$$U_0(r) = g_<(r) \tag{2.30}$$

Substituting $U(r)$ in Eq. (2.29) into Eq. (2.20a) gives for $P\Psi$ after some manipulation

$$P\Psi\rangle = P\Psi_0\rangle + \frac{G\mathscr{V}_sP\Psi\rangle}{E-\mathscr{E}_s}, \tag{2.20b}$$

where it is understood that integration in the last term does *not* involve all variables

$$\begin{aligned} \langle G\mathscr{V}_sP\Psi\rangle &= \langle G\mathscr{V}_s[U(r_1)Y_{L0}(\Omega_1)\phi_0(r_2) \pm U(r_2)Y_{L0}(\Omega_2)\phi_0 r_1]\rangle \\ &= \langle G(r_1,r')\mathscr{V}_s(r',r'')U(r'')\rangle Y_{L0}(\Omega_1)\phi_0(r_2) \\ &\quad \pm \langle G(r_2,r')\mathscr{V}_s(r',r'')U(r'')\rangle Y_{L0}(\Omega_2)\phi_0(r_1) \end{aligned} \tag{2.31}$$

In Eq. (2.20b), we also have used

$$P\Psi_0\rangle = U_0(r_1)Y_{L0}(\Omega_1)\phi_0(r_2) \pm U_0(r_2)Y_{L0}(\Omega_2)\phi_0(r_1) \tag{2.32}$$

To repeat, $P\Psi_0\rangle$ satisfies

$$\langle \phi_0 Y_{L0}[H'-E]P\Psi_0\rangle = 0 \tag{2.33}$$

We remind the reader that we are seeking a solution for $P\Psi$, Eq. (2.18), which we have converted into Eq. (2.20b). Multiply Eq. (2.20b) by $\langle \Phi_s QHP$; using Eq. (2.22) and the exchange property of $\Phi_s(1,2) = \pm\Phi_s(2,1)$ to obtain

$$2\langle v_{0s}U\rangle = 2\langle v_{0s}U_0\rangle + \frac{\langle \Phi_s QHPG\mathscr{V}_sP\Psi\rangle}{E-\mathscr{E}_s} \tag{2.34}$$

The same properties and Eq. (2.25) allow us to simplify the last term in Eq. (2.34). In detail,

$$\begin{aligned} \langle \Phi_s QHPG\mathscr{V}_sP\Psi\rangle = 2\langle \Phi_s(r_1,r_2)QHP\langle G(r_1,r_1')v_{0s}(r_1')\rangle_{r_1'} Y_{L0}(\Omega_1)\phi_0(r_2)\rangle_{r_1,r_2} \\ \cdot\langle v_{0s}(r_1'')U(r_1'')\rangle_{r_1''} \end{aligned} \tag{2.35}$$

where the subscripts on the kets are the coordinates over which we are integrating. Substituting Eq. (2.35) into Eq. (2.34) and reverting to our more abbreviated integration notation gives

$$\langle v_{0s}U\rangle = \langle v_{0s}U_0\rangle + \frac{\langle v_{0s}U\rangle\langle\Phi_s QHPGv_{0s}Y_{L0}\phi_0\rangle}{E-\mathscr{E}_s} \tag{2.36a}$$

which is manifestly solvable for $\langle v_{0s}U\rangle$

$$\langle v_{0s}U\rangle = \frac{\langle v_{0s}U_0\rangle}{1-\langle\Phi_s QHPGv_{0s}Y_{L0}\phi_0\rangle/(E-\mathscr{E}_s)} \tag{2.36b}$$

Equation (2.36b) is the essential part of the desired solution, since substituting into Eq. (2.20b) without the premultiplication by $\langle\Phi_s$ yields

$$\frac{G\mathscr{V}_s P\Psi\rangle}{E-\mathscr{E}_s} = \frac{G\mathscr{V}_s P\Psi_0\rangle}{E-\mathscr{E}_s-\Delta_s} \tag{2.37}$$

where Δ_s is the shift, which, from (Eq. 2.36b), is explicitly written as follows:

$$\Delta_s = \langle\Phi_s(r_1,r_2)QHP\langle G(r_1,r_1')v_{0s}(r_1')\rangle_{r_1'}Y_{L0}\phi_0(r_2)\rangle_{r_1,r_2} \tag{2.38}$$

Finally, substituting Eq. (2.37) into Eq. (2.20b), we obtain our derived and explicit expression for

$$P\Psi\rangle = P\Psi_0\rangle + \frac{G\mathscr{V}_s P\Psi_0\rangle}{E-\mathscr{E}_s-\Delta_s} \tag{2.20c}$$

The asymptotic form of $P\Psi_0\rangle$ is from Eqs. (2.27b), (2.30), and (2.32)

$$\lim_{r_1\to\infty} P\Psi_0\rangle = \frac{\sin(kr_1 - L\pi/2+\eta_0)}{kr_1}Y_{L0}(\Omega_1)\phi_0(r_2) \tag{2.39}$$

However, from Eq. (2.20c) we can deduce that $P\Psi\rangle$ has a different normalization (hence, the subscript u) as well as phase shift

$$\lim_{r_1\to\infty} P\Psi_u\rangle = \frac{\sin(kr_1 - L\pi/2+\eta_0+\eta_r)}{(\cos\eta_r)\,kr_1}Y_{L0}(\Omega_1)\phi_0(r_2) \tag{2.40}$$

where we find

$$\tan\eta_r = \frac{-\frac{1}{2}\Gamma_s}{E-\mathscr{E}_s-\Delta_s} \tag{2.41}$$

$$\Gamma_s = 2k|\langle\Phi_s QHP\Psi_0(E=E_s)\rangle|^2 \tag{2.42}$$

We see from Eq. (2.41) that the actual position of the resonance (assumed isolated) occurs at

$$E = E_s = \mathscr{E}_s + \Delta_s \tag{2.43}$$

i.e., E_s is shifted from the eigenvalues $\mathscr{E}_s$ of QHQ by an amount Δ_s given in Eq. (2.38), which now can legitimately be called the shift. Furthermore, the shape of a hypothetical cross section associated with the purely resonant part of the phase shift η_r (i.e., $\sin^2\eta_r$) is from Eq. (2.41) purely Lorentzian about $E = E_s$ with full width Γ_s at half-maximum (which is usually called the half-width) given by Eq. (2.42).

We can express $P\Psi$ in terms of the resonant phase shift by using Eq. (2.20c) in Eq. (2.41), and the definition of $\mathscr{V}_s$, Eq. (2.19c), without premultiplying by $\langle \phi_0 Y_{L0}$,

$$P\Psi_u = P\Psi_0 - \frac{\tan\eta_r}{\frac{1}{2}\Gamma_s} GPHQ\Phi_s\rangle\langle\Phi_s QHP\Psi_0\rangle \tag{2.20d}$$

In order for $P\Psi$ to have the usual asymptotic form, we normalize Eq. (2.20d) by multiplying by $\cos\eta_r$, i.e.,

$$P\hat{\Psi}\rangle = \cos\eta_r\left[P\Psi_0\rangle - \frac{\tan\eta_r}{\frac{1}{2}\Gamma_s} GPHQ\Phi_s\rangle\langle\Phi_s QHP\Psi_0\rangle\right] \tag{2.44}$$

We can rewrite this as

$$P\hat{\Psi}\rangle = \cos\eta_r P\Psi_0\rangle - \sin\eta_r P\Psi_1\rangle \tag{2.45}$$

where, using the definition in Eq. (2.42) of Γ_s in Eq. (2.44), $P\Psi_1$ can be written as

$$P\Psi_1\rangle = \frac{1}{k}\frac{GPHQ\Phi_s\rangle}{\langle\Phi_s QHP\Psi_0\rangle} \tag{2.46}$$

$P\Psi_1$ like $P\Psi_0$ is also a slowly varying function of energy E. To obtain the total wave function Ψ, we have to know $Q\Psi$. It is now hopefully obvious from Eq. (2.6) that we can write

$$Q\hat{\Psi}\rangle = \sum_n \frac{1}{E - \mathscr{E}_n} \Phi_n\rangle\langle\Phi_n QHP\hat{\Psi}\rangle \tag{2.6$'$}$$

It is convenient to define Q-space Green's functions in addition to the P-space Green's function defined in Eq. (2.28)

$$G_Q \equiv \sum_n \frac{\Phi_n\rangle\langle\Phi_n}{E - \mathscr{E}_n} \tag{2.47}$$

$$G_Q^{(s)} \equiv \sum_{n(\neq s)} \frac{\Phi_n\rangle\langle\Phi_n}{E - \mathscr{E}_n} \tag{2.48}$$

so that we can write

$$Q\hat{\Psi}\rangle = G_Q QHP\hat{\Psi}\rangle \tag{2.6$''$}$$

Inserting $P\hat{\Psi}$ from Eq. (2.45),

$$Q\hat{\Psi} = G_Q^{(s)} QHP[\cos\eta_r P\Psi_0\rangle - \sin\eta_r P\Psi_1\rangle] + \frac{1}{E-\mathscr{E}_s}\Phi_s\rangle[\cos\eta_r\langle\Phi_s QHP\Psi_0\rangle - \sin\eta_r\langle\Phi_s QHP\Psi_1\rangle] \qquad (2.49)$$

Let us note the last square bracket can be simplified in terms of Γ_s and Δ_s: using Eqs. (2.41) and (2.36b), we find

$$[\cos\eta_r\langle\Phi_s QHP\Psi_0\rangle - \sin\eta_r\langle\Phi_s QHP\Psi_1\rangle] = -\frac{\sin\eta_r(E-\mathscr{E}_s)}{(\Gamma_s k/2)^{1/2}} \qquad (2.50)$$

so that

$$Q\hat{\Psi}\rangle = G_Q^{(s)} QHP\cos\eta_r P\Psi_0\rangle - G_Q^{(s)} QHP\sin\eta_r P\Psi_1\rangle - \frac{\sin\eta_r\Phi_s\rangle}{(\Gamma_s k/2)^{1/2}}$$
$$= \cos\eta_r Q\Psi_0\rangle - \sin\eta_r Q\Psi_1\rangle \qquad (2.49')$$

Therefore, using $\Psi = P\Psi + Q\Psi$ we have from Eqs. (2.45) and (2.49′)

$$\Psi\rangle = \cos\eta_r\Psi_0\rangle - \sin\eta_r\Psi_1\rangle \qquad (2.51)$$

where

$$\Psi_0 = P\Psi_0\rangle + G_Q^{(s)} QHP\Psi_0\rangle \qquad (2.52)$$

$$\Psi_1 = P\Psi_1\rangle + Q\Psi_1\rangle \qquad (2.53)$$

The $P\Psi_1$ is given in Eq. (2.46) and $Q\Psi_1$ can be inferred from Eq. (2.49′)

$$Q\Psi_1\rangle = \frac{1}{k}\frac{\Phi_s\rangle}{\langle\Psi_0 PHQ\Phi_s\rangle} + G_Q^{(s)} QHP\Psi_1\rangle \qquad (2.54)$$

These expressions will be used in deriving equations for the line shape parameter and radiation absorption profile discussed in the next section. [The preceding derivation is given in abbreviated form in Ref. 26]

We conclude this section with some observations on the nonresonant function Ψ_0. It is already explicit in Eqs. (2.49′) and (2.52) that Ψ_0 has a $Q\Psi_0$ as well as a $P\Psi_0$ part. This is at first surprising, because Ψ_0 is often referred to as the "nonresonant" continuum function: the word *nonresonant* seems by definition to imply that Ψ_0 is strictly in P space. However, as is clear from the definition in Eq. (2.48) of $G_Q^{(s)}$, $Q\Psi_0$ contains the effects of all resonances other than the sth resonance. It is clear that from the point of view of the sth resonance, other resonances are indeed a part of its background. However to group them as part of the nonresonant background represents an alternative approach from the traditional one that is embodied in the Breit–Wigner formalism and also incorporated in the Fano approach[2] as a generalization of his single-level results. We shall discuss this interesting difference more at the end of the next section; here, we conclude with a presentation of the nonresonant equations satisfied by Ψ_0 as distinct to the well-known equation satisfied by $P\Psi_0$.

We assert that the complete nonresonant function

$$\Psi_0 = P\Psi_0 + Q\Psi_0 \tag{2.52'}$$

satisfies

$$[H - \mathscr{E}_s Q\Phi_s\rangle\langle Q\Phi_s - E]\Psi_0 = 0 \tag{2.55}$$

To see this, we solve the auxiliary eigenvalue equation

$$[QHQ - \mathscr{E}_s Q\Phi_s\rangle\langle Q\Phi_s]Q\Phi_n = \mathscr{E}_n \Phi_n \tag{2.56}$$

This latter equation has the eigenspectrum

$$\mathscr{E}_n, \Phi_n(= Q\Phi_n) \qquad n = 1, 2, 3, \ldots s-1, s+1, \ldots \tag{2.57}$$

where the $\mathscr{E}_n$ and Φ_n are precisely those of the original QHQ Eq. (2.15). But from Eq. (2.57), we see that Eq. (2.56) lacks the sth states of QHQ. This is precisely what we desire, because when we construct the optical-potential equation associated with the $P\Psi_0$ part of Eq. (2.55)

$$\left[PHP + PHQ\frac{1}{E - QHQ + \mathscr{E}_s Q\Phi_s\rangle\langle\Phi_s Q}QHP - E\right]P\Psi_0 = 0 \tag{2.58}$$

and expand in terms of the eigenfunctions of Eq. (2.56), we see from Eq. (2.57) that

$$PHQ\frac{1}{E - (QHQ - \mathscr{E}_s Q\Phi_s\rangle\langle\Phi_s Q)}QHP = \sum_{n(\neq s)} \frac{PHQ\Phi_n\rangle\langle\Phi_n QHP}{E - \mathscr{E}_n} \tag{2.59}$$

The RHS of Eq. (2.59) is exactly the nonresonant potential defining $P\Psi_0$, Eq. (2.18). In addition, using the appropriate relationship between $P\Psi_0$ and $Q\Psi_0$,

$$Q\Psi_0 = \frac{1}{E - (QHQ - \mathscr{E}_s\Phi_s\rangle\langle\Phi_s)}QHP\Psi_0\rangle \tag{2.60a}$$

We see on expanding in terms of Eq. (2.57) that this is identical to the second term in Eq. (2.52)

$$Q\Psi_0\rangle = \sum_{n(\neq s)} \frac{\Phi_n\rangle\langle\Phi_n QHP\Psi_0\rangle}{E - \mathscr{E}_n} \tag{2.60b}$$

3. THE PROJECTION-OPERATOR FORMALISM OF THE LINE-SHAPE PARAMETER q

In addition to their occurrence as resonances in (electron) scattering, autoionization states may also manifest themselves by radiative decay to, or absorption from, a truly bound state. In that case, the line profile of the radiation becomes another characteristic parameter of the states involved, the autoionization state in particular. This is because, as we shall see, the line profile is largely determined by the fact the resonant state can truly autoionize in competition with its radiative decay.

A classic example of autoionization-state line profiles were those measured by Madden and Codling[6] in vacuum UV photoabsorption of noble gases. A corresponding classic theoretical description of such line profiles was given by Fano[2], where, for an isolated resonance, he showed that the ratio of resonant to nonresonant radiative cross sections can be written as

$$\frac{|\langle\Psi_E|T|\Psi_g\rangle|^2}{|\langle\psi_E|T|\Psi_g\rangle|^2}=\frac{(\varepsilon+q)^2}{1+\varepsilon^2} \tag{3.1}$$

where ψ_g is the ground state, ε is a dimensionless energy

$$\varepsilon=\frac{E-E_s}{\frac{1}{2}\Gamma_s} \tag{3.2}$$

and q is a ratio of radiative transition matrix elements of an electromagnetic transition operator T between resonant and nonresonant functions. For the purposes of this section, we can let T be the dipole length operator

$$T=z_1+z_2 \tag{3.3}$$

where z_i is the z coordinate of the ith electron, and here we take $i=1,2$. The quantity q can have either sign, so that from Eq. (3.1), we see that the radiative line will be enhanced on one side and diminished on the other side of the resonance at $E=E_s$; q is accordingly called the line profile parameter.

One of our purposes in this section is to derive an explicit expression for q in terms of the Feshbach formalism. We shall find that the expression is somewhat different from that given by Fano.[2] The difference is due to a subtle difference in the mode of description inherent in the Feshbach formalism from that used by Fano or, more precisely, in the underlying Breit–Wigner approach,[7] to which we shall return at the end of this section. First, let us derive q.

To do this, we note that the total wave functions labelled Ψ_E and ψ_E by Fano are clearly Ψ and Ψ_0, respectively. Hence, using Eqs. (2.51) and (2.52) for the above in the LHS of Eq. (3.1), we find that

$$\left|\frac{\langle\Psi|T|\Psi_g\rangle}{\langle\Psi_0|T|\Psi_g\rangle}\right|^2 = |\cos\eta_r - \sin\eta_r q|^2 \tag{3.4}$$

where the desired q is

$$q = \frac{\langle\Psi_1|T|\Psi_g\rangle}{\langle\Psi_0|T|\Psi_g\rangle} \tag{3.5a}$$

and Ψ_1, Ψ_0 are given by Eqs. (2.52) and (2.53), respectively. That q has been defined consistently with Eq. (3.1) can readily be checked by using Eq. (2.41) in dimensionless form [cf. Eq. (3.2)]

$$\tan\eta_r = -1/\varepsilon \tag{2.41$'$}$$

and noting then that the RHS of Eq. (3.4) does, indeed, reduce to the RHS of Eq. (3.1)!

Substitution for Ψ_1 in Eq. (3.5) gives our final expression for q, which is essentially that used for calculations

$$q = \frac{k^{-1}}{\langle\Psi_0|T|\Psi_g\rangle\langle\Psi_0 PHQ\Phi_s\rangle}\left[\langle\Phi_s|T|\Psi_g\rangle + \langle\Phi_s QHPG|T|\Psi_g\rangle + \sum_{n(\neq s)} \frac{\langle\Phi_n|T|\Psi_g\rangle\langle\Phi_n QHPGPHQ\Phi_s\rangle}{E - \mathscr{E}_n}\right]$$

We write

$$q = q_0 + \delta q_b + \delta q_c + \delta q_r \tag{3.5b}$$

then

$$q_0 = \frac{\langle\Phi_s|T|\Psi_g\rangle}{v_T} \tag{3.5c}$$

$$\delta q_b = \frac{1}{v_T}\sum_{v}\frac{\langle\Phi_s Q|H|P\Psi_v\rangle\langle P\Psi_v|T|\Psi_g\rangle}{E_s - E_v} \tag{3.5d}$$

$$\delta q_c = \frac{1}{\pi v_T}\mathscr{P}\int\frac{\langle\Phi_s QHP\Psi_0(E')\rangle\langle P\Psi_0(E')|T|\Psi_g\rangle}{E_s - E'}E^{1/2}dE' \tag{3.5e}$$

$$\delta q_r = \frac{1}{v_T}\sum_{n(\neq s)}\frac{\langle\Phi_n|T|\Psi_g\rangle\langle\Phi_n QHPGPHQ\Phi_s\rangle}{E_s - E_n} \tag{3.5f}$$

where

$$v_T \equiv k\langle\Psi_0|T|\Psi_g\rangle\langle\Psi_0 PHQ\Phi_s\rangle \tag{3.5g}$$

In the derivation, Eq. (3.5), we used (Ref. 20)

$$G = \sum \frac{P\Psi_\nu\rangle\langle P\Psi_\nu}{E_s - E_\nu} + \frac{\mathscr{P}}{\pi} \int \frac{P\Psi_0(E')\rangle\langle P\Psi_0(E')\sqrt{E'}\,dE'}{E_s - E'} \tag{3.6}$$

4. RELATIONSHIP OF BREIT–WIGNER AND FESHBACH RESONANCE PARAMETERS

As we have already indicated the resonance parameters, as they come out of the Feshbach theory, are energy dependent, very weakly energy dependent to be sure, but for the purpose of precision calculations, this variation can be important. However, from the point of experiment, determining resonance parameters usually comes from fitting the data to resonance expressions while assuming energy-independent parameters. Furthermore, there is considerable motivation for having energy-independent parameters from the point of view of traditional resonance theory, where resonances are identified as poles in the complex energy plane whose positions are by definition energy independent. In fact, from such a theoretical background, there has now come, in the form of complex rotation, developments that allow this approach to be a useful tool for resonance calculations (cf. Chapter 4 by Junker and Chapter 5 by McCurdy in this volume).

However, in this chapter we shall strictly confine ourselves to the pragmatic aspects whereby results from a Feshbach calculation are directly compared with an experimental fit based on energy-independent resonance parameters. We shall call the latter the Breit–Wigner parameters. To derive this relationship, we start with the equation for the resonant phase shift in the Feshbach form

$$\eta(E) = \eta_0(E) + \tan^{-1}\left[\frac{\frac{1}{2}\Gamma(E)}{\mathscr{E} + \Delta(E) - E}\right] \tag{4.1}$$

The Feshbach resonance energy, which we here denote as E_{F}, is defined as the solution to the equation

$$E_{\mathrm{F}} = \mathscr{E} + \Delta(E_{\mathrm{F}}) \tag{4.2}$$

But note that $\Delta(E_{\mathrm{F}})$ as such does *not* appear in Eq. (4.1). Let us nevertheless expand around E_{F}

$$\Gamma(E) = \Gamma_{\mathrm{F}} + \Gamma'_{\mathrm{F}}\delta E \tag{4.3a}$$

$$\Delta(E) = \Delta_{\mathrm{F}} + \Delta'_{\mathrm{F}}\delta E \tag{4.3b}$$

where

$$\delta(E) = E - E_{\mathrm{F}} \tag{4.4a}$$

$$\Gamma_F = \Gamma(E_F) \tag{4.4b}$$

$$\Delta_F \equiv \Delta(E_F) \tag{4.4c}$$

$$\Gamma'_F \equiv \left[\frac{\partial \Gamma}{\partial E}\right]_{E=E_F} \tag{4.4d}$$

$$\Delta'_F \equiv \left[\frac{\partial \Delta}{\partial E}\right]_{E=E_F} \tag{4.4e}$$

The first derivative of $\eta(E)$ from Eq. (4.1) can then be written

$$\eta(E) = \eta_0(E) + \tan^{-1}\left\{\frac{1}{2}\left[\frac{\Gamma_F/(1-\Delta'_F)}{E_F - E} - \frac{\Gamma'_F}{1-\Delta'_F}\right]\right\} \tag{4.5}$$

$$\tan^{-1}\left\{\frac{1}{2}\left[\frac{\Gamma_F/(1-\Delta'_F)}{E_F - E} - \frac{\Gamma'_F}{1-\Delta'_F}\right]\right\} = \tan^{-1}\left\{\frac{1}{2}\left[\frac{\Gamma_F/(1-\Delta'_F)}{E_F - E}\right]\right\} + 0(\delta E)^2 \tag{4.6}$$

Using Eq. (4.6) in Eq. (4.1) gives

$$\eta(E) = \eta_0(E) + \tan^{-1}\left[\frac{\frac{1}{2}\Gamma_F/(1-\Delta'_F)}{E_F - E}\right] + 0(\delta E)^2 \tag{4.7}$$

Observe that all the parameters appearing in Eq. (4.7) are now independent of E, and therefore, can now be directly compared with the Breit–Wigner form, i.e.,

$$\tan^{-1}\left[\frac{\frac{1}{2}\Gamma_{BW}}{E_{BW} - E}\right] = \tan^{-1}\left[\frac{\frac{1}{2}\Gamma_F/(1-\Delta'_F)}{E_F - E}\right] \tag{4.8}$$

from which we derive[8]

$$\Gamma_{BW} = \frac{\Gamma_F}{1-\Delta'_F} + \text{second-order correction} \tag{4.9}$$

$$E_{BW} = E_F + \text{second-order correction} \tag{4.10a}$$

It is emphasized that these relationships are only of first order, as stated in Eq. (4.10a). In Eq. (4.9), the first-order correction is explicitly given; the apparent absence of a first-order correction in Eq. (4.10a) is due to the fact that E_F itself already contains a first-order correction $\Delta(E_F)$ [cf. Eq. (4.2)]. Nevertheless, the question of the second-order correction for E_{BW} is of interest. Treating the resonance as an isolated pole in the complex energy plane,

Drachman[9] has derived the result

$$E_{\mathrm{BW}} = E_{\mathrm{F}} - \tfrac{1}{4}\Gamma_{\mathrm{F}}\Gamma'_{\mathrm{F}} \tag{4.10b}$$

Drachman has also pointed out,[9] and now we may verify, that E_{BW} as given in Eq. (4.10b) is the energy at which $d^2\eta_r/dE^2 = 0$, where η_r is the resonant (second) term in Eq. (4.5). The shift from E_{F} is due to the asymmetry of the $\tan^{-1}$ around its singularity at $E = E_{\mathrm{F}}$. The fact that the resonance does *not* occur at the singularity of $\tan \eta_r$ comes about as follows: if one defines $E_{\mathrm{BW}} = E_{\mathrm{F}}$, then the resultant nonresonant phase shift η_0 will not be a smooth function of E, as Drachman has confirmed for us in terms of the example given in his paper.[9] However, from a practical point of view, since the correction in Eq. (4.9) to Γ_{F} is first order, this correction is much more important than the correction in Eq. (4.10b) to E_{F} (but not $\mathscr{E}_{\mathrm{F}}$), which is second order. We shall show a specific example of that in our precision result for $H^-(2s^2; {}^1S)$.

5. VARIATIONAL CALCULATION OF $\mathscr{E}_n$

In this section, we discuss the calculation of the resonance position $\mathscr{E}_n$.[10] From the point of view of the Feshbach theory, the calculation of $\mathscr{E}_n$ constitutes the heart of the resonance problem. For if $\mathscr{E}_n$ and its associated eigenfunction $Q\Phi_n$ are known accurately, then all other resonance parameters and corrections can be calculated from it effectively by perturbation theory. And in addition, $\mathscr{E}_n$ and Φ_n can be calculated from the Rayleigh–Ritz variational principle, and thus very accurately $\delta[\mathscr{E}_n] = 0$ where

$$[\mathscr{E}_n] = \frac{\langle Q\Phi_{LS} H Q\Phi_{LS}\rangle}{\langle Q\Phi_{LS} Q\Phi_{LS}\rangle} = \frac{\langle \Phi_{LS} Q H Q \Phi_{LS}\rangle}{\langle \Phi_{LS} Q \Phi_{LS}\rangle} \tag{5.1}$$

since $Q^2 = Q$.

Our chief contribution to this field has been the introduction and actual calculation of $\mathscr{E}_n$ using Hylleraas expansions for the Φ_{LS}. The use of Hylleraas coordinates with projection operators requires a nontrivial integration problem that we solved as outlined in Ref. 10, but which we shall treat in a more general way here. For doing actual calculations, we have also consistently used the symmetric Euler angle decomposition of the two-electron fixed-nucleus problem.[11] The reader is referred to our review paper[11] for details; we shall repeat here only the most salient aspects. Equation (5.1) is stationary with respect to variations in the wave function. This gives rise to a set of secular equations that can be solved for the eigenvalues for each set of nonlinear parameters. The eigenvalues are minimized with respect to the nonlinear parameters in the wave functions Φ_{LS}. The Φ_{LS} is the spatial part of the wave function of angular momentum L, parity

$(-1)^\kappa$, and spin S of the system, where $S=0$ for singlet states and $S=1$ for triplet states. The Φ_{LS} can be written as[24]

$$\Phi_{LS}=\sum_{\kappa}{}'' \left\{ \cos\left(\frac{\kappa\theta_{12}}{2}\right)[f_{L\kappa}+(-1)^S \tilde{f}_{L\kappa}]\mathscr{D}_L^{\kappa +} + \sin\left(\frac{\kappa\theta_{12}}{2}\right)[f_{L\kappa}-(-1)^S \tilde{f}_{L\kappa}]\mathscr{D}_L^{\kappa -} \right\} \tag{5.2a}$$

$$\Phi_{LS}(\mathbf{r}_1,\mathbf{r}_2)=(-1)^S\Phi_{LS}(\mathbf{r}_2,\mathbf{r}_1) \tag{5.2b}$$

where $\mathscr{D}_L^{\kappa +}$ are the exchange rotational harmonics[11] whose arguments are our symmetric Euler angles, which are usually written θ, ϕ, ψ.[11] The $f_{L\kappa}$ are functions of the residual internal coordinates that in one way or another define the triangle formed by electrons 1 and 2 and the nucleus. For Hylleraas expansions, the $f_{L\kappa}$ are

$$f_{L\kappa}(r_1,r_2,r_{12})=e^{-\alpha r_1-\beta r_2}r_1^{(L+\kappa)/2}r_2^{(L-\kappa)/2}\sum_{l,m,n} C_{lmn}r_1^l r_2^m r_{12}^n \tag{5.3a}$$

and by definition

$$\tilde{f}_{L\kappa}(r_1,r_2,r_{12})\equiv f_{L\kappa}(r_2,r_1,r_{12}) \tag{5.3b}$$

which are such that $\Phi_{LS}(r_1,r_2)$ has the correct symmetry under exchange $(r_1 \rightleftarrows r_2)$.

In Eq. (5.3), the angular momentum L is assumed to be of the same parity as κ[i.e., $(-1)^L=(-1)^\kappa$]. If $(-1)^{L+1}=(-1)^\kappa$, then L is replaced by $L+1$ in Eq. (5.3a). The double prime on the sum in Eq. (5.2a) indicates that only κ's of a given parity are included; also $\kappa \leqslant L$. The C_{lmn} are the linear parameters and α and β are the nonlinear parameters.

We can simplify[10] the expectation value in Eq. (5.1). Using Eq. (2.13a), we write

$$\begin{aligned}\langle\Phi_{LS}QHQ\Phi_{LS}\rangle &= \langle\Phi_{LS}H\Phi_{LS}\rangle-\sum_{i=1}^{2}[\langle\Phi_{LS}P_iH\Phi_{LS}\rangle+\langle\Phi_{LS}HP_i\Phi_{LS}\rangle]\\ &\quad+\sum_{i=1}^{2}\sum_{j=1}^{2}\langle\Phi_{LS}P_iHP_j\Phi_{LS}\rangle+\langle\Phi_{LS}HP_1P_2\Phi_{LS}\rangle\\ &\quad+\langle\Phi_{LS}P_1P_2H\Phi_{LS}\rangle+\langle\Phi_{LS}P_1P_2HP_1P_2\Phi_{LS}\rangle\\ &\quad-\sum_{i=1}^{2}[\langle\Phi_{LS}P_1P_2HP_i\Phi_{LS}\rangle+\langle\Phi_{LS}P_iHP_1P_2\Phi_{LS}\rangle]\end{aligned} \tag{5.4}$$

Given the (anti) symmetry of the Φ_{LS}, Eq. (5.2b), it can be seen that the following terms are equal ($i \neq j = 1,2$):

$$\langle \Phi_{LS} P_i P_j H P_j \Phi_{LS} \rangle = \langle \Phi_{LS} P_i P_j H P_i \Phi_{LS} \rangle \tag{5.5a}$$

$$\langle \Phi_{LS} P_i H P_j \Phi_{LS} \rangle = \langle \Phi_{LS} P_j H P_i \Phi_{LS} \rangle \tag{5.5b}$$

Using these equations in Eq. (5.4) gives

$$\begin{aligned} \langle \Phi_{LS} Q H Q \Phi_{LS} \rangle = \langle \Phi_{LS} H \Phi_{LS} \rangle &- 2[\langle \Phi_{LS} P_1 H \Phi_{LS} \rangle + \langle \Phi_{LS} H P_1 \Phi_{LS} \rangle] \\ &+ 2\langle \Phi_{LS} P_1 H P_1 \Phi_{LS} \rangle + 2\langle \Phi_{LS} P_1 H P_2 \Phi_{LS} \rangle \\ &+ \langle \Phi_{LS} H P_1 P_2 \Phi_{LS} \rangle + \langle \Phi_{LS} P_1 P_2 H \Phi_{LS} \rangle \\ &- 2[\langle \Phi_{LS} P_1 P_2 H P_1 \Phi_{LS} \rangle + \langle \Phi_{LS} P_1 H P_1 P_2 \Phi_{LS} \rangle] \\ &+ \langle \Phi_{LS} P_1 P_2 H P_1 P_2 \Phi_{LS} \rangle \end{aligned} \tag{5.6}$$

The two quantities in each of the preceding square brackets are the transpose of each others, i.e., for any terms $i\rangle$ and $j\rangle$ in the expansion of $\Phi_{LS}\rangle$, we have for the matrix elements

$$\langle i|P_1 H|j\rangle = \langle j|HP_1|i\rangle \tag{5.7}$$

There are three basic kinds of terms in Eq. (5.6). The first is $\langle \Phi_{LS} H \Phi_{LS} \rangle$, which occurs in ordinary Hylleraas-type variational calculations of bound states, this term need not be discussed. The second is when $P_1 P_2$ occurs but not an individual P_i; an example is $\langle \Phi_{LS} H P_1 P_2 \Phi_{LS} \rangle$. To demonstrate the calculation of this term, let us write Φ_{LS} in symbolic form

$$\Phi_{LS} = f_L(r_1, r_2, r_{12}) \mathscr{D}_L(\boldsymbol{\beta}) \tag{5.2c}$$

where $\boldsymbol{\beta}$ represents the Euler angles.[11] Using the definition of the projection operators P_1 and P_2, we find

$$P_1 P_2 \Phi_{LS} = \phi_0(\mathbf{r}_1)\phi_0(\mathbf{r}_2) \int d\mathbf{r}_1 d\mathbf{r}_2 \phi_0(\mathbf{r}_1)\phi_0(\mathbf{r}_2) f_L(r_1, r_2, r_{12}) \mathscr{D}_L(\boldsymbol{\beta}) \tag{5.8}$$

The volume element can be written as[11]

$$d\mathbf{r}_1 d\mathbf{r}_2 = r_1 dr_1 r_2 dr_2 r_{12} dr_{12} d\boldsymbol{\beta} \tag{5.9}$$

where

$$d\boldsymbol{\beta} = \sin\theta \, d\theta \, d\phi \, d\psi \tag{5.10}$$

Also using

$$\int \mathscr{D}_L d\boldsymbol{\beta} = \delta_{L0} (8\pi^2)^{1/2} \tag{5.11}$$

we can reduce Eq. (5.8) to the form

$$P_1 P_2 \Phi_{LS} = \delta_{L0} (8\pi^2)^{1/2} \phi_0(\mathbf{r}_1)\phi_0(\mathbf{r}_2) \int d\tau \phi_0(\mathbf{r}_1)\phi_0(\mathbf{r}_2) f_L(r_1, r_2, r_{12}) \tag{5.12}$$

We see that this term is zero unless $L = 0$, since the $\phi_0(r)$ is independent of angles. The remaining integral in Eq. (5.10) can be carried out easily. Thus, the integral $\langle \Phi_{LS} H P_1 P_2 \Phi_{LS} \rangle$ can be readily obtained and is only nonzero for S states ($L=0$).

The third type of terms containing only one P_i is more difficult, and the calculation of such terms was the major technical accomplishment that allowed us to initiate these types of calculations.[10] Let us enlarge the definition in Eq. (2.11) of projectors to include the angles

$$P_1 = \phi_0(\mathbf{r}_1) Y_{L0}(\Omega_2) \rangle \langle \phi_0(\mathbf{r}_1) Y_{L0}(\Omega_2) \tag{5.13a}$$

and

$$P_2 = \phi_0(\mathbf{r}_2) Y_{L0}(\Omega_1) \rangle \langle \phi_0(\mathbf{r}_2) Y_{L0}(\Omega_1) \tag{5.13b}$$

It can be verified that with the definition of the projection operators, the basic properties, such as Eq. (2.3), remain unchanged, $P_1 \Phi_{LS}$ is therefore

$$P_1 \Phi_{LS} = \phi_0(\mathbf{r}_1) Y_{L0}(\Omega_2) \int d\mathbf{r}_1 d\Omega_2 \phi_0(\mathbf{r}_1) Y^*_{L0}(\Omega_2) \Phi_{LS} \tag{5.14}$$

The $Y_{L0}(\Omega_2)$ can be expanded[11] in terms of $\mathscr{D}_L(\boldsymbol{\beta})$ functions (only L is involved, because both are eigenfunctions of L^2 with eigenvalues $L(L+1)$. Suppressing summation over other indices, we can write

$$Y_{L0}(\Omega_2) = \alpha_L(\theta_{12}) \mathscr{D}_L(\boldsymbol{\beta}) \tag{5.15}$$

Here, θ_{12} is the angle between $\mathbf{r}_1$ and $\mathbf{r}_2$. Let us note

$$\begin{aligned} d\mathbf{r}_1 d\Omega_2 &= r_1^2 dr_1 d\Omega_1 d\Omega_2 \\ &= r_1^2 dr_1 \sin\theta_{12} d\theta_{12} d\boldsymbol{\beta} \\ &= r_1^2 dr_1 \frac{r_{12} dr_{12}}{r_1 r_2} d\boldsymbol{\beta} \\ &= \frac{r_1}{r_2} dr_1 r_{12} dr_{12} d\boldsymbol{\beta} \end{aligned} \tag{5.16}$$

Substituting Eqs. (5.15) and (5.16) into Eq. (5.14), we obtain

$$P_1 \Phi_{LS} = \frac{\phi_0(\mathbf{r}_1) Y_{L0}(\Omega_2)}{r_2} \int\int r_1 dr_1 r_{12} dr_{12} \alpha_L(\theta_{12}) f_L(r_1, r_2, r_{12}) \tag{5.17}$$

For calculational purposes, it is convenient to redefine the f_L functions

$$f_L(r_1, r_2, r_{12}) \equiv \alpha_L(\theta_{12}) g_L(r_1, r_2, r_{12}) \tag{5.18}$$

In that way, the integral in Eq. (5.17) can always be reduced to the form of the LHS of the following equation:

$$\begin{aligned} &\int\int r_1 dr_1 r_{12} dr_{12} \cos\theta_{12} g_L(r_1, r_2, r_{12}) \\ &\quad = \int\int r_1 dr_1 r_{12} dr_{12} \left(\frac{r_1^2 + r_2^2 - r_{12}^2}{2 r_1 r_2} \right) g_L(r_1, r_2, r_{12}) \end{aligned} \tag{5.19}$$

The RHS of Eq. (5.19) can straightforwardly be integrated if the function $g_L(r_1, r_2, r_{12})$ is of conventional Hylleraas form. Calling this result $F(r_2)/r_2$, we have

$$F(r_2) \equiv \int\int r_1 dr_1 r_{12} dr_{12} \left(\frac{r_1^2 + r_2^2 - r_{12}^2}{2r_1}\right) g_L(r_1, r_2, r_{12}) \tag{5.20}$$

Thus finally

$$P_1 \Phi_{LS} = \phi_0(\mathbf{r}_1) Y_{L0}(\Omega_2) \frac{F(r_2)}{r_2} \tag{5.21}$$

Therefore, referring back to Eq. (5.6), we can infer that

$$\langle \Phi_{LS} H P_1 \Phi_{LS} \rangle \propto \int d\mathbf{r}_1 d\mathbf{r}_2 \Phi_{LS} H \phi_0(\mathbf{r}_1) Y_{L0}(\Omega_2) \frac{F_L(r_2)}{r_2} \tag{5.22}$$

Once again, this integral is of the same form as that occurring in bound-state calculations. Other terms involving P_i can be similarly evaluated.

Relevant precision calculations of $\mathscr{E}_n$ will be given in the following sections.

6. PRECISION CALCULATION OF THE LOWEST ^1S RESONANCE IN ELECTRON–HYDROGEN SCATTERING

The lowest ^1S resonance in electron–hydrogen (e–H) scattering is the most basic resonance in atomic physics. The first (complete) calculation of this state, which was a close coupling treatment by Burke and Schey,[13] has been followed by a large number of subsequent calculations. It is not our intention to review these calculations, including even our own, except for the one calculation[8] that we believe deserves the appellation of precision.

Since we can very accurately evaluate $\mathscr{E}$ using our Hylleraas expansion described in previous sections, what is required is a very accurate evaluation of Δ [cf. Eq. (2.43)]. However, Δ is not easily directly evaluated; thus, the key to the calculation is the observation by Chung and Chen,[14] which avoids having to calculate Δ directly. In fact, we are now in a position to derive very quickly the basic relationship of Chung and Chen.[14] From our foregoing derivations, particularly in Eq. (2.6), we can infer that

$$\langle \Phi_s Q \Psi_u \rangle = \frac{\langle \Phi_s Q H P \Psi_u \rangle}{E - \mathscr{E}_s} \tag{6.1}$$

Substitute for $P\Psi_u$ in Eq. (2.20d) to obtain

$$\langle\Phi_s Q\Psi_u\rangle = \frac{1}{E-\mathscr{E}_s}\left[\langle\Phi_s QHP\Psi_0\rangle - \frac{\tan\eta_r}{\frac{1}{2}\Gamma_s}\langle\Phi_s QHPGPHQ\Phi_s\rangle\langle\Phi_s QHP\Psi_0\rangle\right] \tag{6.2}$$

Using the fact the integrals on the RHS are related to the width or shift, we rewrite the RHS as

$$\langle\Phi_s Q\Psi_u\rangle = \frac{1}{E-\mathscr{E}_s}\left(\frac{\Gamma_s}{2k}\right)^{1/2}\left(1+\frac{\Delta_s}{E-E_s}\right) \tag{6.3a}$$

$$= \left(\frac{\Gamma_s}{2k}\right)^{1/2}\frac{1}{E-E_s} \tag{6.3b}$$

Now transpose Eq. (6.3b) to read

$$E_s = E - \frac{[\Gamma_s(E)/2k]}{\langle\Phi_s Q\Psi_u\rangle} \tag{6.3c}$$

Observe that the second term on the right is a function of E and all the quantities can be calculated in precision fashion. The resonance position E_s is obtained at the energy where the second term is zero. In order to determine all the quantities in Eq. (6.3c), $\mathscr{E}$ and Φ, precisely (henceforth dropping the subscript s), Ho *et al.*[8] obtained the minimized $\mathscr{E}$ from a Rayleigh–Ritz variational principle by using a Hylleraas-type trial function given in Eq. (5.2a). Specifically, for a ^1S state, it is given by

$$\Phi(\mathbf{r}_1,\mathbf{r}_2) = e^{-\alpha r_1-\beta r_2}\sum_{l,m,n} C_{lmn} r_1^l r_2^m r_{12}^n + (1\leftrightarrow 2) \tag{6.4}$$

(This problem was discussed in Section 5.) The authors found that the minimum value of $\mathscr{E}$ is obtained when $\alpha=\beta$, an expected result, since the lowest resonance is dominated by configurations $2s^2+2p^2$. The results of minimizing QHQ for various numbers of terms are shown in Table 1. They correspond to the number of terms such that

$$1+m+n=\omega \tag{6.5}$$

where $\omega = 4, 5, 6, 7$, and 8. The corresponding number of terms are $N = 22, 34$, 50, 70, and 95 in the expansion in Eq. (6.4). We obtained the lowest eigenvalue for $N = 95$ when $\alpha=\beta=0.49$, and this has converged to six significant figures.

To solve for $P\Psi$, the optical potential in Eq. (2.16)

$$\mathscr{V}_{\text{op}} = \sum_n \frac{PHQ\Phi_n\rangle\langle\Phi_n QHP}{E-\mathscr{E}_n} \tag{6.6}$$

TABLE 1
Optimization of the Lowest $H^-(^1S)$ Eigenvalue of QHQ^a

	N				
$\alpha=\beta$	22	34	50	70	95
0.39	−0.29748096	−0.29752739	−0.29755590	−0.29756211	−0.29756491
0.40	−0.29748709	−0.29753143	−0.29755779	−0.29756301	−0.29756536
0.41	−0.29747690	−0.29753635	−0.29756032	−0.29756429	−0.29756603
0.435	−0.29744880	−0.29753725	−0.29756125	−0.29756490	−0.29756638
0.45	−0.29739988	−0.29753509	−0.29756132	−0.29756527	−0.29756663
0.46	−0.29735282	−0.29753132	−0.29756081	−0.29756539	−0.29756675
0.47	−0.29729131	−0.29752502	−0.29755973	−0.29756539	−0.29756684
0.48	−0.29721227	−0.29751539	−0.29755796	−0.29756525	−0.29756689
0.49	−0.29711215	−0.29750145	−0.29755529	−0.29756494	−0.29756689
0.50	−0.29698708	−0.29748199	−0.29755146	−0.29756440	−0.29756685

[a]Results are in Ry.

was constructed by Ho *et al.*[14] from the spectrum of eigenvalues and eigenfunctions obtained by calculating QHQ. Because the electrons are indistinguishable, we must normalize $P\Psi$ as follows

$$P\Psi = \frac{1}{\sqrt{2}}[U_l(\mathbf{r}_1)\phi_0(\mathbf{r}_2) + U_l(\mathbf{r}_2)\phi_0(\mathbf{r}_1)] \tag{6.7}$$

The optical-potential equation thus reduces to the integro differential equation for $u_l(r)$

$$\left[\frac{d^2}{dr^2} - \frac{l(l+1)}{r^2} + k^2 + V_{\text{st}}\right]u_l(r) + \int [V_{\text{ex}}(r,r') + V_{\text{op}}(r,r')]u_l(r')dr' = 0 \tag{6.8}$$

where $U_l(r) = (1/r)u_l(r)Y_{l0}(\Omega)$.

The static and exchange potentials V_{st} and V_{ex} constitute the well-known (static) exchange approximation. They are given in many places, starting with Morse and Allis.[5] The optical-potential V_{op} in Eq. (6.8) is obtained by premultiplying Eq. (6.6) by $<\phi_0 Y_{l0}$ to obtain

$$\int V_{\text{op}}(r,r')u_l(r')dr' = r\sum_{n=1}^{N} \frac{\langle\phi_0 Y_{l0} PHQ\Phi_n\rangle\langle\Phi_n QHP\Psi\rangle}{E - \mathscr{E}_n} \tag{6.9}$$

The equation for the nonresonant continuum $P\Psi_0$ is obtained by omitting the $n=s=1$ term V_{op}; using the form

$$P\Psi_0^{(s)} = \frac{1}{\sqrt{2}}[\tilde{U}_l(\mathbf{r}_1)\phi_0(\mathbf{r}_2) + \tilde{u}_l(\mathbf{r}_2)\phi_0(\mathbf{r}_1)] \tag{6.10}$$

Thus, we obtain an analogous equation for $\tilde{u}(l=0)$

$$\left[\frac{d^2}{dr^2}+k^2+V_{\rm st}(r)\right]\tilde{u}(r)+\int[V_{\rm ex}(r,r')+\tilde{V}_{\rm op}(r,r')]\tilde{u}(r')dr'=0 \tag{6.11}$$

where

$$\int \tilde{V}_{\rm op}(r,r')\tilde{u}(r')dr' = r\sum_{n(\neq s)}\frac{\langle\phi_0 PHQ\Phi_n\rangle\langle\Phi_n QHP\Psi_0\rangle}{E-\mathscr{E}_n} \tag{6.12}$$

Equations (6.8) and (6.11) were solved by converting them to integral equations and numerically solving the resulting equations.[8,15] This is done by using the Green's function associated with the plane-wave part of those equations. The Green's function is, in analogy to Eq. (2.28) ($r_>, r_<$ are the greater and lesser of r and r')

$$g(r,r') = -\frac{1}{k}\sin kr_< \cos kr_> \tag{6.13}$$

so that the integral equation for the solution of Eq. (6.8) is

$$u(r) = \sin kr + \int g(r,r')U(r,r')u(r'')dr'dr'' \tag{6.14}$$

where U is the symbolic sum of the potentials $V_{\rm st}$, $V_{\rm ex}$, and $V_{\rm op}$. A similar equation is obtained for $\tilde{u}(r)$ by replacing $V_{\rm opt}$ by $\tilde{V}_{\rm op}$ in U on the RHS of Eq. (6.14)

$$\tilde{u}(r) = \sin kr + \int g(r,r')\tilde{U}(r,r')\tilde{u}(r'')dr'dr'' \tag{6.15}$$

As radial counterparts of the functions $P\Psi_0\rangle$ and $P\Psi_{\rm u}\rangle$ of Eqs. (2.39) and (2.40), the functions $\tilde{u}(r)$ and $u(r)$ are normalized as

$$\lim_{r\to\infty}\tilde{u}(r) = \sin(kr+\eta_0) \tag{6.16a}$$

$$\lim_{r\to\infty}u(r) = \frac{\sin(kr+\eta)}{\cos\eta_r} \tag{6.16b}$$

Here, the nonresonant and total phase shifts η_0 and η actually come from solutions of the respective integral in Eqs. (6.14) and (6.15). The resonant phase η_r in Eq. (6.16b) is

$$\eta_r = \eta - \eta_0 \tag{6.17}$$

Ho *et al.*[8] obtained $u(r)$ and $\tilde{u}(r)$ at $N'=48$ Gaussian integration points by solving a system of approximating algebraic equations. They used 16 Gauss–Legendre points for the range from the origin to $r=R$ and 32 Gauss–Laguerre points to cover the remaining range R to $111.75+R$. The results were checked by changing R from 3.5 to 4.5 and 5.5. From the solutions of $u(r)$

and $\tilde{u}(r)$ for each E, $P\Psi$ and $P\Psi_0$ can be calculated as well as the shift Δ as a function of E.

Referring to Eq. (6.3c), we plot

$$y_1 = E - \frac{[\Gamma(E)/2k]}{\langle \Phi Q \Psi \rangle} = \mathscr{E} + \Delta(E) \tag{6.18a}$$

$$y_2 = E \tag{6.18b}$$

as a function of E. In Fig. 1, these curves are shown with the nonlinear parameters $\alpha = \beta = 0.49$. The intersection of $y_1(E)$ and $y_2(E)$ defines the resonant energy E_r. It is noted that the specific values of $y_1(E)$ form a straight line to high precision and E_r is obtained to the desired accuracy. From E_r and $\mathscr{E}$, $\Delta(E_r)$ is

$$\Delta(E_r) \equiv E_r - \mathscr{E} \tag{6.19}$$

However, it should be stressed that $\Delta(E_r)$ is essentially only of theoretical interest, but by knowing E_r, we have from y_1 the Feshbach width

$$\Gamma(E_r) = \Gamma_{\rm F}$$

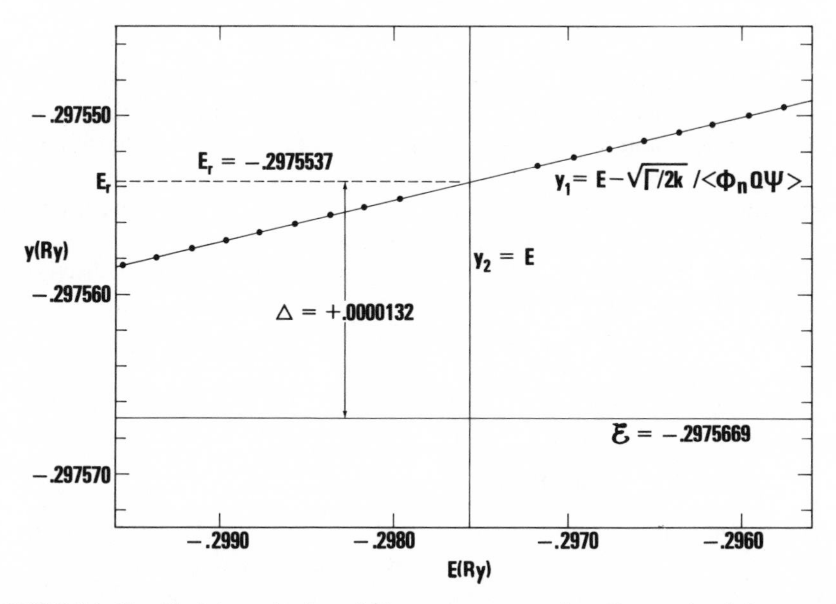

FIGURE 1. Graphical determination of ^1S resonance energy in e-H scattering [cf. text and Ref. 8].

TABLE 2
Convergence Behavior of Resonant Parameters and Test of Numerical Accuracy[a]

	N	34 ($\omega=5$)	50 ($\omega=6$)	70 ($\omega=7$)	95 ($\omega=8$)	
	$\mathscr{E}$	9.5580730	9.5573405	9.5572093	9.5571826	9.5571731[b] 9.5571758[c]
$N'=40$	E_r	9.5582611	9.5575476	9.5573679	9.5573617	
	Δ	+0.0001881	+0.0002071	+0.0001586	+0.0001791	
$R=4.5$	Γ	0.0476415	0.0470261	0.0470590	0.0470605	
$N'=48$	E_r	9.5582636	9.5575501	9.5573703	9.5573642	
	Δ	+0.0001906	+0.0002096	+0.0001610	+0.0001816	
$R=3.5$	Γ	0.0476482	0.0470327	0.0470656	0.0470670	
$N'=48$	E_r	9.5582615	0.5575480	0.5573683	9.5573621	
	Δ	+0.0001885	+0.0002076	+0.0001590	+0.0001795	
$R=4.5$	Γ	0.0476418	0.0470263	0.0470593	0.0470608	
$N'=48$	E_r	9.5582580	9.5575446	9.5573648	9.5573587	
	Δ	+0.0001850	+0.0002041	+0.0001555	+0.0001704	
$R=5.5$	Γ	0.0476327	0.0470173	0.0470502	0.0470517	

[a]Results are based on $\alpha=\beta=0{\cdot}49$. See text for definitions of N' and R. Units are in eV (1 Ry = 13.605826 eV).
[b]From $\delta_\omega - \delta_{\omega-1} = c\omega^P$.
[c]From $\delta_\omega - \delta_{\omega-1} = c'a^\omega$.

Results taken from our previous calculation are presented in Tables 2 and 3. The reader is referred to our paper[8] for a detailed discussion of extrapolations and other tests of the accuracy of the results. Suffice it here to restate that we believe that the calculation presently represents the most accurate calculation of a resonance. We would like only to reemphasize the point that this calculation also includes the correction between Feshbach and Breit–Wigner resonance parameters as discussed at the end of Section 4. Specifically, we find

$$\left.\frac{\partial\Delta}{\partial E}\right|_{E=E_r} = 0.002328\,\text{eV} \tag{6.20}$$

Thus, noting from Table 2 that $\Gamma_F(E_r) = 0.0470605$, we find from Eq. (4.9)

$$\Gamma_{BW} = 0.04717 \pm 0.00002\,\text{eV}$$

The second-order correction to Δ, Eq. (4.10b), however, as implicit in Section 4, is of higher order in E_r if $E_r \gg \Gamma$, which generally is the case. Thus, we find that

$$E_{BW} = E_F = 9.55735 \pm 0.00005\,\text{eV}$$

TABLE 3
^1S Phase Shifts and Widths in the Vicinity of the ^1S Resonance[a]

E (eV)	k (a.u.)	$\eta(E)$ (rad)	$\eta_0(E)$ (rad)	$\Delta(E)$ (eV)	$\Gamma(E)$ (eV)
9.529972	0.836919	1.56723	0.85618	1.162(−4)	0.0470794
9.532693	0.837038	1.61929	0.85621	1.225(−4)	0.0470776
9.535415	0.837158	1.67771	0.85625	1.287(−4)	0.0470757
9.538136	0.837277	1.74323	0.85629	1.350(−4)	0.0470739
9.540857	0.837397	1.81658	0.85633	1.413(−4)	0.0470720
9.543578	0.837516	1.89829	0.85636	1.476(−4)	0.0470702
9.546299	0.837635	1.98861	0.85640	1.538(−4)	0.0470683
9.549020	0.837755	2.08726	0.85644	1.601(−4)	0.0470665
9.551742*	0.837874	2.19329	0.85648	1.664(−4)	0.0470646
9.562626*	0.838351	2.64701	0.85665	1.918(−4)	0.0470572
9.565347	0.838471	2.75394	0.85669	1.981(−4)	0.0470553
9.568069	0.838590	2.85368	0.85673	2.045(−4)	0.0470534
9.570790	0.838709	2.94519	0.85678	2.109(−4)	0.0470515
9.573511	0.838828	3.02808	0.85682	2.172(−4)	0.0470496
9.576232	0.838948	3.10256	0.85686	2.236(−4)	0.0470477
9.578953	0.839067	3.16913	0.85691	2.301(−4)	0.0470458
9.581674	0.839186	3.22848	0.85695	2.365(−4)	0.0470439
9.584396	0.839305	3.28139	0.85700	2.429(−4)	0.0470420
9.587117	0.839424	3.32860	0.85705	2.493(−4)	0.0470401

[a]Mesh is 0.0002 Ry from $\mathscr{E}$ except those marked *, which are for 0.0004 Ry. The width in this table is the Feshbach width, which is to be distinguished from the Breit–Wigner width. See, for example, Table 4 and discussion. These values of E correspond to the points in Fig. 1 ($\alpha=\beta=0.49$; $N=95$ terms).

TABLE 4
Lowest ^1S resonance of H^-[a]

Author[b]	Reference	Method	E_r(eV)	(eV)
Burke and Taylor	12	Scattering	9.5603	0.0475
Shimamura	18	Kohn variational	9.5574	0.0472
Bardsley and Junker	16	Complex rotation	9.5572	0.0474
Bhatia	21	Stabilization	9.5570	0.0458
Doolen *et al.*[c]	15	Complex rotation	9.5570	
Bhatia and Temkin	7	QHQ and polarized orbital	9.5564 ± 0.002	0.0472
Chung and Chen	5	Uncorrelated Kohn–Feshbach	9.5569	0.04725
Present work		This calculation	9.55735 ± 0.00005	0.04717 ± 0.00002
		($\mathscr{E} = 9.557175 \pm 0.2 \times 10^{-5}$)		
		($\Delta = +0.00018 \pm 0.5 \times 10^{-4}$)		
Experiment				
Sanche and Burrow	20		9.558 ± 0.010	
Williams[d]			9.557 ± 0.01	0.045 ± 0.005

[a]Results are in eV (1 Ry = 13.605826 eV; see Ref. 24).
[b]Reference numbers are those given in Ref. 8.
[c]G. D. Doolan, J. Nuttal, and R. W. Stagat, *Phys. Rev. A* **10**, 1612 (1974).
[d]J. F. Williams, *J. Phys. B* **1**, L56 (1974).

The results compare excellently with the most accurate experiments[16,17] and well serve as a standard of comparison for other calculations (Table 4).

7. REVIEW OF SEMIPRECISION CALCULATIONS OF OTHER FUNDAMENTAL AUTOIONIZATION STATES OF He AND H^-

We have previously discussed the significance of the ^1P autoionization states of He**(^1P) as fundamental checks of the continuum solutions of the Schrödinger equation. In this section, we discuss our calculation of these plus D and S states of He and H^-. Although these calculations are not quite at the level of rigor and precision of the calculation of the ^1S resonance of H^- discussed in the Section 6, they still represent a very high level of accuracy that has very recently been strikingly confirmed by a new measurement of the lowest and most fundamental He**($2s2p$) autoionization state.[28] Specifically, we shall see that this new measurement confirms a whole series of calculations that had persistently deviated from a previous measurement whose central value had long been believed canonical, without sufficient attention having been paid to the concomitant experimental error.

In this connection, let us note at the outset that a radiative-absorption profile described by the RHS of Eq. (3.1) has its maximum absorption at $\varepsilon = q^{-1}$, which can be derived by simple differentiation. This corresponds to

$$E_{\max\,\mathrm{abs}} = E_s + \frac{\Gamma_s}{2q} \tag{7.1}$$

a fact that must be included when accurate comparison with experiment is made of E_s.

The results to be described were all based on Hylleraas variational calculations of $\Phi_{L,S}$ and yielded $\mathscr{E}$ to generally high precision. However, for the nonresonant continuum, Ψ_0 or $P\Psi_0$, as the case may be, we have used various approximations, which we discuss later, in integral expressions for the small quantities Γ_s, Eq. (2.42); Δ_s, Eq. (2.38); and q, Eq. (3.5a) et seq.

The nonresonant continua we have used correspond to three increasingly elaborate approximations of $P\Psi_0$ of Eq. (2.32): (1) the exchange approximation[5]

$$\Psi_0 \to \Psi^{\mathrm{ex}} = \frac{1}{\sqrt{2}}\left[\frac{u(r_1)}{r_1} Y_{l0}(\Omega_1)\phi_0(r_2) \pm (1 \leftrightarrow 2)\right] \tag{7.2}$$

(2) the polarized orbital approximation[18,19]

$$\Psi_0 \to \Psi_0^{(\mathrm{PO})} = \frac{1}{\sqrt{2}}\left\{\frac{\tilde{u}(r_1)}{r_1}[\phi_0(r_2) + \phi^{(pol)}(\mathbf{r}_1, \mathbf{r}_2)]\, Y_{l0}(\Omega_1) \pm (1 \leftrightarrow 2)\right\} \tag{7.3}$$

where

$$\phi^{(\text{pol})} = -\frac{\varepsilon(r_1,r_2)}{r_1^2}\frac{u_{1s\to p}(r_2)}{r_2}\frac{\cos\theta_{12}}{\sqrt{\pi}} \tag{7.4}$$

and

$$u_{1s\to p}(r_2) = r_2 e^{-Zr_2}(r_2 + \tfrac{1}{2}Zr_2^2) \tag{7.5}$$

(Z is the nuclear charge); (3) the (nonresonant) polarized pseudo-state approximation[20]

$$\Psi_0 \to \Psi_0^{(\text{pss})} = \frac{1}{\sqrt{2}}\Bigg[\frac{F_1(r_1)}{r_1} Y_{10}(\Omega_1)\phi_{1s}(r_2) + \sum_{l=0,2}\frac{F_l(r_2)}{r_2}\bar{\phi}_{2p}(r_2) y_{1l,1}(\hat{r}_1,\hat{r}_2) + (1 \leftrightarrow 2)\Bigg] \tag{7.6a}$$

$\Psi_0^{(\text{pss})}$ has been written explicitly for total P waves, which is the only case to which it has been applied; $\bar{\phi}_{2p}$ is defined and discussed below. The angular factor $y_{1l,1}$ in Eq. (7.6a) represents two ways a p state can be coupled to a p-scattered wave to form a total P wave

$$y_{1l,1} \equiv \sum_m (1lm - m|10)\, Y_{1,m}(\Omega_1)\, Y_{1,-m}(\Omega_2) \tag{7.6b}$$

The justifications of these approximations should, at this point be clear; however, we shall discuss them briefly. The exchange approximation in Eq. (7.2) is clearly the first nontrivial approximation; it satisfies the equation

$$(PHP - E)\Psi^{\text{ex}} = 0 \tag{7.7a}$$

and it is automatically in P space

$$P\Psi^{\text{ex}} = \Psi^{\text{ex}} \tag{7.8}$$

The only nonobvious fact about this approximation is that has a discrete as well as a continuous spectrum.[20] In fact, for charged targets ($Z \geqslant 2$), the discrete spectrum is (convergingly) infinite; the discrete functions satisfy

$$(PHP - E_\nu^{\text{ex}})\Psi_\nu^{\text{ex}} = 0, \qquad E_\nu^{\text{ex}} < -Z^2 \tag{7.7b}$$

The $\Psi_\nu^{(\text{ex})}$ are of the form in Eq. (7.2) where, however, the functions u become the exponentially damped function, $u \to u_\nu(r)$, which are orthonormal

$$\langle \Psi_\mu^{\text{ex}} \Psi_\nu^{\text{ex}} \rangle = \langle u_\nu u_\mu \rangle = \delta_{\mu\nu} \tag{7.9}$$

These discrete terms contribute, for example, to the shift; if we make a spectral expansion of the Green's function, then[20]

$$\Delta = \frac{1}{2\pi}\left\{\sum_\nu \frac{\Gamma_\nu}{E_s - E_\nu} + \mathscr{P}\int \frac{\Gamma(E')dE'}{E - E'}\right\} = \Delta_b + \Delta_c \tag{7.10}$$

where $\mathscr{P}$ represents a principal value integral and Γ_v are the discrete counterparts of the width function in Eq. (2.42)

$$\Gamma_v = 2\pi |\langle P\Psi_v | H | Q\Phi_s \rangle|^2 \tag{2.42$'$}$$

In all our calculations where a discrete contribution to Δ has been included, the exchange approximation for Γ_v has been used irrespective of what approximation has been used for $\Gamma(E')$.

As compared to the exact equation for $P\Psi_0$, which in a slightly more symbolic form than Eq. (2.19a) satisfies

$$\left[PHP + \sum_{n(\neq s)} \frac{PHQ\Phi_n \rangle \langle \Phi_n QHP}{E - \mathscr{E}_n} - E \right] P\Psi_0 = 0 \tag{7.11}$$

the exchange approximation in Eq. (7.7a) lacks the nonresonant-optical-potential terms

$$\mathscr{V}^{(s)} = \sum_{n(\neq s)} \frac{PHQ\Phi_n \rangle \langle \Phi_n QHP}{E - \mathscr{E}_n} \tag{7.12}$$

The additional approximations we have considered attempt in an increasingly incisive manner to incorporate the effect of $\mathscr{V}^{(s)}$. The polarized orbital approximation[18,19] identifies this contribution with induced polarization and incorporates it via Eq. (7.3) in well-known ways. The nonresonant polarized pseudo state is adapted from the approximation of Burke *et al.*[21], which is itself a variant of the closely coupled polarized orbital approximation of Damburg and Karule.[22] We have discussed that variation in a previous article;[23] suffice it here to state that the nonresonant character of Eq. (7.6) comes from identifying the resonant state with the ϕ_{2p}, ϕ_{2s} states of the target and taking $\tilde{\phi}_{2p}$ as a polarized (orbital) pseudostate that is orthogonal to them. This is done by letting

$$\tilde{\phi}_{2p}(r) = \frac{\tilde{u}_{1s\to p}(r)}{r} \tag{7.13}$$

where

$$\tilde{u}_{1s\to p}(r) = a u_{1s\to p}(r) + b R_{2p}(r) \tag{7.14}$$

and a and b are such that $\langle \tilde{u}_{1s\to p}(r) R_{2p}(r) \rangle = 0$. With regard to polarized orbital and polarized pseudostate approximations, the functions $\Psi_0^{(\mathrm{PO})}$ and $\Psi_0^{(pss)}$ are what come out of the calculation, but $P\Psi(\neq \Psi)$ is obtained by the operation of P, Eq. (2.13b), on them, and $P\Psi$ is what is used in calculating the specific-resonant quantities.

We shall not go into detail about the numerical solutions nor the reduction of the formulas to actual calculational form, since that has been discussed in our previous papers.[20,24] Rather, we shall give a compilation of

TABLE 5
^3P and ^1D Resonances of H^-

		State		
Our calculations		^3P		^1D
$\mathscr{E}$(Ry)		-0.2851969		-0.256174
$\mathscr{E}$(eV)[a]		9.72549		10.12037
Nonresonant approx.		Pol. orb.		Pol. orb.
Δ_c(eV)		0.0130		0.0040
Δ_b		—		—
$\Delta = \Delta_c + \Delta_b$		0.0130		0.0040
E(eV)		9.7385		10.1244
Γ(eV)		0.0063		0.010
Other results	$E_{^3P}$	$\Gamma_{^3P}$	$E_{^1D}$	$\Gamma_{^1D}$
Scattering				
Burke, Ref. 13	9.7417	0.0059	10.1267	0.0088
Stabilization[b]	9.7403	0.0049		
Sanche and Burrow,	9.738	0.0056	10.128	0.0073
Ref. 6	± 0.010	± 0.0005	± 0.010	± 0.002
Williams[c]	9.735	0.0060		
		± 0.0005		

[a]Result relative to ground state of H using $R = 13.605826$ eV.
[b]A. K. Bhatia, *Phys. Rev. A* **9**, 9 (1974); **10**, 729 (1974).
[c]J. F. Williams, *J. Phys. B* **7**, L56 (1974).

our final results for the states mentioned at the beginning of this section, taking the opportunity to correct various typographical and other errors in one or two of those entries.

In Table 5, we give our results for two higher ($^3P, {}^1D$) resonances of H^-(e − H). The nonresonant continuum used then was the polarized orbital approximation. Note that H^- has no bound-state spectrum in these symmetries, so that the shift has only a continuum part. The agreement with the two experiments cited,(16,17) which are generally conceded to be the most accurate experiments, is very satisfactory.

Coming now to helium, we show in Table 6 our results for the $^{1,3}S$, ^3P, and $^{1,3}D$ states.(24) The nonresonant continua used are as noted in footnote *c*, and the agreement with experiment [cf. in particular the compilation of Martin(25)] are in all cases within the experimental error, which is quoted by Martin as being of the order ± 0.05 eV. It is only important to note that to the extent these states, being dipole forbidden from the ground state, are observed through electron impact, it is appropriate to use the rydberg of infinite mass

TABLE 6

Summary of First Four Autoionization States of He of S, P, and D Angular Momentum (Excluding ^1P)

	λ	$\mathscr{E}_\lambda$(Ry)	$\mathscr{E}$(eV)[a]	Γ(eV)	Δ_c(eV)	Δ_b(eV)	Δ(eV)	E_∞(eV)
^1S	1	−1.5576265	57.8223	1.25×10^{-1}	1.177×10^{-2}	9.346×10^{-3}	2.112×10^{-2}	57.8435
	2	−1.2454971	62.0691	6.67×10^{-3}	5.307×10^{-3}	1.664×10^{-2}	2.195×10^{-2}	62.0911
	3	−1.1801597	62.9581	3.87×10^{-2}	3.355×10^{-3}	9.959×10^{-4}	4.351×10^{-3}	62.9624
	4	−1.0964678	64.0968	2.412×10^{-3}	1.478×10^{-3}	2.748×10^{-3}	4.226×10^{-3}	64.1010
^3S	1	−1.2052107	62.6173	4.21×10^{-5}	4.73×10^{-8}	2.061×10^{-5}	2.066×10^{-5}	62.6173
	2	−1.1195332	63.7830	7.09×10^{-6}	1.97×10^{-5}	8.294×10^{-5}	1.026×10^{-4}	63.7831
	3	−1.0976990	64.0800	1.22×10^{-4}	6.60×10^{-5}	2.353×10^{-4}	3.013×10^{-4}	64.0803
	4	−1.0650259	64.5246	1.32×10^{-4}	3.24×10^{-6}	1.594×10^{-4}	1.626×10^{-4}	64.5247
^3P	1	−1.522983	58.2937	8.90×10^{-3}	1.455×10^{-2}	1.267×10^{-2}	2.720×10^{-2}	58.3209
	2	−1.169776[b]	63.0994	2.61×10^{-3}	5.013×10^{-3}	2.219×10^{-3}	7.232×10^{-2}	63.1066
	3	−1.158012[b]	63.2594	4.88×10^{-5}	4.730×10^{-6}	6.825×10^{-5}	7.298×10^{-5}	63.2595
^1D	1	−1.405634	59.8903	0.0729	0.0220[c]	0.0023[c]	0.0243	59.9146
	2	−1.138752	63.5215	0.0187	0.0044	0.65×10^{-4}	0.0045	63.5259
	3	−1.112855	63.8738	5.81×10^{-4}	1.815×10^{-4}	0.10×10^{-4}	1.915×10^{-4}	63.8740
	4	−1.073241	64.4128	7.12×10^{-3}	1.667×10^{-3}	0.56×10^{-5}	1.673×10^{-3}	64.4145
^3D	1	−1.167611	63.1288	2.72×10^{6}[c]	2.60×10^{-4}[c]	9.91×10^{-4}	1.25×10^{-3}	63.1301
	2	−1.121361	63.7581	1.92×10^{-4}	9.17×10^{-5}	3.62×10^{-6}	9.53×10^{-5}	63.7582
	3	−1.083336	64.2755	3.31×10^{-6}	1.20×10^{-4}	2.78×10^{-4}	3.98×10^{-4}	64.2759
	4	−1.066831	64.5000	1.36×10^{-4}	5.02×10^{-5}	5.50×10^{-6}	5.57×10^{-5}	64.5001

[a] Results in eV relative to the ground state of He −5.80744875 Ry [C. L. Pekeris, *Phys. Rev.* **146**, 48 (1966)] using $R = 13.605826$ eV from B. N. Taylor, W. H. Parker, D. N. Langenberg, *Rev. Mod. Phys.* **41**, 375 (1969).

[b] From Chen and Chung, Ref. 14.

[c] 1, 3, D widths and Δ_c have been calculated with polarized orbital nonresonant functions. All other results use the exchange approximation.

(R_∞) to effect the conversion to electron volts. Only for photo-excitation experiments should R_M be used; this has been shown in the appendix of Ref. 24.

We now come to the final set of calculations that we wish to discuss here—the ^1P autoionization states of He (Table 7). Here, we also calculated q in view of the fact that the experiment, being one of photoabsorption, also determines q. Our most recent calculation[26] has now included all the different contributions (bound, continuum, and other resonant) to q that we discussed in Section 3. These contributions are small for the wide resonances ($s = 1,3$), but note the relative size of δq_b for the narrow resonance ($s = 2$). All these calculations used the exchange approximation for the nonresonant continuum after we were convinced of the reliability of that approximation. This was done by comparing the calculation of Δ with one involving the nonresonant polarized pseudostate continuum.[20] The value of Δ given in Table 7 uses the latter but differs from the exchange approximation value by

TABLE 7
Final Results for Resonance Parameters

	He		
Parameter	$s=1$	$s=2$	$s=3$
q_0	−3.139	−3.556	−2.849
δq_c	3.677(−1)	1.678	3.635(−1)
δq_b	−7.631(−2)	−2.636	−1.129(−1)
δq_r	−9.892(−4)	−9.387(−2)	−4.595(−3)
q	−2.849	−4.606	−2.604
E(eV)[a]	60.1450	62.7594	63.6610
Γ(eV)[a]	0.0363	1.06(−4)	0.009
Δ(eV)[a]	−0.00729		
		Ref. 27	
q	-2.80 ± 0.25		-2.0 ± 1.0
E(eV)	60.133 ± 0.015	62.756 ± 0.010	63.656 ± 0.007
Γ(eV)	0.038 ± 0.004		0.008 ± 0.004
		Ref. 28	
q	-2.55 ± 0.16		-2.5 ± 0.5
E(eV)	60.151 ± 0.010		63.655 ± 0.01
Γ(eV)	0.038 ± 0.002		0.0083 ± 0.002

[a]Results for E and Γ for He from our previous paper [Ref. 24].

less than 0.15%. That may be anomalously small, but the corresponding widths differ by about 1.5%, which is probably a more realistic estimate of the accuracy of the exchange approximation.

The reason we went to the trouble (with the valuable collaboration of Prof. Burke) of using the pseudostate approximation(20) for Δ was the persistent deviation of our previous results for the position E_r from the central value of the classic experiment of Madden and Codling.(27) However, as we have indicated, the calculated value has persisted. We are, therefore, very gratified that in a recent experiment by Morgan and Ederer,(28) done at the National Bureau of Standards in a division headed by Dr. Madden, a new value of E_r has been found that is completely in accord with our calculations of that quantity over the years. While trying to avoid any hubris and emphasizing that the assessment of the comparative inherent experimental accuracy is not within our purview, we must nevertheless conclude that the agreement with the recent experiment does, indeed, provide strong support for the essential correctness and cogency of the Feshbach formalism, which it has been the purpose of this chapter to elucidate.

REFERENCES

1. H. Feshbach, *Ann. Phys. (N.Y.)* **19**, 287 (1962).
2. U. Fano, *Phys. Rev.* **124**, 1866 (1961).
3. Y. Hahn, T. F. O'Malley, L. Spruch, *Phys. Rev.* **128**, 932 (1962).
4. T. F. O'Malley and S. Geltman, *Phys. Rev.* **137**, A1344 (1965).
5. P. M. Morse and W. P. Allis, *Phys. Rev.* **44**, 269 (1933).
6. R. P. Madden and K. Codling, *Astrophys. J.* **141**, 364 (1965).
7. G. Breit and E. P. Wigner, *Phys. Rev.* **49**, 519, 642 (1936).
8. Y. K. Ho, A. K. Bhatia, and A. Temkin, *Phys. Rev. A* **15**, 1432 (1977).
9. R. J. Drachman, *Phys. Rev. A* **15**, 1430 (1977).
10. A. K. Bhatia, A. Temkin, and J. F. Perkins, *Phys. Rev.* **153**, 177 (1967).
11. A. K. Bhatia and A. Temkin, *Rev. Mod. Phys.* **36**, 1050 (1964).
12. P. G. Burke and H. M. Schey, *Phys. Rev.* **126**, 149 (1962).
13. P. G. Burke, in Invited Papers of the Fifth International Conference on the Physics of Electronic and Atomic Collisions 1967 edited by P. M. Branscomb, p. 128, Univ. of Colorado, Boulder 1968.
14. K. T. Chung and J. C. Y. Chen, *Phys. Rev. A* **13**, 1655 (1976).
15. Y. K. Ho and P. A. Fraser, *J. Phys. B* **9**, 3213 (1976).
16. L. Sanche and P. D. Burrow, *Phys. Rev. Lett.* **29**, 1639 (1972).
17. J. F. Williams, *J. Phys. B* **7**, L 56 (1974); *Electron and Photon Interactions with Atoms* (edited by H. Kleinpoppen and M. R. C. McDowell, p. 309, Plenum, New York 1976).
18. A. Temkin, *Phys. Rev.* **107**, 1004 (1957).
19. A. Temkin and J. C. Lamkin, *Phys. Rev.* **121**, 788 (1961).
20. A. K. Bhatia, P. G. Burke, and A. Temkin, *Phys. Rev. A* **8**, 21 (1973); **10**, 459 (1974).
21. P. G. Burke, D. F. Gallaher, and S. Geltman, *J. Phys. B* **2**, 1142 (1969).
22. R. J. Damburg and E. Karule, *Proc. Phys. Soc. (London)* **90** 637 (1967).
23. R. J. Drachman and A. Temkin, in *Case Studies in Atomic Collision Physics*, vol. 2 (edited by E. McDaniel and M. R. C. McDowell, p. 399, North-Holland, Amsterdam 1972).
24. A. K. Bhatia and A. Temkin, *Phys. Rev. A* **11**, 2018 (1975).
25. W. C. Martin, *J. of Phys. and Chen. Ref. Data* **2**, 257 (1973).
26. A. K. Bhatia and A. Temkin, *Phys. Rev. A* **29**, 1875 (1984).
27. R. P. Madden and K. Codling, *Astrophys. J.* **141**, 364 (1965).
28. H. D. Morgan and D. Ederer, *Phys. Rev. A* **29**, 1901 (1984).

CHAPTER TWO

PROJECTION AND QUASI-PROJECTION OPERATORS FOR ELECTRON IMPACT RESONANCES ON MANY-ELECTRON ATOMIC TARGETS

A. Temkin and A. K. Bhatia

1. INTRODUCTION

In Chapter 1, we dealt with the theory and applications of the projection-operator formalism to two-electron systems. The reason for singling out two-electron autoionization systems (implying a one-electron target) was given in Chapter 1; to repeat briefly, the fact that the target eigenfunctions are known analytically and the special character of a one-electron target allow exact projection operators in the Feshbach sense[1] to be written down, which thereby allows precision calculations of the composite (two-electron) system approaching the accuracy of ordinary two-electron bound-state systems to be carried out [cf., for example, Ref. 2]. In the case of autoionization of two- or more electron target systems, the inability of achieving comparable precision arises from two causes: not only can we not give the exact eigenfunctions analytically, but we cannot even supply a formal, explicit expression for the projection operators that we need at the outset.

The problem of writing down a projection operator even formally arises because of the requirement of antisymmetry: the operator must preserve the indistinguishability of scattered and orbital electrons. It might be thought that we could simply symmetrize the obvious asymmetric operator, but such a procedure would destroy idempotency, which is a necessary property in the original Feshbach theory (we shall see later, however, that it can be dispensed with to some extent).

A. TEMKIN and A. K. BHATIA ■ Atomic Physics Office, Laboratory for Astronomy and Solar Physics, Goddard Space Flight Center, National Aeronautics and Space Administration, Greenbelt, MD 20771.

Feshbach[1] also dealt with the problem of antisymmetry; unfortunately, his formal solution to the problem was incomplete or, more accurately, heuristically intended, in that he did not include spin coordinates at all, and angular-momentum considerations are at best implicit. In addition, the final form of his projection operators contains a partial antisymmetrizer on both sides of a projector, which is very confusing until we note that the operator itself can be put into a manifestly symmetric form.

One object then of this chapter is to put Feshbach's derivation and notation into a complete form, so that we can directly construct and apply the projection operator in an actual calculation. Of course, to do so will involve approximating the eigenfunction of the target and, having done that, solving for the eigensolutions of an ensuing homogeneous equation, in terms of whose solutions the projection operator is expressed. The practicality of this procedure will be demonstrated for the open-shell approximation of the He-like targets. This, hopefully, will provide not only a valuable pedagogical example of how the construction works in practice but also a useful (and first) idempotent operator of the simplest nonseparable approximation of the He-like (ground-state) target (Section 2).

We shall also derive explicitly the kernel of the auxiliary integral equation using a Hylleraas-type of target-state wave function. Although we have not been able to solve the eigenvalue equation analytically, we can readily write down a variational principle for the eigensolutions. Since they are one-dimensional functions, such variational (or numerical) solutions may be more convenient to work with for calculational purposes.

In Section 3, we shall drop the question of idempotency and review the ideas associated with quasi-projection operators (QPO). This review will be brief, since most of the basic material has already been covered in the literature. One interesting sidelight is the fact that our new notation for projection operators (above) leads naturally to the same form of QPOs that we had originally introduced quite independently of Feshbach's theory.

In the case of QPOs, the formalism can be explicitly generalized to treat autoionization states that lie in the region of inelastic scattering, and this is also reviewed. Special attention is given to a recent calculation of the lowest ${}^2P^0$ resonance of He^-. This state is of interest because it was thought to be a shape resonance. We shall, therefore, take this opportunity to discuss the difference between these kinds of resonances from a novel point of view (in terms of the literature; we have, in fact, discussed this for many years). In the case of the above ${}^2P^0$ resonance, although Feshbach (i.e., core excited) in nature, it is a special type that can occur only in many-electron targets. It is suggested that some other resonances, heretofore considered shape resonances, may be of this special Feshbach type.

From the spectrum of these quasi-projection or projection-operator calculations, we can readily construct a (quasi-) optical potential that we would expect to lead to a convergent method of calculating nonresonant phase shifts. In practice, this scheme (and various modifications) have never worked out. We conclude this chapter with a preliminary explanation of this phenomenon (Section 4).

2. IDEMPOTENT RESONANCE PROJECTION OPERATIONS

2.1. *Notation and Preliminaries*

We start by considering the scattering of an electron from an atomic target (atom or ion) consisting of N electrons. The total (time-independent) wave function Ψ has the asymptotic form

$$\lim_{r_i \to \infty} \Psi = (-1)^{p_i} \frac{\sin(kr_i - \pi l/2 + \eta_l)}{kr_i} \psi_0(r^{(i)}). \tag{2.1.1}$$

The Ψ is a fully antisymmetric function of the coordinates of all $i = 1, 2, \ldots, N+1$ electrons

$$\Psi = \Psi(x_1, \ldots, x_i, \ldots, x_j, \ldots, x_{N+1}) = -\Psi(x, \ldots, x_j, \ldots, x_i, \ldots, x_{N+1}),$$

where x_i are the totality of coordinates (spin and space) of the ith electron

$$x_i \equiv (\mathbf{r}_i, s_i) \tag{2.1.2}$$

A *cyclic* permutation of the particle labels i that brings x_i into the first position is denoted by p_i

$$p_i \equiv \begin{pmatrix} 1, 2, \ldots, & i, \ldots, & N+1 \\ i, i+1, \ldots, & N+1, 1, \ldots, & i-1 \end{pmatrix} \tag{2.1.3}$$

[This implies that $(-1)^{p_i}$ is the parity of p_i; in particular, $(-1)^{p_1} = 1$ always; cf. below.] The channel functions ψ_0 in Eq. (2.1.1) is defined as the target function ϕ_0 coupled to the orbital angular and spin of the scattered particle

$$\psi_0(r^{(i)}) = \sum_{\text{mag q.n.}} (l_i L_0 m_i M_0 | L M_L)(\tfrac{1}{2} S_0 m_s M_{S_0} | S M_S) Y_{l_i m_i}(\Omega_i) \chi^{(i)}_{1/2 m_s} \phi_0(x^{(i)}) \tag{2.1.4}$$

Rotation invariance and spin independence of the Hamiltonian will guarantee that the physical results we shall be concerned with here are independent of the total magnetic quantum numbers M_L and M_S. In general, we shall take them to be the most convenient values commensurate with the magnetic quantum numbers of the target M_0 and M_{S0}.

In Eq. (2.1.4), $x^{(i)}$ is the collection of the coordinates of the N electrons of the target *with* x_i *missing*

$$x^{(i)} \equiv (x_{i+1}, x_{i+2}, \ldots, x_{N+1}, x_1, \ldots, x_{i-1}) \tag{2.1.5}$$

In a similar fashion, we define $r^{(i)}$ as the collection of coordinates with only the radial coordinate r_i of the ith electron missing

$$r^{(i)} \equiv (\Omega_i s_i, x_{i+1}, \ldots, x_{N+1}, x_1, \ldots, x_{i-1}) \tag{2.1.6}$$

so that

$$dx^{(i)} = \frac{1}{dx_i} \prod_{j=1}^{N+1} dx_j \tag{2.1.7a}$$

and

$$dr^{(i)} = d\Omega_i \, ds_i \, dx^{(i)} \tag{2.1.7b}$$

For later use, we also define (for $i \neq 1$)

$$dx^{(1i)} = \frac{1}{dx_1 dx_i} \prod_{j=1}^{N+1} dx_j \tag{2.1.7c}$$

2.2. *Derivation of the Idempotent Projection Operator*

The basic projection operator of the Feshbach theory[1] is the P operator, whose basic property is that it does not alter the asymptotic form of the wave function; i.e., for any i,

$$\lim_{r_i \to \infty} P\Psi = \lim_{r_i \to \infty} \Psi \tag{2.2.1}$$

This condition by itself does not determine a unique P; indeed, P is physically not uniquely determinable[1]. For example, the operator

$$P \equiv \begin{cases} 0 & r_j < R \quad \text{for all } j = 1, \ldots, N+1 \\ 1 & r_i > R \quad \text{for any } i \end{cases}$$

where R is some large radius is not only perfectly acceptable (clearly $P^2 = P$) but, in fact, very useful for nuclear physics, where the forces and the nuclei really do have a rather well-defined boundary characterized by R. However, for atomic physics, such a definition is very unrealistic, and the essential point of Feshbach's main definition of P is that it is truly independent of R. That P is derived from two independent conditions

$$P\Psi \equiv \sum_{i=1}^{N+1} (-1)^{p_i} u(r_i) \psi_0(r^{(i)}), \qquad i = 1, 2, \ldots, N+1 \tag{2.2.2}$$

and

$$\langle \psi_0(r^{(1)}) | (1 - P)\Psi \rangle_{r_1} = 0 \tag{2.2.3a}$$

The subscript r_1 indicates the coordinate not integrated over, i.e.,

$$\langle \psi_0(r^{(1)})\Psi \rangle_{r_1} \equiv \int \psi_0(r^{(1)})\Psi \, dr^{(1)} \tag{2.2.4}$$

The function $u(r_i)$ has the asymptotic form

$$\lim_{r_i \to \infty} u(r_i) = \frac{\sin(kr_i - \pi l/2 + \eta_l)}{kr_i} \tag{2.2.5}$$

(Note that in this section, radial functions are *not* multiplied by r, even though we use lower-case letters for them.) In connection with Eq. (2.2.2), it is important to observe that *the definition of $P\Psi$ is more than a statement of the asymptotic form of Ψ: Equation (2.2.2) defines $P\Psi$ for all values of all r_i once the functions $u(r_i)$ have been defined.* In fact, Eq. (2.2.2) corresponds to the form of the well-known static-exchange approximation;[3] in general, $u(r_i)$ is the solution of an integrodifferential equation that may even go beyond the exchange approximation, and it is not simply a sinusoidal function for small values of r_i.

The object of this exercise is to obtain a specific form of P that will be similarly valid for all values of r_i. To this end, we shall substitute Eq. (2.2.1) into Eq. (2.2.3a) and manipulate the resulting expression in such a way that $\Psi\rangle$ appears as the right-most factor. We can then read everything to its left as Q. Note, again, that Eq. (2.2.3a) is a stronger condition than (but includes) the asymptotic requirement

$$\lim_{r_i \to \infty} Q\Psi\rangle = 0 \tag{2.2.3b}$$

Having identified Q, we can trivially find

$$P = 1 - Q \tag{2.2.3c}$$

Let us proceed with this manipulation. Equation (2.2.3a) can be rewritten as

$$\langle \psi_0(r^{(1)})\Psi \rangle_{r_1} = \langle \psi_0(r^{(1)})P\Psi \rangle_{r_1} \tag{2.2.3d}$$

where the superscript again emphasizes that r_1 is not integrated over. Define the left-hand side (for general i) to be

$$w(r_i) \equiv (-1)^{p_i} \langle \psi_0(r^{(i)})\Psi \rangle_{r_i} \tag{2.2.6}$$

and substitute Eq. (2.2.2) into the right-hand side of Eq. (2.2.3d). This gives

$$w(r_1) = u(r_1) + \sum_{i=2}^{N+1} (-1)^{p_i} \int u(r_i) \langle \psi_0(r^{(1)})\psi_0(r^{(i)}) \rangle_{r_1 r_i} r_i^2 \, dr_i \tag{2.2.7a}$$

where $(i \neq 1)$

$$\langle \psi_0(r^{(1)})\psi_0(r^{(i)}) \rangle_{r_1 r_i} = \int \psi_0^*(r^{(1)})\psi_0(r^{(i)}) \, dx^{(1i)} \, d\Omega_1 \, ds_1 \, d\Omega_i \, ds_i \tag{2.2.8}$$

In deriving Eq. (2.2.7a), we have used $\langle \psi_0(r^{(1)})\psi_0(r^{(1)})\rangle_{r_1} = 1$ and the definition of $dx^{(1i)}$ in Eq. (2.1.7c). Keep in mind that in $\langle \psi_0(r^{(1)})\psi_0(r^{(i)})\rangle_{r_1 r_i}$, the subscripts mean that the object is a function of r_1 and r_i.

By virtue of the antisymmetry of ψ_0 in the indices $x_2, \ldots, x_{N+1}$ every term of the sum in Eq. (2.2.7a) gives the same contribution, so that we can rewrite $w(r_1)$

$$w(r_1) = u(r_1) - \int K(r_1 | r_2) u(r_2) r_2^2 dr_2 \tag{2.2.7b}$$

Here, we have used the definition of the kernel (for general indices i and j)

$$K(r_i | r_j) \equiv (-1)^{p_i + p_j + 1} N \langle \psi_0(r^{(i)})\psi_0(r^{(j)})\rangle_{r_i r_j}. \tag{2.2.9}$$

The next step in the manipulating Eq. (2.2.3a) is to diagonalize K, toward that end, we define the integral equation eigenvalue problem

$$v_\alpha(r_1) = \lambda_\alpha \int_0^\infty K(r_1 | r_2) v_\alpha(r_2) r_2^2 dr_2 \tag{2.2.10a}$$

Assume that Eq. (2.2.10a) can be solved for its eigenvalues λ_α and its eigenfunctions $v_\alpha(r)$. By using the hermiticity of K, it is easy to show in the usual way that the eigenfunctions can be made orthonormal

$$\langle v_\alpha v_\beta \rangle = \int_0^\infty v_\alpha(r) v_\beta(r) r^2 dr = \delta_{\alpha\beta} \tag{2.2.10b}$$

Comparing the last two equations shows that K can be written as

$$K(r_1 | r_2) = \sum_{\beta=1}^{n_\lambda} \frac{v_\beta(r_1) v_\beta(r_2)}{\lambda_\beta} \tag{2.2.11}$$

where n_λ is the number of eigenvalues (which rigorously may be infinite if K is nonseparable; we shall discuss that problem in Section 2.4).

We digress for a moment to point out that solutions of the integral equation (2.2.10a) have played an important role in quantum chemistry, where they were introduced by Löwdin[4] and are known as natural orbitals. It is not our purpose to go into that area, although it may well be that some of those developments may be of use for the specific problems that arise here. As far as we know, the quantum chemical literature does not deal explicitly with the role of natural orbitals in constructing projection operators.

Before proceeding further, let us also note the following sum rule concerning the sum of the inverses of the eigenvalues λ_α

$$\sum_{\alpha=1}^{n_\lambda} \frac{1}{\lambda_\alpha} = (-1)^{p_2+1} N \langle \chi_S(1; 2, \ldots, N+1) \chi_S(2; 3, \ldots, N+1) \rangle \tag{2.2.12}$$

The LHS may be derived from Eq. (2.2.11) and the RHS from Eq. (2.2.9). The spin inner product is to be interpreted literally for only $N=1$ and $N=2$ electron targets, but the generalization for $N>2$ may be readily derived from Eq. (2.2.9) using the general expression for ψ_0 in Eq. (2.1.4). We shall verify this sum rule in specific cases later.

Returning to the derivation of P, we substitute Eq. (2.2.11) into Eq. (2.2.7b) to obtain

$$w(r_1)=u(r_1)-v_1(r_1)\langle v_1 u\rangle-\sum_{\lambda_\beta\neq 1}\frac{v_\beta(r_1)\langle v_\beta u\rangle}{\lambda_\beta} \tag{2.2.13}$$

where $v_1(r_1)$ is the eigenfunction corresponding to $\lambda_\beta=1$ (if any). Premultiply Eq. (2.2.13) by each specific v_β, using Eq. (2.2.10b) to obtain

$$\langle v_\beta u\rangle=\frac{\lambda_\beta}{\lambda_\beta-1}\langle v_\beta w\rangle, \qquad \lambda_\beta\neq 1 \tag{2.2.14}$$

so that by substituting into Eq. (2.2.13), we have

$$u(r_1)=w(r_1)+v_1(r_1)\langle v_1 u\rangle+\sum_{\lambda_\beta\neq 1}\frac{v_\beta(r_1)\langle v_\beta w\rangle}{\lambda_\beta-1} \tag{2.2.15}$$

Keeping in mind our aim of replacing $u(r_1)$ by $w(r_1)$ [because $w(r_1)$ directly involves Ψ in Eq. (2.2.6)], we see this is effectively done in Eq. (2.2.15), since the one term (v_1) that does not involve w makes no contribution; viz.,

$$\left\langle \psi_0(r^{(1)})\left|\sum_{i=1}^{N+1}(-1)^{p_i}v_1(r_i)\psi_0(r^{(i)})\right.\right\rangle_{r_1}$$

$$=v_1(r_1)-\int_0^\infty K(r_1|r_2)v_1(r_2)r_2^2dr_2=0 \tag{2.2.16}$$

where the fact that the RHS of Eq. (2.2.16) vanishes follows from Eq. (2.2.10), because by definition, $\lambda_\alpha=1$. Thus, substituting the remaining part of Eq. (2.2.15)

$$u(r_1)\to w(r_1)+\sum_{\lambda_\alpha\neq 1}\frac{v_\alpha(r_1)\langle v_\alpha w\rangle}{\lambda_\alpha-1} \tag{2.2.17}$$

into $\langle\psi_0 P\Psi\rangle_{r_1}$ gives

$$\langle\psi_0(r^{(1)})P\Psi\rangle_{r_1}$$

$$=\left\langle \psi_0(r^{(1)})\left|\sum_{i=1}^{N+1}(-1)^{p_i}\left[w(r_i)+\sum_{\lambda_\alpha\neq 1}^{n_\alpha}\frac{v_\alpha(r_i)\langle v_\alpha w\rangle}{\lambda_\alpha-1}\right]\right|\psi_0(r^{(i)})\right\rangle \tag{2.2.18a}$$

Now, to exhibit the explicit dependence of w as an inner product involving

$\Psi\rangle$, we use Eq. (2.2.6) in Eq. (2.2.18a)

$$\langle\psi_0(r^{(1)})|P\Psi\rangle = \left\langle \psi_0(r^{(1)})\middle|\sum_i(-1)^{p_i}\left[(-1)^{p_i}\langle\psi_0(r^{(i)})|\Psi\rangle\right.\right.$$

$$\left.\left.+\sum_{\lambda_\alpha\neq 1} v_\alpha(r_i)\frac{\langle v_\alpha(-1)^{p_i}\langle\psi_0(r^{(i)})\Psi\rangle\rangle}{\lambda_\alpha - 1}\right]\psi_0(r^{(i)})\right\rangle \quad (2.2.18b)$$

And, finally, rearranging factors on the RHS of Eq. (2.2.18b) so that $\Psi\rangle$ appears last, we have [noting that $(-1)^{2p_i} = 1$]

$$\langle\psi_0(r^{(1)})|P\Psi\rangle_{r_i} = \left\langle \psi_0(r^{(1)})\middle|\sum_i\left[\psi_0(r^{(i)})\rangle\langle\psi_0(r^{(i)})\right.\right.$$

$$\left.\left.+\sum_{\lambda_\alpha\neq 1}\frac{v_\alpha(r_i)\psi_0(r^{(i)})\rangle\langle v_\alpha(r_i)\psi_0(r^{(i)})}{\lambda_\alpha - 1}\right]\Psi\right\rangle \quad (2.2.18c)$$

Comparing the LHS and RHS of Eq. (2.2.18c), we have our desired explicit representation of the P operator

$$P = \sum_{i=1}^{N+1}\left[\psi_0(r^{(i)})\rangle\langle\psi_0(r^{(i)}) + \sum_{\lambda_\alpha\neq 1}\frac{\psi_0(r^{(i)})v_\alpha(r_i)\rangle\langle v_\alpha(r_i)\psi_0(r^{(i)})}{\lambda_\alpha - 1}\right] \quad (2.2.19a)$$

We can write Eq. (2.2.19a) in a somewhat more efficient form by factoring out the $\psi_0\rangle\langle\psi_0$ part

$$P = \sum_{i=1}^{N+1}\psi_0(r^{(i)})\rangle\left[1 + \sum_{\lambda_\alpha\neq 1}\frac{v_\alpha(r_i)\rangle\langle v_\alpha(r_i)}{\lambda_\alpha - 1}\right]\langle\psi_0(r^{(i)}) \quad (2.2.19b)$$

Let it first be noted that P is completely symmetric in the particle labels $i = 1, \ldots, N+1$ and therefore will not alter the symmetry of any function Ψ that is completely antisymmetric. For example, if Ψ is derived from another function Ψ' of the same symmetry (but otherwise arbitrary)

$$\Psi = P\Psi' \quad (2.2.20)$$

then from Eq. (2.2.3d), we obtain

$$\langle\psi_0(r^{(1)})P\Psi'\rangle_{r_1} = \langle\psi_0(r^{(1)})P^2\Psi'\rangle_{r_1} \quad (2.2.21)$$

The property expressed in Eq. (2.2.21) is *not* sufficient to show that P is effectively idempotent for *scattering calculations*. To obtain that condition, we require the satisfaction of the well-known optical-potential scattering equation

$$\langle\psi_0(H_{PP} + V_{\text{op}} - E)P\Psi\rangle = \langle\psi_0(H_{PP} + V_{\text{op}} - E)P^2\Psi\rangle \quad (2.2.22)$$

This requirement is equivalent to the condition that arbitrary matrix elements of the operators P and Q, where

$$Q = 1 - P \tag{2.2.3c}$$

and P are given by Eq. (2.2.19), are such that arbitrary *matrix elements* of P^2 and Q^2 are equal to the same elements of P and Q

$$\langle \Phi_1 P^2 \Phi_2 \rangle = \langle \Phi_1 P \Phi_2 \rangle \tag{2.2.23a}$$

$$\langle \Phi_1 Q^2 \Phi_2 \rangle = \langle \Phi_1 Q \Phi_2 \rangle \tag{2.2.23b}$$

The only requirement is that the functions Φ_1 and Φ_2 be completely (anti)symmetric in their $(N+1)$ particle indices. We have demonstrated elsewhere[5] that these forms of P and Q have that property. But it is to be emphasized that Eqs. (2.2.23a) and (2.2.23b) will not hold if the functions Φ_1 and Φ_2 are not totally antisymmetric. It is for this reason that the operators P and Q are not formally idempotent as operator identities.

The net result is that the variational solutions of QHQ written in its conventional Schrödinger-like form

$$QHQ\Phi_n = \mathscr{E}_n \Phi_n \tag{2.2.23b}$$

will be idempotent with respect to Q and ultimately the same will be true for $\mathscr{V}_{00}$.

Finally, we discuss the relationship of our "idempotent" P operator to that derived by Feshbach[1]. His P operator is

$$P = \frac{\mathfrak{A}}{(N+1)^{1/2}} \left\{ \phi_0(x^{(j)}) \rangle \langle \phi_0(x^{(j)}) \right. \\ \left. + \sum_{\lambda_\alpha \neq 1} \frac{v_\alpha(x_j)\phi_0(x^{(j)}) \rangle \langle v_\alpha(x_j)\phi_0(x^{(j)})}{\lambda_\alpha - 1} \right\} \frac{\mathfrak{A}}{(N+1)^{1/2}} \tag{2.2.24}$$

In Eq. (2.2.24), the target-wave functions ϕ_0 are seen to have their arguments in one specific order $[x^{(j)} = x_{j+1}, x_{j+2}, \ldots, x_{N+1}, x_1, \ldots, x_{j-1}]$, although the ϕ_0 are assumed to be antisymmetric in those (N) particle coordinates. The $\mathfrak{A}$ is the antisymmetrizer of an $(N+1)$ electron wave function in the form of a scattering wave function

$$\mathfrak{A}\{u(x_i)\phi(x^{(i)})\} \equiv \sum_{j=1}^{N+1} (-1)^{p_j} u(x_j)\phi(x^{(j)}) \tag{2.2.25}$$

To show the formal equivalence to our form of P, we note that if Ψ is already antisymmetric in all $N+1$ electron coordinates, then

$$\mathfrak{A}\Psi \equiv \sum_{i=1}^{N+1} (-1)^{p_i} \Psi(x_i, x^{(i)}) = (N+1)\Psi(x_1, \ldots, x_{N+1}) \tag{2.2.26}$$

since by antisymmetry

$$\Psi(x_i, x^{(i)}) = (-1)^{p_i}\Psi(x_1, x_2, \ldots, x_{N+1}) \tag{2.2.27}$$

Thus, in operating with Feshbach's P,

$$P\Psi = \frac{\mathfrak{A}}{(N+1)^{1/2}}\{\Pi_j\}\frac{\mathfrak{A}}{(N+1)^{1/2}}\Psi = \mathfrak{A}\{\Pi_j\Psi(x_1, x_2, \ldots, x_{N+1})\} \tag{2.2.28}$$

where Π_j is the operator in curly bracket in Eq. (2.2.25)

$$\Pi_j = \phi_0(x^{(j)})\rangle\langle\phi_0(x^{(j)}) + \sum_{\lambda_\alpha \neq 1} \frac{\phi_0(x^{(j)})v_\alpha(r_j)\rangle\langle v_\alpha(r_j)\phi_0(x^{(j)})}{\lambda_\alpha - 1} \tag{2.2.29}$$

Using the definition of $\mathfrak{A}$, we can develop Eq. (2.2.28) further

$$P\Psi = \sum_{i=1}^{N+1} (-1)^{p_i}\Pi_i\Psi(x_i, x^{(i)}) \tag{2.2.30}$$

But now, we use the antisymmetry of Ψ again to obtain finally [with $(-1)^{2p_i} = 1$]

$$P\Psi = \left[\sum_{i=1}^{N+1} \Pi_i\right]\Psi(x_1, x_2, \ldots, x_{N+1})\rangle \tag{2.2.31}$$

The operator in square brackets is formally identical to our P operator in Eq. (2.2.19).

The difference in the two ways of writing P in Eq. (2.2.19) or (2.2.22) amounts to symmetrizing an operator in calculating matrix elements of symmetric operators between unsymmetrized wave functions or the reverse. The more important practical difference is the fact that our operators are defined in terms of channel functions ψ_0 where the spin and angular parts of the incident partial wave are explicitly and appropriately coupled to target state eigenfunctions ϕ_0. This allows explicit operators to be deduced in specific cases, as we shall now proceed to do.

2.3. *Realization of P for the One-Electron Target*

We deal here with the case $N = 1\,(N + 1 = 2)$. Let $i \neq j = 1, 2$; the channel function in Eq. (2.1.4) reduces to (ϕ_0 is assumed to be the $1s$ ground state)

$$\begin{aligned} \psi_0(r^{(1)}) &= Y_{l0}(\Omega_1)\chi_S(2,1)\phi_0(r_2) \\ \psi_0(r^{(2)}) &= Y_{l0}(\Omega_2)\chi_S(1,2)\phi_0(r_1) \end{aligned} \tag{2.3.1}$$

where χ_S are the spin functions

$$\chi_S(i, j) = \begin{cases} 2^{-1/2}(\alpha_i\beta_j - \alpha_j\beta_i) & S = 0 \quad \text{(singlet)} \\ 2^{-1/2}(\alpha_i\beta_j + \beta_i\alpha_j) & S = 1, \quad M_S = 0 \quad \text{(triplet)} \end{cases} \tag{2.3.2}$$

Note that

$$\chi_S(2,1) = (-1)^{S+1}\chi_S(1,2) \tag{2.3.3}$$

from which $P\Psi$ in Eq. (2.2.2) assumes the conventional form of the exchange approximation

$$P\Psi = [u(r_1)\phi_0(r_2)Y_{l0}(\Omega_1) + (-1)^S u(r_2)\phi_0(r_1)Y_{l0}(\Omega_2)]\chi_S(1,2) \tag{2.3.4}$$

In deriving Eq. (2.3.4), we have also used the two permutations

$$p_1 = \begin{pmatrix} 1 & 2 \\ 1 & 2 \end{pmatrix} \qquad p_2 = \begin{pmatrix} 1 & 2 \\ 2 & 1 \end{pmatrix} \tag{2.3.5a}$$

so that

$$(-1)^{p_1} = 1 \qquad (-1)^{p_2} = -1 \tag{2.3.5b}$$

We next evaluate the kernel in Eq. (2.2.9), whose key element is

$$\begin{aligned}\langle \psi_0(r^{(1)})\psi_0(r^{(2)})\rangle_{r_1r_2} &= \phi_0(r_2)\phi_0(r_1)\int Y_{l0}^*(\Omega_1)\chi_S^*(2,1)\, Y_{l0}(\Omega_2)\chi_S(1,2) \\ &\quad \cdot d\Omega_1\, d\Omega_2\, ds_1\, ds_2 \\ &= \delta_{l0}\phi_0(r_2)\phi_0(r_1)(-1)^{S+1}(4\pi) \\ &= \delta_{l0}(-1)^{S+1}R_{1s}(r_1)R_{1s}(r_2)\end{aligned}$$

Thus,

$$K(r_1|r_2) = \begin{cases} (-1)^{S+1}4\pi\phi_0(r_1)\phi_0(r_2)\delta_{l0} \\ (-1)^{S+1}R_{1s}(r_1)R_{1s}(r_2)\delta_{l0} \end{cases} \tag{2.3.6}$$

where ϕ_0 is the normalized hydrogenic ground state (nuclear charge Z)

$$\phi_0(r) = \frac{1}{(4\pi)^{1/2}} R_{1s}(r) \tag{2.3.7a}$$

and R_{1s} is the radially normalized function

$$R_{1s}(r) = 2Z^{3/2}e^{-Zr} \tag{2.3.7b}$$

Before proceeding further, let us note that because $K \neq 0$ only for $l = 0$, so, too, is $v_\alpha \neq 0$ only for $l = 0$, thus, the second sum in P, Eq. (2.2.19), enters only for S states. We shall come back to that point later, but here we confine ourselves to $l = 0$.

The next step is to solve the integral eigenvalue equation (2.2.10), which here reduces to

$$v_\alpha(r_1) = \lambda_\alpha R_{1s}(r_1)(-1)^{s+1}\int R_{1s}(r_2)v_\alpha(r_2)r_2^2\,dr_2 \tag{2.3.8a}$$

It can readily be verified that this equation is satisfied by

$$v_\alpha(r) = R_{1s}(r)$$
$$\lambda_\alpha = (-1)^{S+1} \tag{2.3.8b}$$

Thus, the only $\lambda_\alpha \neq 1$ eigenvector we must worry about occurs for $S = 0$ (i.e., singlet states) for which—from the definition of the channel function in Eq. (2.3.1)—we have for the general form of the P operator in Eq. (2.2.19b)

$$P = (4\pi^{-2})\{\chi_0(2,1)\rangle[p_2 - \tfrac{1}{2}p_1p_2]\langle\chi_0(2,1) + \chi_0(1,2)\rangle[p_1 - \tfrac{1}{2}p_2p_1]\langle\chi_0(1,2)\} \tag{2.3.8c}$$

where we have used the proportionality of v_α to ϕ_0 in this case. In Eq. (2.3.8c), the p_j are purely spatial projectors

$$p_j = v_\alpha(r_j)\rangle\langle v_\alpha(r_j)$$

Using the fact that $p_1p_2 = p_2p_1$ and $\chi_0(1,2)\rangle\langle\chi_0(1,2) = \chi_0(2,1)\rangle\langle\chi_0(2,1)$, we can simplify P even further

$$P = (4\pi)^{-2}[p_1 + p_2 - p_1p_2]\chi_0(1,2)\rangle\langle\chi_0(1,2) \tag{2.3.9}$$

In the form of Eq. (2.3.9), P is essentially the well-known operator given by Hahn, O'Malley, and Spruch.[6] Their P lacks on the spin factor $\chi_0\rangle\langle\chi_0$, however, if P operates on an eigenfunction of good total S and L, where L means total angular momentum (and $L = l$ for scattering from the ground state of the hydrogenic target), then $\chi_0\rangle\langle\chi_0$ is redundant. Most important, their P_i is defined to mean integration over coordinates of the ith electron, whereas our p_i includes integration over $d\Omega_j$ as well. This accounts for the extra factor of $(4\pi)^{-1}$ in Eq. (2.3.9) as compared to their well-known form

$$P = P_1 + P_2 - P_1P_2 \tag{2.3.10}$$

The calculations and methodology of our previous article as well as a host of others have been explicitly or implicitly based on these projection operators. Historically, it is of interest to note that the form of Eq. (2.3.10) (and its complement Q) was given[6] without being derived from the general form in Eq. (2.2.19). Rather, it was incisively postulated by the authors[6] as satisfying all the requirements of the Feshbach theory including symmetry, idempotency, and the asymptotic properties. Subsequently, several researchers realized that this was a special case of Feshbach's general operator [cf. discussion following Eq. (2.2.3d)], in particular Hahn[7].

But as we shall see in the next section, although the Feshbach procedure for constructing P and Q for many-electron (specifically two-electron) targets can to varying extents be carried out, in no other case can P and Q have all the important properties and explicitness it does in the one-electron target.

2.4. *Determination of P for the Two-Electron Target*

2.4.1. Realizations for the Open-Shell Approximation. We derive here the projection operators for the two-electron (He-like) target, where the latter are described in the open-shell approximation

$$\phi_0(x_1, x_2) = N_{\mu\nu}^{-1}[e^{-\mu r_1 - \nu r_2} + e^{-\nu r_1 - \mu r_2}]\chi_0(1, 2) \tag{2.4.1}$$

The nonlinear parameters μ and ν are chosen to minimize the energy of the He-like target (ground-state) wave function. For He itself, the parameters are well-known: $\mu = 2.1832$, $\nu = 1.1886$; $N_{\mu\nu}$ is the associated normalization factor

$$N_{\mu\nu}^2 = (4\pi)^2[(2\mu\nu)^{-3} + 8(\mu + \nu)^{-6}] \tag{2.4.2}$$

The channel wave functions from Eq. (2.1.4) are readily constructed

$$\psi_0(r^{(1)}) = Y_{l0}(\Omega_1)\alpha_1\phi_0(x_2, x_3) \tag{2.4.3}$$

In Eq. (2.4.3), we have taken $M_L = 0$ ($\Rightarrow m_i = 0$, since $L_0 = M_0 = 0$), and we have let the total $m_s = +\frac{1}{2}$, represented by α_1 in Eq. (2.4.3). For the $N + 1 = 3$ electron system, the cyclic permutations p_1, p_2, p_3 of (123), (231), (312) are all of even parity

$$(-1)^{p_1} = (-1)^{p_2} = (-1)^{p_3} = 1 \tag{2.4.4}$$

thus, $P\Psi$ in Eq. (2.2.2) is

$$P\Psi = u(r_1)\psi_0(r^{(1)}) + u(r_2)\psi_0(r^{(2)}) + u(r_3)\psi_0(r^{(3)}) \tag{2.4.5}$$

where $\psi_0(r^{(i)})$ are obtained from $\psi_0(r^{(1)})$ by appropriate cyclic permutation of the subscripts 1, 2, 3 on the RHS of Eq. (2.4.3).

We must next derive and solve the homogeneous integral equations (2.2.10a) and (2.2.10b) for the auxiliary eigenfunctions $v_\alpha(r)$ from which P is constructed. The kernel of the integral equation is

$$K(r_1|r_2) = -2\langle\psi_0(r^{(1)})\psi_0(r^{(2)})\rangle \tag{2.4.6}$$

which is readily determined from Eqs. (2.4.1) and (2.4.3) to be

$$K(r_1|r_2) = \frac{2(4\pi)^2}{N_{\mu\nu}^2}\delta_{l0}\left[\frac{e^{-\mu(r_1+r_2)}}{(2\nu)^3} + \frac{e^{-\nu(r_1+r_2)}}{(2\mu)^3} + \frac{e^{-(\mu r_1 + \nu r_2)} + e^{-(\mu r_2 + \nu r_1)}}{(\mu + \nu)^3}\right] \tag{2.4.7}$$

For the open-shell target. The eigenfunctions $v_\alpha(r)$ of the fundamental integral equation (2.2.18) may then be readily ascertained to be

$$v_\alpha(r) = C_{\alpha\mu}e^{-\mu r} + C_{\alpha\nu}e^{-\nu r} \tag{2.4.8}$$

with eigenvalues

$$\frac{1}{\lambda_\alpha} = \tfrac{1}{2}(1 \pm 4(K_{\nu\nu}K_{\mu\mu})^{1/2} I_{\mu\nu}) \tag{2.4.9}$$

where $K_{\gamma\delta}$ are the coefficients of the $e^{-(\gamma r_1 + \delta r_2)}$ terms in $K(r_1|r_2)$ in Eq. (2.4.7) $(\gamma, \delta = \mu$ or $\nu)$ and

$$I_{\gamma\delta} = \int e^{-(\gamma+\delta)r} r^2\, dr = \frac{2}{(\gamma+\delta)^3} \tag{2.4.10}$$

[$\alpha = (1, 2)$ corresponds to $(+, -)$ in Eq. (2.4.9)]. It can again be verified that the eigenvalues satisfy the sum rule in Eq. (2.2.12). The coefficients of each eigenvector have the relationship, also determined by substituting into Eq. (2.2.10),

$$C_{\alpha\nu} = \pm \left(\frac{\mu}{\nu}\right)^{3/2} C_{\alpha\mu} \tag{2.4.11}$$

and the absolute normalization from $\langle v_\alpha v_\alpha \rangle = 1$ requires

$$C_{\alpha\mu}^2 I_{\mu\mu} + 2C_{\alpha\mu}C_{\alpha\nu}I_{\mu\nu} + C_{\alpha\nu}^2 I_{\nu\nu} = 1 \tag{2.4.12}$$

Equation (2.4.12) is a complete and explicit solution for the auxiliary eigenvectors and eigenvalues needed to construct the idempotent projection operator for the open-shell target

$$P = \sum_{i=1}^{3} \left[\psi_0(r^{(i)})\rangle\langle\psi_0(r^{(i)}) + \sum_{\alpha=1(\lambda_\alpha \neq 1)}^{2} \frac{v_\alpha(r_i)\psi_0(r^{(i)})\rangle\langle\psi_0(r^{(i)})v_\alpha(r_i)}{\lambda_\alpha - 1} \right]$$

where it is clear that as long as $\mu \neq \nu$ there is no $\lambda_\alpha = 1$ eigenvalue. (When $\mu = \nu$, then the v_α collapse into one eigenfunction for which $\lambda_\alpha = 1$, and there is no second sum in P. *But this must be considered carefully*.)

It is clear that the same procedure that we used in Section 2.4 can be extended to any configuration-interaction-type approximation for two-electron targets.

$$\phi_0(x_1, x_2) = \left[\sum_\lambda \tilde{\phi}_\lambda(r_1, r_2) P_\lambda(\cos\theta_{12}) \right] \chi_0(1, 2) \tag{2.4.13}$$

In such a function, only the $\lambda = 0$ relative partial wave will survive in the kernel in Eq. (2.2.9). If each $\tilde{\phi}_\lambda$ component can be analytically expressed as an exponential times powers

$$\tilde{\phi}_\lambda(r_1, r_2) = \frac{e^{-(\mu r_1 + \nu r_2)}}{(16\pi^2)^{1/2}} \sum_{m,n} C_{m,n}^{(\lambda)} r_1^m r_2^n + (1 \leftrightarrow 2) \tag{2.4.14}$$

then it is clear the resultant eigenvalue equation for $v_\alpha(r)$ and λ_α can be solved analytically in essentially the same way it is done for the open-shell target in Eq. (2.4.7).

Whether such a process would be worthwhile, particularly for the ^1S ground state of He or its isoelectronic ions—where the orbital electrons may be very close together—is less clear, because, as is well known, a configuration interaction expansion cannot properly describe the cusp in the wave function (Kato's theorem) engendered by the singularity in the Coulomb potential at $r_{12}=0$

$$\left(\frac{\partial\phi_0}{\partial r_{12}}=\frac{1}{2}\phi_0\right)_{r_{12}=0} \tag{2.4.15}$$

We therefore describe some developments for Hylleraas-type functions in the following section.

2.4.2. Developments for Hylleraas-Type Target Approximations. The most natural wave function that can describe such cusp behavior is a Hylleraas expansion, whose spatial part,

$$\phi_0(r_1,r_2,r_{12})=\frac{e^{-\gamma(r_1+r_2)}}{(16\pi^2)^{1/2}}\sum_{l,m,n} C_{lmn}(r_1^l r_2^m + r_1^m r_2^l)r_{12}^n \tag{2.4.16}$$

is seen to contain the coordinate r_{12} explicitly and can therefore obey the cusp condition.

It is of interest then to inquire whether an idempotent P operator can be constructed for such a ϕ_0, where the sum contains odd powers (n) of r_{12}. That is a problem of ongoing research by the authors and has not yet been completely solved. We shall here give a précis of developments we have thus far made, leaving further developments to future articles.

The first thing we have found is that the kernel can be evaluated in closed form. Specifically, Eq. (2.2.9) reduces to

$$K(r_1|r_2)=\left(\frac{1}{4\pi}\right)\int\phi_0(\mathbf{r}_2,\mathbf{r}_3)\phi_0(\mathbf{r}_3,\mathbf{r}_1)\,d\Omega_1\,d\Omega_2\,d^3r_3 \tag{2.4.17}$$

When we insert the Hylleraas expansions in Eq. (2.4.16) and carry out the angular expressions, we obtain

$$\begin{aligned}K(r_1|r_2)=4\pi e^{-\gamma(r_1+r_2)}\sum C_{lmn}C_{l'm'n'}[&r_2^l r_1^{m'}\mathscr{I}_{n,n'}^{(m+l')}(r_1|r_2)\\ &+r_2^l r_1^{l'}\mathscr{I}_{n,n'}^{(m+m')}(r_1|r_2)+r_2^m r_1^{m'}\mathscr{I}_{n,n'}^{(l+l')}(r_1|r_2)\\ &+r_2^m r_1^{l'}\mathscr{I}_{n,n'}^{(l+m')}(r_1|r_2)]\end{aligned} \tag{2.4.18}$$

where

$$\mathscr{I}_{n,n'}^{(p)}(r_1|r_2)=\frac{1}{(4\pi)^3}\int e^{-2\gamma r_3}r_{23}^n r_{13}^{n'} r_3^p\,d\Omega_1\,d\Omega_2\,d^3r_3 \tag{2.4.19a}$$

When either or both n, n' are odd, it may appear that the integral cannot be carried out in closed form, but it can—the trick is to use the expansion of Sack[8]

$$r_{ij}^n = \sum_\lambda F_\lambda^{(n)}(r_i, r_j) P_\lambda(\cos\theta_{ij}) \tag{2.4.20}$$

Although this expansion function $F_\lambda^{(n)}$ for $n =$ odd is infinite, when we carry out the angular integrals in Eq. (2.4.19a), only the $\lambda = \lambda' = 0$ components survive, thus,

$$\mathscr{I}_{n,n'}^{(p)}(r_1|r_2) = \int_0^\infty e^{-2\gamma r_3} F_0^{(n)}(r_2, r_3) F_0^{(n')}(r_1, r_3) r_3^{p+2} dr_3 \tag{2.4.19b}$$

The $F_0^{(n)}$ are finite polynomials that are a terminating hypergeometric series

$$F_0^{(n)}(r_i, r_j) = r_>^n F\left[\frac{-n}{2}, \frac{-(n+1)}{2}, \frac{3}{2}, \frac{r_<^2}{r_>^2}\right] \tag{2.4.21}$$

where as usual $r_<$ and $r_>$ are the lesser and greater of r_i, r_j. This means that the integral in Eq. (2.4.19b) divides itself into three parts: $r_3 < r_<, r_< < r_3 < r_>$, $r_< < r_> < r_3$; each part can be done in closed form. An example of one specific result is

$$\mathscr{I}_{01}^{(0)} = \frac{1}{4\gamma^5 r_>} \{[1 + (\gamma r_>)^2] - e^{-2\gamma r_<}[1 + \tfrac{1}{2}\gamma r_>]\} \tag{2.4.22}$$

That specific integral depends on only the greater of r_1 and r_2, but, in general, $r_<$ as well as $r_>$ will enter the result. However, we always have the property

$$\mathscr{I}_{n,n'}^{(p)}(r_1|r_2) = \mathscr{I}_{n',n}^{(p)}(r_2|r_1) \tag{2.4.23a}$$

Thus in Eq. (2.4.18), we can restrict the n, n' sum to $n > n'$ in the form

$$\begin{aligned}\sum_n \sum_{n'} & C_{lmn} C_{l'm'n'} \mathscr{I}_{n,n'}^{(p)}(r_1|r_2) \\ &= \sum_{n \geqslant n'} \sum (1 - \tfrac{1}{2}\delta_{nn'}) C_{lmn} C_{l'm'n'} [\mathscr{I}_{nn'}^{(p)}(r_1|r_2) + \mathscr{I}_{nn'}^{(p)}(r_2|r_1)]\end{aligned} \tag{2.4.23b}$$

where the $(1 - \frac{1}{2}\delta_{nn'})$ factor on the RHS of Eq. (2.4.23b) prevents double counting the $n = n'$ terms.

The factor in square brackets on the RHS of Eq. (2.4.23b) is now manifestly symmetric in r_1 and r_2 and can be explicitly evaluated; they are finite polynomials for any (n, n').

What is not clear at this point is whether the resultant kernel in Eq. (2.4.18) allows the homogeneous equation

$$v_\alpha(r_1) = \lambda_\alpha \int_0^\infty K(r_1 | r_2) v_\alpha(r_2) r_2^2 dr_2 \tag{2.2.10a}$$

to be solved in closed analytic form. Clearly, if it can, then resonances in electron scattering from two-electron target can be calculated with accuracy approaching those of the one-electron target, which, in view, of the special place of He as a target, would be very valuable for experimental purposes.

2.4.3. Variational Approach for Auxiliary Integral Equations. If Eq. (2.2.10a) cannot be analytically solved for a Hylleraas-generated kernel, then it is still possible to solve it numerically by solving Eq. (2.2.10a) over a discrete grid as a straightforward matrix eigenvalue problem. More generally, we can use analytic approximations of $v_\alpha(r)$ to determine the coefficients and the eigenvalues from the Rayleigh–Ritz variational principle

$$\delta\left[\frac{1}{\lambda_\alpha}\right] = 0 \tag{2.4.24a}$$

where from Eq. (2.2.10a), the functional is

$$\left[\frac{1}{\lambda_\alpha}\right] = \frac{\langle v_\alpha K v_\alpha \rangle}{\langle v_\alpha v_\alpha \rangle} \tag{2.2.24b}$$

Indeed, the variational principle in Eq. (2.4.24) can serve as a variational basis for solving the auxiliary functions v_α for more than two-electron targets—even in cases where they might in principle be solved analytically. The investigation of the utility and practicalities of this approach would suggest a fertile area for further study!

3. QUASI-PROJECTION OPERATORS

This section will be rather brief in view of the fact that the major ideas and some calculated results are already in the literature. We shall confine ourselves here to a few introductory remarks. In Sections 3.2, 3.3, and 3.4, we review the major aspects of the approach, as always trying to stress points that were not emphasized in our original articles. Section 3.5 concentrates more heavily on the $He^-(^2P^0)$ resonance in electron-helium (e-He) scattering not only because it was thought to be a shape resonance by those who first considered it, whereas it turned out to be a Feshbach resonance, but because it is a different kind of Feshbach resonance, which deserves special discussion in its own right. More important, because it is wide, it may very well be a prototype of a class of resonances that is of general importance in, say, its astrophysical consequences.

3.1. *Justification and Representations*

In Section 2, we saw that the derivation of idempotent projection operators that preserve exchange between incoming and orbital electrons are not a trivial matter. (As a practical matter, we do believe that as a result of the developments described in Section 2, we can now realistically approach the problem of calculating projection operators and autoionization states of three-electron systems in the elastic domain as accurately as in true bound states.) Beyond that, however, idempotent projectors are not yet likely to provide a useful calculational tool for the immediate future.

We would, nevertheless, like to have some formalism that would preserve the most important properties of projection operators and yet would not be so difficult to construct. Let us ask the question: What are the most important properties that we want to retain? From the point of resonances, our answer is that we want to have a mathematical object $\hat{Q}$ such that the eigenspectrum of $\hat{Q}H\hat{Q}$ is discrete, with eigenvalues $\mathscr{E}_n$ that are (close to) the positions of the resonances. For the two-electron system, the Feshbach theory[1], as initiated in the calculation of eigenvalues of QHQ by O'Malley and Geltman[9] and their indubitable proximity to resonances of the e-H, e-He^+ systems, the ideal object is clearly the idempotent projection operator Q, which, when operating on the exact (continuum) solution of the Schrödinger equation, makes it vanish asymptotically

$$\lim_{r_i \to \infty} Q\Psi = 0 \qquad i = 1, 2, \ldots, N+1 \tag{3.1.1}$$

The idea that underlies quasi-projection operators [10] is the observation that the asymptotic property in Eq. (3.1.1) alone (rather than the idempotency $Q^2 = Q$) guarantees a discrete spectrum of $\hat{Q}H\hat{Q}$. Once this point has been realized, we can readily construct an object $\hat{Q}$ (in fact, many forms of $\hat{Q}$) for which

$$\lim_{r_i \to \infty} \hat{Q}\Psi = 0 \qquad i = 1, 2, \ldots, N+1 \tag{3.1.2}$$

and that also respects exchange but for which idempotency, even in the Feshbach sense, does not hold, i.e. where

$$\langle \psi_0 \hat{Q}\Psi \rangle \neq 0 \tag{3.1.3}$$

A heuristic proof of the discreteness of the spectrum of $\hat{Q}H\hat{Q}$ is given in Ref. (10). To paraphrase it in a few words, we show that if the spectrum of $\hat{Q}H\hat{Q}$ were continuous, then that would contradict the orthonormality of the eigenfunctions $\langle \hat{Q}\Phi_n, \hat{Q}\Phi_m \rangle = \delta_{nm}$, quadratic integrability being a strong version of the condition in Eq. (3.1.2).

Having recognized the falling away of the requirement of idempotency, we can easily generate operators [which we have named[10] quasi-projection operators] with the requisite property in Eq. (3.1.2). Two such operators (in the elastic domain) are[10]

$$\hat{P}_b = \sum_{i=1}^{N+1} P_i, \qquad \hat{Q}_b = 1 - \hat{P}_b \tag{3.1.4}$$

and

$$\hat{Q}_a = \prod_{i=1}^{N+1} (1 - P_i), \qquad \hat{P}_a = 1 - \hat{Q}_a \tag{3.1.5}$$

where the projector P_i is defined by

$$P_i \equiv \psi_0(r^{(i)})\rangle\langle\psi_0(r^{(i)}) \tag{3.1.6}$$

Observe that Eq. (3.1.4) is the first part of the projection operator in Eq. (2.2.19). However, neither $\hat{Q}_a$ nor $\hat{Q}_b$ is idempotent even in the projected sense in Eq. (2.2.21).

The lower form of the operators $\hat{Q}_a$ and $\hat{P}_a$ is an obvious generalization of the operators [$Q = (1 - P_1)(1 - P_2)$, $P = 1 - Q$, in the preceding article] for $N = 1$ electron targets. In the $N = 1$ case, Q and P are the idempotent projection operators introduced by Hahn, O'Malley, and Spruch[6] and first used calculationally by O'Malley and Geltman[9]. However, for $N > 1$, they *are no longer idempotent*. [This fact was apparently not realized by O'Malley, who introduced Q_a and P_a independently and, in fact, earlier.[11]] It was also erroneously thought by us that $\hat{Q}_a$ would eliminate the ground $\phi_0(x^{(j)})$ in any subset of the N-electron coordinates [cf. Appendix C of Ref. (10)]. The point is that although $P\Psi$ in Eq. (2.2.2) appears to contain only the general state of the target, it does so in an explicitly antisymmetric way. Mathematically, however, this is somewhat misleading, as we shall briefly indicate. Because the set of eigenfunctions Φ_n (including the continuum) is complete, the function $P_a\Psi$ can be expanded in a formally *unantisymmetrized* form (with respect to any specific coordinate—call it 1)

$$\hat{P}_a\Psi = \sum_{n=0} \chi_n(r_1)\psi_n(r^{(1)}) \tag{2.2.2$'$}$$

The same can be done for the exact wave function itself

$$\Psi = \sum_{n=0} \theta_n(r_1)\psi_n(r^{(1)})$$

But in both cases, because the expansions are not explicitly antisymmetrized, both Ψ and $P\Psi$ will contain more than just the ground states ψ_0. And in addition, the sets $\chi_n(r_1)$ and $\theta_n(r_1)$ will be different from each other, and they will not be orthonormal. Thus, in general,

$$\langle\psi_0(r^{(1)})(1 - \hat{P}_a)\Psi\rangle = \chi_0(r_1) - \theta_0(r_1) \neq 0$$

and the operators P_a and Q_a will not be idempotent, even in the Feshbach sense. Only when we construct operators P and Q as in Section 2 will the functions $\chi_0(r_i)$ and $\theta_0(r_i)$ be, in effect, the same, so that $\langle \psi_0 Q\Psi \rangle = 0$.

What then are the advantages of quasi-projection operators? The main advantage of the quasi-projection operators we have introduced is that they do *not* require the (solution for the) auxiliary functions $v_\alpha(r_1)$—not to mention the complication of projecting with them, if they were known. A more practical advantage, however, is the fact that the quasi-projection operators $\hat{Q}_b$ and $\hat{P}_b$ (we shall henceforth drop the subscript b) can readily be generalized to calculate resonances in the inelastic domain,[12] and for more than two-electron targets. We shall complete this section with a review of inelastic quasi-projection operators.[12]

Define the generalized P operator

$$\hat{P}_{(\nu)} = \sum_{n=0}^{\nu} \sum_{i=1}^{N+1} P_i^{(n)} \equiv \sum_{n=0}^{\nu} P^{(n)} \tag{3.1.7a}$$

and its complement

$$Q_{(\nu)} \equiv 1 - \hat{P}_{(\nu)} \tag{3.1.7b}$$

The parenthesized subscript (ν) here gives the highest threshold of the target *above* which the eigenvalues of $Q_{(\nu)}HQ_{(\nu)}$ correspond to resonances. In other words, if an eigenvalue of $Q_{(\nu)}HQ_{(\nu)}$ lies above $\varepsilon_{\nu+1}$ [energy of the ($\nu+1$) state of target], then it has *no* meaning as a resonance. In Eq. (3.1.7), the generalized projectors $P_i^{(n)}$ are

$$P_i^{(n)} = \psi_n(r^{(i)})\rangle\langle\psi_n(r^{(i)}) \tag{3.1.8}$$

where ψ_n is the channel function associated with nth excited state of the target

$$\psi_n(r^{(i)}) = \sum (L_n l_i M_L m_i | L\bar{M}_L)(S_n \tfrac{1}{2} M_S m_s | S\bar{M}_S)\phi_n(x^{(i)}) Y_{l_i m_i}(\Omega_i)\chi^{(i)}_{1/2\, m_s} \tag{3.1.9}$$

With these definitions, we can proceed with the demonstration of Eq. (3.1.2); that limit is a sufficient condition for the existence of a discrete spectrum of $Q_{(\nu)}HQ_{(\nu)}$ below the ($\nu+1$) threshold. Let us rewrite Eq. (3.1.2) in the form

$$\lim_{r_i\to\infty} \hat{P}_\nu \Psi = \lim_{r_i\to\infty} \Psi \tag{3.1.2'}$$

The RHS of Eq. (3.1.2′) is the asymptotic form of the full (scattering) wave function; it can conveniently be expressed in terms of the channel functions ψ_n in Eq. (3.1.9)

$$\lim_{r_i\to\infty} \Psi = (-1)^{p_i} \sum_{n=0}^{\nu} \frac{\sin(k_n r_i + \sigma_n)}{k_n r_i} \psi_n(r^{(i)}) \tag{3.1.10}$$

The phase term σ_n is the sum of phase shift, angular momentum term, and

Coulomb piece if the target has a net charge. We want to show that the LHS of Eq. (3.1.2′) is equal to Eq. (3.1.10). Observe that

$$\lim_{r_j \to \infty} \psi_n(r^{(i)}) = \delta_{ij} \psi_n(r^{(j)}) \tag{3.1.11}$$

which stems from $\phi_n(x^{(i)})$ vanishing as $r_j \to \infty$ if r_j is among its arguments. This will be the case *unless* $r^{(i)} = r^{(j)}$ (i.e. $i = j$). In the latter case, the limit will have no effect, since the specific r_j is not contained in $\psi_n(r^{(i)})_{i=j}$. We also use the orthonormality of the target eigenfunctions (assuming, of course, they have the same arguments)

$$\langle \phi_n(x^{(j)}) \phi_{n'}(x^{(j)}) \rangle = \delta_{nn'} \tag{3.1.12}$$

With these properties, it is now straightforward to evaluate the limit on the LHS of Eq. (3.1.2) using the definition of the quasi-projection operator. The demonstration is left as an exercise for the reader; it is given explicitly in Ref. (12) [Eq. (2.2)].

3.2. *Spurious States*

Although the spectrum QHQ is discrete, it may contain a finite number of eigenvalues that do not correspond to Feshbach resonances. This section considers why such spurious states arise, how to infer their number before calculation, and how to identify them. After a calculation has been made, we shall also consider the three-electron systems in detail and include an interesting table from which much useful information about elements of the theory and calculation for three-electron systems can be deduced.

The basic reason that spurious states arise is the same reason that eigenvalues of $\langle \Psi H \Psi \rangle$ can arise anywhere in the continuous spectrum of H if we use an arbitrary (but quadratically integrable) ansatz for Ψ. For convenience, we shall first define an antisymmetric sum (indicated by subscript A) involving the product of N-particle antisymmetric functions with a one-particle function

$$[\psi(r^{(j)})\theta(r_j)]_A \equiv \sum_{j=1}^{N+1} (-1)^{p_j} \psi(r^{(j)})\theta(r_j) \tag{3.2.1}$$

Let us now expand such a Ψ

$$\Psi = \left\{ \sum_{n=0}^{v} \sum_{m} C_{nm} \psi_n(r^{(j)}) \theta_m(r_j) + \sum_{n=v+1} \sum_{m} C_{nm} \psi_n(r^{(j)}) \theta_m(r_j) \right\}_A \tag{3.2.2}$$

We have broken down the sum into two parts: one that contains and one that does not contain bound states up to the vth state of target, which we take to be the highest energy-accessible state, since we are here thinking in terms of calculating resonances in electron scattering. If we were to associate scattering wave functions with eigenfunctions of Ψ coming from the diagonalization of $\langle \Psi H \Psi \rangle$ (subject to the normalization condition $\langle \Psi\Psi \rangle = 1$), *those where the* C_{nm} *are dominant for* $n < v+1$ would, in fact, be approximations of ordinary scattering wave functions. The reason is that by definition ordinary scattering dominantly involves states that are energetically accessible (note that this includes shape resonances also). Depending on the completeness of Ψ, there could be an arbitrarily large number of such states, since the single-particle orbitals [summation $\theta_m(r_j)$] in Eq. (3.2.1) is unrestricted. However, if we could devise an operator $Q_{(v)}$ such that

$$Q_{(v)}\Psi = \left[\sum_{n=v+1} \sum_{m} \bar{C}_{nm} \psi_n(r^{(j)}) \theta_m(r_j) \right]_A \tag{3.2.3}$$

then diagonalization of $\langle \Psi Q_{(v)} H Q_{(v)} \Psi \rangle$ would eliminate all of these nonresonant eigenvalues, and any eigenstate that remained would necessarily be associated with a target state $\Psi_{n>v}$. If the energy E of that diagonalized eigenfunction were less than ε_{v+1} (i.e., $< \varepsilon_{v+1}$), then such a function would correspond to a real (but temporary) disturbance of the composite system, i.e., a Feshbach resonance! Mathematically, then, the property represented by Eq. (3.2.3.) is the ideal of the Feshbach theory; for that very reason, only the Feshbach Q operator will have that property. What in fact quasi projectors $\hat{Q}_{(v)}$ do is delimit the number of $n < v$ terms in Eq. (3.2.2) without eliminating them completely. And that is the number of spurious states [i.e., eigenfunctions of $Q_{(v)} H Q_{(v)}$ that do *not* correspond to resonances below the $(v+1)$ threshold of the target system].

The simplest example corresponds to the ^1S symmetry of the two-electron system. Our quasi projector $Q(=\hat{Q}_b)$ in Eq. (3.1.4) in that case is (suppressing the subscript of Q)

$$\hat{Q} = 1 - P_1 - P_2$$

[In that one case, our second form of the quasi projector $\hat{Q}_a$ in Eq. (3.1.5) is the idempotent projection operator of Hahn *et al.*,[6] which we have shown in Section 2.3 to be equivalent to the Feshbach form $Q = 1 - P_1 - P_2 + P_1 P_2$.] The difference between Q and $\hat{Q}$ then is the absence of the product $P_1 P_2$, and it requires precisely this product to yield the formal identity $Q^2 = Q$. But more important from the present point of view, if we expand an arbitrary two-electron function of ^1S symmetry

$$\Psi = \sum_{n \geqslant m=0} \sum C_{nm} [\phi_n(1)\phi_m(2) + \phi_m(1)\phi_n(2)]$$

where the $\phi_n(i)$ are the eigenfunctions of the one-electron target of electron i [the total spin factor $\chi_0(1,2)$ being suppressed], then it is straightforward to calculate that

$$Q\Psi = \sum_{n \geqslant m = 1}\sum C_{nm}[\phi_n(1)\phi_m(2) + \phi_m(1)\phi_n(2)] \tag{3.2.4}$$

whereas

$$\hat{Q}\Psi = -C_{00}\phi_0(1)\phi_0(2) + \sum_{n \geqslant m = 1}\sum C_{nm}[\phi_n(1)\phi_m(2) + \phi_m(1)\phi_n(2)] \tag{3.2.5}$$

The comparison clearly shows that $\hat{Q}\Psi$ has allowed one (nondoubly excited) state to survive, $\phi_0(1)\ \phi_0(2)$, whereas $Q\Psi$ has completely eliminated it. The result is that $\hat{Q}H\hat{Q}$ in this case has one and only one spurious eigenvalue [a poor approximation of the ground $(1s)^2$ state but assuredly not a doubly excited state!]

To demonstrate, however, that the operator is nontrivial, we show in Table 1 the results of the Rayleigh–Ritz variational calculations for QHQ and $\hat{Q}H\hat{Q}$ using a Hylleraas form of total wave function [i.e., the form in Eq. (2.4.16) is associated with Ψ and the target state $\phi_0 \propto e^{-Zr}$]. Note that the Hylleraas form of Ψ (in particular, terms containing r_{12} to an odd power) precludes a trivial association of any specific term with any particular configuration (indeed, the mathematical problem of projecting is a nontrivial one[13]). What we see is that the lowest root is, as stated, a poor approximation of the $(1s)^2$ state of the composite system (in the case of H^-, the energy does not even correspond to binding). As an autoionization state, however, that eigenvalue is meaningless, i.e., spurious as a resonance. But beyond that, each eigenvalue below the first excited threshold does correspond to a resonance, and, in fact, the respective eigenvalues are very close approximations of QHQ.

Let us turn quickly to the more general case of the $N = 2$ target. Consider

TABLE 1

Quasi-Projection $\langle \hat{Q}H\hat{Q} \rangle$ Results for ^1S Autoionization State of He and H^-

		$\mathscr{E}_n - \varepsilon_0$(eV)				
		$e - H(H^-)$		$e - He^+(He)$		
No. of terms	$n =$	1^a	2	1^a	2	3
13		2.4920	9.5607		33.2415	37.506
22		2.4157	9.5403	−21.4771	33.2290	37.4825
34		2.3367	9.5410	−21.5358	33.2281	33.4785
50		2.3134	9.5406	−21.5629	33.2278	33.478
$\langle QHQ \rangle_{50\ \text{term}}$			9.5387		33.2267	33.471

[a] Entries are energies (in eV) relative to the ground state of the respective (H, He^+) target. The exact values to which the $n = 1$ columns are to be compared are −0.7558 and −24.5917 [C. L. Pekeris, *Phys. Rev.* **112**, 569 (1958)].

first $\hat{Q}_{(0)}$ and $\hat{P}_{(0)}$, the quasi projectors for resonances below the first excited threshold

$$\hat{P}_{(0)} = \sum_{i=1}^{3} P_i, \qquad \hat{Q}_{(0)} = 1 - \hat{P}_{(0)}$$

We want to determine the number of spurious states associated with $Q_{(0)}$. Expand a wave function in terms of eigenstates as in Eq. (3.2.2)., it is now straightforward to work out $\hat{Q}_0\Psi$

$$\hat{Q}_{(0)}\Psi = \sum_{n=1} \sum_{m} [C_{nm}\psi_n(r^{(j)})\theta_m(r_j)]_A \tag{3.2.6}$$

For demonstration purposes, we shall consider only ^2S states of the three-electron system. The thing to notice about Eq. (3.2.6) is that there is no $n=0$ term in the sum, even though there was in Ψ in Eq. (3.2.2). In other words, our quasi-projector operator has eliminated the one state ψ_0 that survives (asymptotically) in elastic scattering: *there are no spurious states associated with* $\hat{Q}_{(0)}$. In deriving Eq. (3.2.6), we have used the fact that the single-particle orbitals $\theta_m(r_j)$ can always be so chosen that they are orthogonal to the target ground state.[9]

$$\int \theta_m(\mathbf{r}_i)\phi_0(\mathbf{r}_1, \mathbf{r}_2)d^3r_i = 0 \qquad i = 1, 2 \text{ for all } m \tag{3.2.7}$$

This is eminently reasonable because the ground state is a closed-shell configuration $(1s)^2$ and the only kind of orbital for which Eq. (3.2.7) could not be imposed would be a $(1s)$-like orbital; but from the exclusion principle, a $(1s)^3$ configuration is forbidden.

This immediately leads to the expectation that there will be spurious states associated with inelastic quasi-projectors, since the excited states of He are not closed shells [i.e., their configurations are $(1snl)$ with $nl \geqslant 2s$]; thus, they contain only singly occupied orbitals. To see this in detail, consider Ψ in Eq. (3.2.2) with two channels open

$$\Psi = \sum_{m=1} C_{0m}(\psi_0\theta_m)_A + \sum_{m=1} C_{1m}(\psi_1\theta_m)_A + \sum_{n=2}\sum_{m=2} C_{nm}(\psi_n\theta_m)_A \tag{3.2.2$'$}$$

Simplifying the notation even further, we write

$$\Psi = \Psi^{(\mathrm{I})} + \Psi^{(\mathrm{II})} + \Psi^{(\mathrm{III})}$$

The relevant $\hat{Q}_{(\nu)}$ operator in this case is form Eq. (3.1.7a)

$$\hat{Q}_{(1)} = 1 - \hat{P}^{(0)} - \hat{P}^{(1)}$$

Operating here with $\hat{P}^{(0)}$ and $\hat{P}^{(1)}$ is straightforward but lengthier; we find

$$\hat{P}^{(0)}\Psi^{(\mathrm{I})} = \Psi^{(\mathrm{I})}, \qquad \hat{P}^{(1)}\Psi^{(\mathrm{I})} = C_{01}(\tfrac{3}{2})^{1/2}(\psi_1\theta_{1s})_A \tag{3.2.8a}$$

$$\hat{P}^{(0)}\Psi^{(\mathrm{II})} = 0, \qquad \hat{P}^{(1)}\Psi^{(\mathrm{II})} = \Psi^{(\mathrm{II})} + \tfrac{1}{2}C_{11}(\psi_1\theta_{2s})_A \tag{3.2.8b}$$

and

$$(\hat{P}^{(0)} + \hat{P}^{(1)})\Psi^{(III)} = 0 \tag{3.2.8c}$$

In deriving Eq. (3.2.8), we have used the fact that the configurations

$$\psi_0(1s^2)\theta_{2s}(r_j) = \psi_1(1s2s\,^3S)\theta_{1s}(r_j) = \psi_2(1s2s\,^1S)\theta_{1s}(r_j)$$

are equivalent (i.e., they would be identical if they were constructed from the same orbital). This leads to the requirement that the minimum $m = 1$ in $\Psi^{(I)}$ and $\Psi^{(II)}$. In $\Psi^{(III)}$, the minimum $m = 2$, because we can also show (remember $\theta_1 = \theta_{2s}$)

$$(\psi_2\theta_{2s})_A \propto \psi_1\theta_{2s}$$

thus, in order for the terms of $\Psi^{(III)}$ to be linearly independent of those of $\Psi^{(II)}$, we must have $\theta_m > \theta_{2s}$ ($\Rightarrow m > 1$). Equation (3.2.8c) then follows from the fact that the single-particle orbitals can be chosen to be orthogonal

$$\langle \theta_n \theta_m \rangle = \delta_{nm}$$

Using Eqs. (3.2.8a)–(3.2.8c), we have

$$\begin{aligned}\hat{Q}_{(1)}\Psi &= \Psi - (\hat{P}^{(0)} + \hat{P}^{(1)})\Psi \\ &= \Psi^{(I)} + \Psi^{(II)} + \Psi^{(III)} - \{[\Psi^{(I)} + (\tfrac{3}{2})^{1/2}C_{01}(\psi_1\theta_{1s})_A] \\ &\quad - [\Psi^{(II)} + \tfrac{1}{2}C_{11}(\psi_1\theta_{2s})_A]\} \\ &= \Psi^{(III)} - (\tfrac{3}{2})^{1/2}C_{01}(\psi_1\theta_{1s})_A - \tfrac{1}{2}C_{11}(\psi_1\theta_{2s})_A\end{aligned} \tag{3.2.9}$$

In other words, we see that $\hat{Q}_{(1)}$ allows two spurious states of (^2S symmetry) corresponding to the two linearly independent terms involving $\Psi_1[=\Psi(1s2s^3S)]$. That is to say, diagonalization of $Q_{(1)}HQ_{(1)}$ may permit two eigenfunctions (of ^2S symmetry) that describe *nonresonant* scattering from the 2^3S state. However, all other eigenfunctions [providing their energy lies below $\varepsilon(2\,^1S)$] would correspond to real resonances!

In Table 2, we summarize the action of the $\hat{P}^{(v)}$ operators on configurations up to the $(1s3s)^3S$ of the two-electron target. From Table 2, the number of spurious states and their configurations allowed by $Q_{(v)}$ can be ascertained. To see that, let us write Eq. (3.1.7b) in the form

$$Q_v = 1 - \sum_{n=0}^{v} P^{(n)} \tag{3.1.7c}$$

The open channels up to the vth have the form

$$\Psi_{\text{open}} = \sum_{(n,m)=0}^{v} (\psi_n\theta_m)_A \tag{3.2.10}$$

(We can set all $C_{nm} = 1$ here and in Eq. (3.2.2) for this discussion.) Consider

TABLE 2

$P^{(n)}\,\Psi$ *for Open-Channel Parts of* $\Psi = C_{nm}(\Psi_n\theta_m)_A$ *of Three-Electron Systems*[a]

Total synon	$\psi_n\theta_m$	$P^{(0)}\Psi$	$P^{(1)}\Psi$	$P^{(2)}\Psi$	$P^{(3)}\Psi$	$P^{(4)}\Psi$
	$(1s)^2(2s) = (1s)(1s2s)(1s)$	$(\psi_0\theta_{2s})_A$	$\frac{1}{2}(\psi_2\theta_{1s})_A$	$\frac{1}{2}(\psi_2\theta_{1s})_A$	0	0
$2S$	$(1s2s)(2s)$	0	$\frac{\sqrt{3}}{2}(\psi_1\theta_{2s})_A$	$\frac{1}{2}(\psi_2\theta_{2s})_A$	0	0
	$(1s2p)(2p)$	0	0	0	$\frac{\sqrt{3}}{2}(\psi_3\theta_{2p})_A$	$\frac{1}{2}(\psi_4\theta_{2p})_A$
$2P^0$	$(1s)^2(2p) = (1s2p)(1s)$	$(\psi_0\theta_{2p})_A$	0	0	$\frac{\sqrt{3}}{2}(\psi_3\theta_{1s})_A$	$\frac{1}{2}(\psi_4\theta_{1s})_A$
	$(1s2s)^3S(2p)$	0	$(\psi_1\theta_{2p})_A$	0	$\frac{1}{2}(\psi_3\theta_{2s})_A$	$\frac{\sqrt{3}}{2}(\psi_4\theta_{2s})_A$
	$(1s2s)^1S(2p)$	0	0	$(\psi_2\theta_{2p})_A$	$\frac{\sqrt{3}}{2}(\psi_3\theta_{2s})_A$	$-\frac{1}{2}(\psi_4\theta_{2s})_A$
	$(1s2p)^3P(2s)$	0	$\frac{1}{2}(\psi_1\theta_{2p})_A$	$\frac{\sqrt{3}}{2}(\psi_2\theta_{2p})_A$	$(\psi_3\theta_{2s})_A$	0
	$(1s2s)^1P(2s)$	0	$\frac{\sqrt{3}}{2}(\psi_1\theta_{2p})_A$	$\frac{1}{2}(\psi_2\theta_{2p})_A$	0	$(\psi_4\theta_{2s})_A$
	$(1s2p)^3P(3s)$	0	0	$(\psi_3\theta_{3s})_A$	$(\psi_3\theta_{3s})_A$	
	$(1s2p)$	0	0	0	0	$(\psi_4\theta_{3s})_A$
$2P^e$	$(1s2p)(2p)$	0	0	0	$\frac{\sqrt{3}}{2}(\psi_3\theta_{2p})_A$	$\frac{1}{2}(\psi_4\theta_{2p})_A$

[a] $\psi_0, \ldots, \psi_4$ correspond to the states $(1s^2)^1S$, $(1s2s)^3S$, $(1s2s)^1$, $S(1s2p)^3P$, $(1s2p)^1P$ of the target. When the result is the same for states that differ only in spin, only the configuration is given.

TABLE 3

Number and Designations of Spurious States in the Quasi-Projection Techniques for Three-Electron Ions[a]

Symmetry of autoionization state	$v+1$ Target state[a]	1 $2^3S(1s2s)$	2 $2^1S(1s2s)$	3 $2^3P(1s2p)$	4 $2^1P(1s2p)$	5 $3^3S(1s3s)$
$2S^0$		0	2	2	3	3
Configurations of spurious states				$(1s^22s)$ $(1s2s^2)$	$(1s^22s)$ $(1s2s^2)$ $(1s2p^2)$	$(1s^22s)(1s^22s)$ $(1s2s^2)(1s2s^2)$ $(1s2s^2)$
$2P^0$		0	0	0	2	2
Configurations of spurious states					$(1s^22p)$ $(1s2s2p)$	$(1s^22p)$ $(1s2s)(2p)$
$2P^0$		0	0	0	1	1
Configurations of spurious states					$(1s2p^2)$	$(1s2p)^2$

[a] This table corrects previous similar tables in Refs. (12) and (14).

${}^2P$ states below the $v + 1 = 4$[i.e., $(1s2p)$ ${}^1P$ state of the target]. This implies that $v = 3$; we want to consider

$$\hat{Q}_{(3)}\Psi_{\text{open}} = \left(1 - \sum_{\mu=0}^{3} \hat{P}^{\mu}\right)\Psi_{\text{open}}$$

where

$$\Psi_{\text{open}} = (\psi_{1\,{}^1S}\theta_{2P} + \psi_{2\,{}^3S}\theta_{2p} + \psi_{2\,{}^1S}\theta_{2p} + \psi_{2\,{}^3P}\theta_{3s})_A$$

Here, the one-particle orbital multiplying ψ_{2^3P} must be taken as θ_{3s} rather θ_{2s} because we can show that

$$(\psi_{2\,{}^3P}\theta_{2s})_A = (a\psi_{2\,{}^1S}\theta_{2p} + b\psi_{2\,{}^3S}\theta_{2p})_A$$

Thus, in order for the 2^3P channel function to be linearly independent of the other two, it must be multiplied by a linearly independent single-particle orbital. The same is true for $(\psi_{2\,{}^1P}\theta_{2s})_A$. Referring to Table 3, we see then that

$$\begin{aligned}\hat{Q}_{(3)}\Psi_{\text{open}} &= C_1(\psi_{1\,{}^1S}\theta_{2p})_A + C_2(\psi_{2\,{}^3S}\theta_{2p})_A + C_3(\psi_{2\,{}^1S}\theta_{2P})_A + C_4(\psi_{2\,{}^3P}\theta_{3s})_A \\ &\quad - \{C_1[\psi_{1\,{}^1S}\theta_{2p} + (\tfrac{1}{2}\sqrt{3})^{1/2}\psi_{2\,{}^3P}\theta_{1s}]_A + C_2(\psi_{2\,{}^3S}\theta_{2p} + \tfrac{1}{2}\psi_{2\,{}^3P}\theta_{2s})_A \\ &\quad + C_3[\psi_{2\,{}^1S}\theta_{2p} + (\tfrac{1}{2}\sqrt{3})^{1/2}\psi_{2\,{}^3P}\theta_{2s}]_A + C_4(\psi_{2\,{}^3P}\theta_{3s})_A\} \\ &= -((\tfrac{1}{2}\sqrt{3})^{1/2}\psi_{2\,{}^3P}\theta_{1s} + \{[\tfrac{1}{2} + (\tfrac{1}{2}\sqrt{3})^{1/2}]\}\psi_{2\,{}^3P}\theta_{2s})_A\end{aligned}$$

That is, two linearly independent open-channel functions survive $\hat{Q}_3$, which implies that $\hat{Q}_{(3)}H\hat{Q}_{(3)}$ will allow two spurious resonant eigenvalues below the $2\,{}^1P$ threshold. These nonresonant eigenvalues are expected to be among the lowest that occur (because they involve the lowest orbitals); however, that need not be the case. Only after calculating and examining the orbital content of each eigenfunction can we unambiguously identify the spurious states. In Table 2, we give a summary of spurious states below the 3^3S threshold. [Table 2 then corrects earlier tables in both Refs. (12) and (14).]

3.3. *Previous Calculations*

In this chapter, we shall not repeat the results of our quasi-projector calculations; rather, we shall make a few comments concerning them, reserving for the next section some further details on the first inelastic resonance of He^- $[(1s2s2p)\,{}^2P]$ for important reasons that we shall give there.

All our calculations have been restricted to three-electron systems[14–16]. The main reason for this restriction is the obvious one of simplicity; however, even with regard to spurious states, the analysis of their number and configurations can reasonably only be done as previously indicated for such cases. As a rule, more general and more powerful methods will be required for $N > 2$ targets, and we would suggest that as a worthwhile topic for the interested researcher.

With regard to the $N = 2$ target, we regard our original calculation[9] for the elastic He$[1s(2s)^2\,{}^2S]$ resonance as fundamental. It was the first calculation to converge to a well-defined value in the continuum $\mathscr{E} = 19.363$, 19.386 eV (based on closed- and open-shell approximations of the He ground state ϕ_0) and a width $\Gamma = 0.0139$ eV based on a closed-shell ground state and an exchange adiabatic nonresonant continuum $\hat{P}\Psi_0$ using the well-known width formula $\Gamma = 2k|\langle \hat{P}\Psi_0|V|\hat{Q}\phi\rangle|^2$ (cf. Chapter 1). Not only are these values quite accurate, but the calculation also *ruled out* additional structure that some experiments had found below the first excited (2^3S) state of He and had attributed to He^- ($^2P^0$). [It is now generally agreed that He^- does not have any additional Feshbach resonances below He(2^3S)]. It should be added, however, that since that time more accurate experiments[17,18] and calculations—in particular, complex rotation calculations[19] (cf. Chapter 4 by B. R. Junker in this volume) have been done for this as well as other (inelastic) resonances of He^-.

Other quasi-projection calculations that we consider to be accurate involve the autoionization states of Li in the elastic domain[20]. Two of our calculations using the inelastic quasi-projection formalism[14,15] are unfortunately less satisfactory. In effect, they were not so well monitored by the present authors as they should have been. Thus, for example, in our first inelastic calculation,[14] had a more careful search of the nonlinear parameter space been made, it would have revealed the lowest inelastic He^-($^2P^0$) resonance (cf. the following section for more details on that resonance). Also, we now consider many of the energy positions and/or identifications to be in doubt. Although as far as we know, there were no errors in the calculations themeselves, it is clear that those calculations will have to be redone using a new program about which we have more knowledge and over which we have more control (cf. next section).

3.4. He^-($^2P^0$): A New Type of Feshbach Resonance

In a recent article, we stated that the enhancement of the threshold excitation cross section of the 2^3S state in e-He scattering provides perhaps the earliest evidence of a resonance in atomic (as opposed to nuclear) physics (cf. Ref. 34).[21] We went on to note that resonance also provided the basis for the first application of the Breit–Wigner theory to an atomic resonance by Baranger and Gerjuoy.[22] The resonance in question [He^-($^2P^0$)] lies between the 2^3S ($E_{2^3S} = 19.82$ eV) and 2^1S ($E_{2^1S} = 20.61$ eV) states of He, and it is wide ($\Gamma \cong 0.5$ eV). It was among the first resonances to emerge, together with the famous elastic Schulz resonance, from a close coupling calculation involving a many-electron target.[23] Because the resonance is wide—nearly as wide as the spacing between the 2^3S and 2^1S

states of the target—it was naturally assumed in this[23] as well as later calculations[24] that the resonance was a shape resonance coming from a potential barrier associated with the open (2^3S) state of He.

Although later experiments and calculations confirmed and refined the shape and position of the resonance, this was the accepted view until, in a calculation by Chung, this resonance unambiguously showed up as an eigenvalue from his hole-projection technique.[25] That immediately shed a new light on this resonance, because Chung's hole-projection technique could, in our opinion, only reveal Feshbach resonances, and Chung so interpreted his result.[25] From our point of view, this potentially important result needed confirmation for two reasons: If it were a Feshbach resonance, then it should come out of our formalism as well. Secondly, from the eigenvector associated with such a resonant eigenvalue, we could also calculate a width; only if such a resonance were suitably wide could we be sure that this was indeed the observed resonance and that its interpretation as a Feshbach resonance was consistent. [The fact that our calculations did confirm these findings indicated to us the substantive merit of Chung's method and led ultimately to its inclusion in this volume (cf. Chapter 3 by K. T. Chung and B. F. Davis).]

We shall briefly describe our calculation. The state should have shown up in our inelastic quasi-projection calculations[14] but as mentioned in the preceding section, those original inelastic calculations encountered difficulties. We thus decided to go back to the original elastic quasi-projection program[10] and extend it to the inelastic domain by including only configurations that were automatically orthogonal to the ^3S symmetry (and, hence, specifically to the 2^3S state of He). Since the original program allowed the use of an open-(as well as a closed-) shell projector for the ground state, it had the great advantage of enabling us to test the (at that time still untested) result of Chung.[25] At the same time, it allowed calculation of partial width to the ground state and (most important and with minimum generalization) to the first excited (2^3S) state.

Without belaboring the point, the results[21] were positive on both accounts; not only did we confirm Chung's position, but we found a very wide width to the 2^3S state. A précis of results is given in Table 4. Some newer results have also been added to Table 4, including importantly a complex rotation calculation by Junker.[26] That result is probably the most accurate, although a recent calculation by Kimninos *et al.*[27] claims comparable accuracy with an analysis that advances the case that the resonance does not have a Lorentzian shape. We do not wish to go into that matter; our point is to emphasize the predictive power of the present projection-operator methods, whose accuracy is concomitantly sufficient for most (astrophysical and laboratory plasma) practical diagnostic analysis for which resonant phenomena are likely to be important.

We shall rather concentrate on another aspect of this resonance: the

TABLE 4
Results from e-He($^2P^0$) Resonance (in eV)[a]

Physical quantity (target ground state, target 2^3S state)	ϕ_0(closed)	ϕ_0(open)	$\Gamma_2\,{}^2S$ ϕ_0(closed) ϕ_1(open)	$\Gamma_1\,{}^1S$ ϕ_0(closed)
No. of radial configurations				
15	20.703 39	20.739 36		
19	20.557 26	20.580 55	0.326	0.0018
23	20.556 74	20.578 59	0.364	0.0021
31	20.526 05	20.561 43	0.424	0.0024
40	20.524 89	20.560 29	0.437	0.0024
Ref. (20)	(20.536, 20.495)[b]			
Ref. (26)	20.33		0.355	
Ref. (27)	20.284		0.575	
Experiment[c]	20.3 ± 0.3		(Total width) 0.5	

[a] The 2^3S and 2^1S thresholds are at 19.8236 and 20.6162 eV, respectively.
[b] The two results correspond to two somewhat different calculations.
[c] The experimental results represent a rough average of results quoted in Ref. (21) (Table 1).

fact that it can be wide, i.e., decay rapidly to a state to which it is (by definition of a Feshbach resonance) orthogonal! We shall conclude this chapter with a brief discussion of this point, preceded by a heuristic explanation of the difference between a Feshbach and a shape resonance and why, in particular, this type of "core-rearranged" Feshbach resonance may be particularly important in many applications. All of these points, in fact, were presented as part of an ICPEAC satellite meeting,[(28)] and it is that discussion that we shall paraphrase here.

3.4.1. Difference between Feshbach and Shape Resonances. From an historical perspective, the origin of what we now call Feshbach resonances goes back to early in the history of quantum theory. (In point of fact, the list of people associated with such developments reads like a virtual "Who's Who" of the great names associated with the development of quantum mechanics.) Of relevance to us are two key ideas

1. Bohr's concept of the compound nucleus (meaning, here, the compound target plus electron system), which itself includes two parts.
   (*a*) The energy of the incident particle is shared by particles within the target.
   (*b*) The intermediate state is so long-lived (compared to the transit time of the scattered particle) that the decay is independent of the mode of formation.
2. The Breit–Wigner analysis and parametrization of resonances.

With regard to 2, the classic paper by Breit and Wigner[29] clearly shows that what they had in mind physically was a compound nucleus in the sense of Bohr; however, to illustrate it mathematically, they chose a shape rather than a "compound-nucleus" resonance. In one sense, that was unfortunate, because the difference between such resonances is the difference between a single-particle and a collective phenomenon. In the former, the physics is that of a particle trapped in a potential containing a barrier. We can easily write down a local potential (an attractive well followed by a repulsive hill) that will exhibit such behavior. In Fig. 1a, we demonstrate a typical radial function [actually $U(r)$] for various $k_i(<k_{i+1})$ as it traverses such a resonance. The phase shift η versus k is given on the right. We see that the shapes of these radial functions undergo a breathing motion analogous to the folds in an accordion as it is pushed and pulled.

This state is contrasted with a Feshbach resonance, which is pictured in Fig. 1b. Here, as k increases, the radial function actually changes sign at the origin; but it does so smoothly in such a way that not only $U(r)$, but its derivatives at the origin are zero

$$[U_{k_{\rm res}}(r)] = \left\{\frac{d}{dr}[U_{k_{\rm res}}(r)]\right\} = 0 \qquad k \cong k_{\rm res}$$

such behavior *cannot* occur if U satisfies a wave equation with a *local* potential

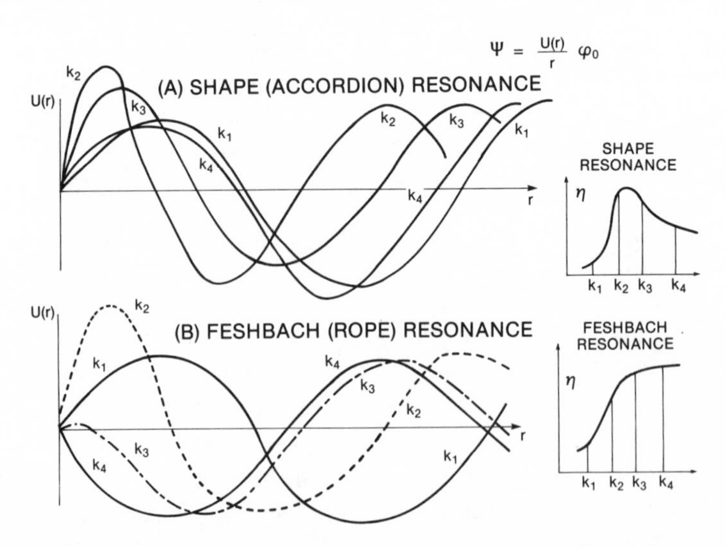

FIGURE 1. Dependence of radial functions on r for various k in the vicinity of (a) a shape resonance; (b) a Feshbach resonance. The associated phase shifts as a function of k are given on the right. Note that for a Feshbach resonance, the radial function $U(r = 0)$ changes sign continuously for some k in the energy range of the resonance (see text).

$V(r)$.

$$\left[-\frac{d^2}{dr^2} + V(r) - k_{\text{res}}^2\right][U_k(r)] = 0$$

Or [assuming $V(r)$ is not more singular than Coulombic] otherwise it would imply that

$$\lim_{r \to 0} \frac{d^2}{dr^2} U_{k_r}(r) = \lim_{r \to 0} [V(r) U_{k_r}(r)] = 0$$

and carrying this further, all derivatives of $[U_{k_r}(r)]_{r=0}$ would be zero. And that, in turn, would mean by Taylor expansion that $U_{k_r}(r)$ is zero for all values of r, i.e., that no physically meaningful wave function would exist for $E = k_r^2$. That is, of course, unacceptable.

The real implication then is that $U_k(r)$ must be the solution of a Schrödinger equation containing a *nonlocal* potential

$$-\left(\frac{d^2}{dr^2} + k^2\right)[U_k(r)] + \int_0^\infty v(r, r')[U_k(r')]dr' = 0$$

With such a potential, it is perfectly possible to have a solution $U_k(r)$ whose value and first derivative are zero at the origin. The behavior of the radial function as a function of k is what we have pictured in Fig. 1b. Notice here that the function changes as a function of k very much as a rope whose end is tied would oscillate if the other end were moved. (We have, therefore, often called this a "rope resonance.") Considerable attention has been given to these and other aspects of separable nonlocal potentials by Mulligan and collaborators [cf. Ref. (30) and subsequent papers].

3.4.2. Wide Feshbach Resonances. The point of the foregoing discussion is not simply to elucidate an interesting but academic difference between two different kinds of resonances. Rather, we have in mind the fact that if resonances are Feshbach in character, then there is a well-defined eigenvalue equation

$$\hat{Q}H\hat{Q}\Phi_n = \mathscr{E}_n \Phi_n$$

that can be defined and calculated, so that the eigenvalues $\mathscr{E}_n$ automatically can be obtained from the calculation. Thus, the resonant enhancements of a phase shift do not have to be hunted for and perhaps missed. Consequently, it is of considerable practical advantage to have such an equation.

The question that naturally arises is under what circumstances will such Feshbach resonances be wide? The hallmark of such compound nucleus is that they are usually very narrow. In the Feshbach formalism, that would appear to be a general characteristic because the projection operator Q (or $\hat{Q}$)

eliminates the energetically allowable target states ϕ_n from Ψ [i.e., recall Eq. (2.2.3b)] $\langle \phi_n | Q\Psi \rangle = 0$). We might very well expect, therefore, that the orthogonality would also force the width integral

$$\Gamma_n^{1/2} \propto \int \mathscr{A}\{U\phi_n\} VQ\Phi d\tau$$

where

$$P\Psi \propto \mathscr{A}\{U\phi_n\}$$

also to be small.

The obvious answer to this paradox is that the integral contains an interaction V, whereas the overlap $\langle \phi_n | Q\Psi \rangle$ does not, and, indeed, the $He^-(^2P^0)$ partial width is a case in point. The trouble is that the specific answer gives no general insight into why it happens. The general answer is that, by a combination of exchange and energetics, the partial width to a close by energetically allowable state can be greatly favored. For example, in the $He(2^3S)$ excitation (which cannot be excited by positron impact), the slow outgoing p-wave, represented by the function U, has essentially no nodes in the interaction region and therefore (mediated by the appropriate

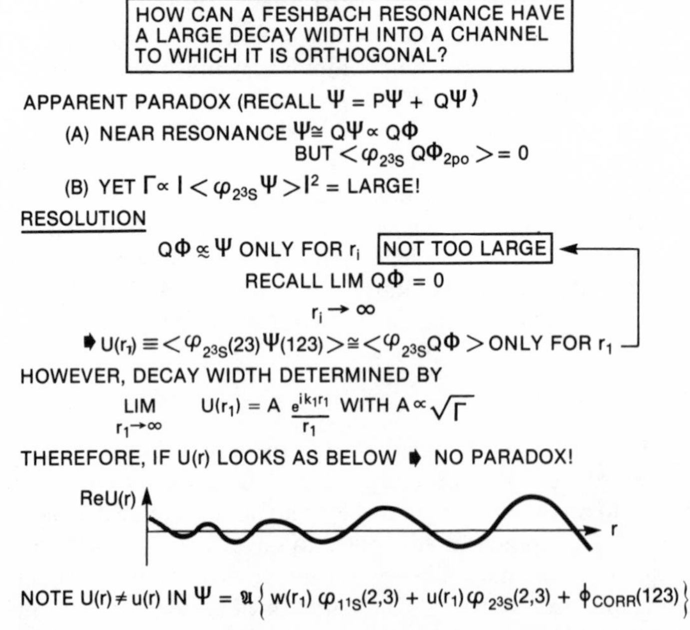

FIGURE 2. Précis of solution of paradox posed by a "wide" Feshbach resonance (see text for further discussion).

r_{ij}^{-1} interaction) overlaps very heavily with the excited orbital ($n = 2s$ here), which is also almost nodeless. In contrast, the scattered orbital entering the partial-width integral to the ground state oscillates rapidly, corresponding to its considerable kinetic energy, and overlaps poorly with any of the orbitals ($1s, 2s$, or $2p$) of the compound state. In fact, we have obtained this kind of enhancement in previous calculations of O VII excitation.[31]

This phenomenon might also appear to be somewhat paradoxical from the point of view of a complete scattering (i.e. to generalize close coupling) calculation. For there, the width is measured by the amplitude of the radial function $U(r)$, which, in effect, is the overlap of ϕ_{2^3S} with Ψ. However, Ψ is dominated by $Q\Phi$ at resonance, and by definition, $\langle \phi_{2^3S} Q\Phi \rangle \cong \langle \phi_{2^3S} \Psi \rangle$. How can Γ be both small and large at the same time?

The resolution of this paradox is outlined in Fig. 2. Basically, it amounts to saying that Ψ is approximately proportional to $Q\Phi$ only when all radial coordinates r_i (including that of the scattered particle) are not large. When a radial coordinate becomes large, this is no longer so (in fact, $\lim r_i \to \infty$, $Q\Phi = 0$), and the effective amplitude must be large there. In effect, then, as pictured in Fig. 2, the function $U(r)$ must have an amplitude $A(r)$ that increases dramatically as a function of r until it levels at its asymptotic value, where it becomes proportional to the (square root of the) width.

4. CONCLUSIONS

4.1. *Resume and Future Work*

In this chapter, we have attempted to establish the framework for constructing and calculating true (in the sense of $\langle \phi_0 Q\Psi \rangle = 0$) projection operators for electron-resonance calculations involving many-electron targets (ions or atoms). The construction is, in principle, complete for two-electron (i.e., He-like) targets described by configuration interaction wave functions, and an explicit form for P (and Q) was derived for the open-shell approximation. Also, considerable progress was made in solving the projection-operator problem using Hylleraas-type target functions. Clearly, further analysis and calculations along these lines will be worthwhile and hopefully will be carried out. Note that this problem is rather specifically devoted to electron scattering from atomic systems. The more general problem of including antisymmetry in an optical-potential formalism of scattering of composite systems that each contain identical particles of its colliding partner (say atom–atom or nucleus–nucleus scattering) has been treated by Kowalski and Picklesimer.[32]

The problem of constructing idempotent projection operators in the inelastic domain is not trivial, but it has recently been carried out by us elsewhere.[35] Whether anyone will want to pursue calculations in the inelastic domain or the extension to more than two-electron targets will depend on the difficulty and success of the elastic projection-operator calculations.

With respect to calculations, the concept of quasi-projection is a particularly useful alternative. Calculations, in fact, in both the elastic and inelastic domain have been carried out, as we have indicated, but they have been confined to two-electron targets, and some of our inelastic calculations should be improved. Nevertheless, the confirmation of the inelastic $He^-(^2P^0)$ resonance has afforded interesting theoretical insight into a new type of Feshbach resonance, which, because it is wide yet close to an energetically accessible threshold, may be a prototype of a process with important practical implications in both astrophysical and laboratory plasmas. We have already speculated[28] that some other well-known resonances [specifically $N^-(^3P^e)$] may, in fact, be of this type and therefore calculable in the projection-operator formalism.

4.2. *Calculation of Nonresonant Phase Shifts*

One problem that has consistently evaded solution is the inability of the optical potential constructed from (many-electron) projections or quasi-projection-operator formalism to yield converged and accurate phase shifts in the nonresonant region. In principle, the application is an obvious one involving the (Q) spectral expansion of the optical potential. That method of implementing the optical-potential formalism was suggested in our first paper on quasi-projection operators.[10] The idea is that the (quasi–optical potential

$$\hat{\mathscr{V}}_{\text{op}} = \hat{P}H\hat{Q}\frac{1}{E-\hat{Q}H\hat{Q}}\hat{Q}H\hat{P}$$

is expanded in a (in principle) complete set of eigenfunctions of $QHQ\Psi_n = \mathscr{E}_n\Phi_n$ to obtain

$$\hat{\mathscr{V}}_{\text{op}} = \sum_{n=1}^{n_{\max}} \hat{P}H\hat{Q}\Phi_n\rangle\frac{1}{E-\hat{\mathscr{E}}_n}\langle\Phi_n\hat{Q}H\hat{P} \tag{4.2.1}$$

If we solve the QHQ (or $\hat{Q}H\hat{Q}$), we automatically obtain a set of eigenvectors and associated eigenvalues ($\hat{\Phi}_n$ and $\hat{\mathscr{E}}_n$, where the carets indicating that target state wave functions may not be exact). We can then substitute these values into the expansion of $\hat{\mathscr{V}}_{\text{op}}$ in Eq. (4.2.1) and solve the resultant optical-potential equation at energies E away from the resonances ($\hat{\mathscr{E}}_n$) to obtain phase shifts as a function of energy. The results that are obtained in scattering electrons from He based on our quasi-projection-operator calculations of He^- were subsequently described[33] and are typical: as the number n_{max} of linear terms in the ansatz for Φ is increased, the optical potential in Eq. (4.2.1) gives nonresonant phase shifts that are too large and appear not to be converging, in spite of the fact that the one (and only) ^2S resonance is accurately given by Φ_1 (and $\mathscr{E}_1$) and both its position and width appear to be excellently convergent.[10]

We believe that at long last we understand the cause of this behavior, and our final paragraph is devoted to an elucidation of this cause.

As we have now often stated, projection- or quasi-projection-operator calculations for many-electron targets always depend on approximate target wave functions. If the target functions ϕ_n were exact, then they would be completely able to describe the complete shielding that is in principle present in the interaction of an electron with, say, a neutral target

$$V_{\text{int}} = -\frac{Z}{r_1} + \sum_{j=2}^{N+1} \frac{1}{r_{ij}}$$

However, because ϕ_0 and, as a result, the autoionization states Φ_n are approximate, they begin to feel false Coulomb effects; that will reflect itself in the spectral expansion of the optical potential [Eq. (4.2.1)]. We can appreciate that most readily in the denominators $(E - \mathscr{E}_n)^{-1}$, because the $\mathscr{E}_n$ become progressively lower than they should be by virtue of this same false Coulomb attraction. Thus, with each dominator too small, the overall optical potential is similarly too attractive. Although for any finite $n_{max}(n < n_{max})$, the result will converge as $n_{max} \to \infty$, the phase shift will inevitably increase indefinitely, since as is well known a Coulomb potential gives rise to a total phase shift that diverges logarithmically.

We might ask if the false Coulomb effects are present in the $\hat{Q}H\hat{Q}$ calculations of the autoionization themselves. The answer is yes, but charged systems (positive ions) also have autoionization states, and their energies for low n need not be overly negative. It is only the multiplicity of such effects as n becomes large that proves the undoing of $\hat{\mathscr{V}}_{op}$. But having recognized the cause of the observed phenomenon, we must still correct it, and we believe recognition is the first step toward the solution.

REFERENCES

1. H. Feshbach, *Ann. Phys. (N.Y.)* **19**, 287 (1962).
2. A. K. Bhatia and A. Temkin, *Phys. Rev. A* **11**, 2018 (1975).
3. P. M. Morse and W. P. Allis, *Phys. Rev.* **44**, 269 (1933).
4. P. O. Löwdin, *Phys. Rev.* **97**, 1474 (1955).
5. A. Temkin and A. K. Bhatia, *Phys. Rev. A* **31**, 1259 (1985).
6. Y. Hahn, T. F. O'Malley, and L. Spruch, *Phy. Rev.* **128**, 932 (1962).

7. Y. Hahn, *Ann. Phys. (N.Y.)* **58**, 137 (1970).
8. R. A. Sack, *J. Math. Phys.* **5**, 245 (1964).
9. T. F. O'Malley and S. Geltman, *Phys. Rev.* **137**, A1344 (1965).
10. A. Temkin, A. K. Bhatia, and J. N. Bardsley, *Phys. Rev. A* **7**, 1663 (1972).
11. T. F. O'Malley, *Phys. Rev.* **150**, 14 (1966).
12. A. Temkin and A. K. Bhatia, *Phys. Rev. A* **18**, 792 (1978).
13. A. K. Bhatia, J. F. Perkins, and A. Temkin, *Phys. Rev.* **153**, 177 (1967).
14. S. Wakid, A. K. Bhatia, and A. Temkin, *Phys. Rev. A* **21**, 496 (1980).
15. S. Wakid, A. K. Bhatia, and A. Temkin, *Phys. Rev. A* **22**, 1395 (1980).
16. A. K. Bhatia, *Phys. Rev. A* **18**, 2523 (1978).
17. S. Cvejanovic, J. Comer, and F. H. Reed, *J. Phys. B.* **7**, 468 (1974).
18. R. Kennerly, R. Van Brunt, and A. Gallagher, *Phys. Rev. A* **23**, 2430 (1981).
19. B. R. Junker and C. Huang, *Phys. Rev. A* **18**, 313 (1978).
20. A. K. Bhatia and A. Temkin, *Phys. Rev. A* **13**, 2322 (1976).
21. A. K. Bhatia and A. Temkin, *Phys. Rev. A* **23**, 3361 (1981).
22. E. Baranger and E. Gerjouy, *Phys. Rev.* **106**, 1182 (1957).
23. P. G. Burke, J. W. Cooper, and S. Ormonde, *Phys. Rev. A* **183**, 245 (1969).
24. R. S. Oberoi and R. K. Nesbet, *Phys. Rev. A* **8**, 2969 (1973).
25. K. T. Chung, *Phys. Rev. A* **23**, 1079 (1981).
26. B. R. Junker, *J. Phys. B: At. Mol. Phys.* **15**, 4495 (1982).
27. Y. Kimninos, G. Aspromallis, and C. A. Nicolaides, *Phys. Rev. A* **27**, 1865 (1983).
28. A. Temkin and A. K. Bhatia, in *Electron Scattering from Atoms and Atomic Ions*, a satellite symposium of the 1981 ICPEAC, Program-Abstract Booklet, unpublished.
29. G. Breit and E. Wigner, *Phys. Rev.* **49**, 519 (1936).
30. B. Mulligan, L. G. Arnold, B. Bagchi, and T. O. Krause, *Phys. Rev. C* **13**, 213 (1976).
31. M. S. Pindzola, A. Temkin, and A. K. Bhatia, *Phys. Rev. A* **19**, 72 (1979).
32. K. L. Kowalski and A. Picklesimer, *Ann. Phys.* (*N.Y.*) **139**, 215 (1982).
33. A. K. Bhatia and A. Temkin, *Phys. Rev. A* **13**, 2322 (1976).
34. H. Maier–Leibnitz, *Z. Phys.* **95**, 499 (1936).

CHAPTER THREE

HOLE-PROJECTION METHOD FOR CALCULATING FESHBACH RESONANCES AND INNER-SHELL VACANCIES

KWONG T. CHUNG AND BRIAN F. DAVIS

1. INTRODUCTION

Inner-shell vacancy states play an important role in understanding quantum systems. Experimentally, they appear in the form of a discrete spectrum that is embedded in the continuum. In most cases, they are coupled to the continuum via the Coulomb interaction, although this coupling may be very weak, so that the energy level and wave function can be obtained approximately by using square-integrable basis functions. This coupling can cause considerable difficulty in the theoretical treatment of these states. These vacancy states can be formed in atomic, molecular, nuclear, and solid systems. They arise in various physical processes; for example, in resonant or nonresonant collisions, photoabsorption, nuclear decay, or elementary particle capture, etc. Once formed, they may decay by inner-shell or outer-shell autoionization, Coster–Kronig transition, x-ray emission, etc.[1] The application of these effects covers a wide range of disciplines in physics and electronics as well as medical and geophysical sciences.

The theoretical study of these quantum states started in the early 1930s;[2] however accurate quantitative results came only when the computer was introduced. The electron–hydrogen resonances obtained by the close-coupling calculation of Burke and Schey[3] has generated considerable excitement. In the meantime, the analyses of Fano[4] and Feshbach[5] placed the theoretical understanding of these states on a firm mathematical basis. Feshbach's analysis, in particular, shows how these states arise in a scattering process, and the name *Feshbach resonance* or *closed-channel resonance* has

---

KWONG T. CHUNG and BRIAN F. DAVIS ■ Department of Physics, North Carolina State University, Raleigh, NC 27650–5367. This work was supported by National Science Foundation Grant No. PHY84–5469.

now been commonly adopted. Many methods have been developed in the intervening time to deal with these states.[6-14] Some of these methods are discussed in other chapters in this volume. In this chapter, we discuss a method recently introduced that is especially well suited for dealing with systems where one or more core electrons are excited. This new approximation method accounts for the excitation of the core by building the vacancy orbitals directly into the wave function.[15] It eliminates the open-channel continua by maximizing the energy with respect to parameters in the vacancy orbital. Thus, the closed-channel resonances are exposed and the minimization procedure of the Rayleigh–Ritz variation method is restored. Although the method has been in existence for only a relatively short period of time, due to its simplicity, a large amount of useful results have been obtained. In this chapter the method and some of these results will be presented and discussed.

2. HOLE-PROJECTION THEOREM

For convenience of discussion, an atomic state is usually characterized by its configuration and symmetry, which is given by the good quantum numbers plus the principal and angular quantum numbers of the individual electrons. In many instances, especially for Feshbach resonances, this description is complicated by degeneracy, but the simple feature of a particular core electron having been excited and thus leaving behind a well-defined vacancy in the inner shell is easily visualized. For example, for the commonly called $(2s2s)^1S$ resonance of helium, $2s2s$ contributes about 72% to the normalization and $2p2p$ about 28%. Hence, the description $(2s2s)^1S$ is only approximate, but it is essentially correct that it is the lowest ^1S resonance with two $1s$ holes. Therefore, if we can build these holes into the wave function, the closed-channel resonances should come out naturally as a discrete spectrum of the Hamiltonian.

Let us assume $\psi(1, 2, 3, \ldots, n)$ is a configuration interaction wave function with the proper angular and spin symmetries of interest. This wave function is not antisymmetrized; hence, each particle in ψ has a well-defined angular momentum. An antisymmetrized wave function with a vacancy can be given by

$$\Psi = \mathscr{A}[1 - |\phi_0(\mathbf{r}_j)\rangle\langle\phi_0(\mathbf{r}_j)|]\psi(1, 2, 3, \ldots, n) \tag{1}$$

where $\mathscr{A}$ is an antisymmetrization operator. We are assuming that $\phi_0(\mathbf{r})$ is the vacancy orbital function for which electron j has the same symmetry and is therefore the only particle that may fill this vacancy. Note that $\phi_0(\mathbf{r})$ does not contain any spin, the projection operator $|\phi_0(r_j)\rangle\langle\phi_0(\mathbf{r}_j)|$ does not effect the spin part of $\psi(1, 2, \ldots, n)$.

The problem is how to determine $\phi_0(\mathbf{r})$. In this regard, let us consider the following theorem[15] for a one-particle system.

Theorem. Let $H(\mathbf{r})$ be a Hermitian operator with normalized eigenfunctions $\psi_0(\mathbf{r}), \psi_1(\mathbf{r}), \ldots, \psi_i(\mathbf{r})$ and corresponding nondegenerate eigenvalues $E_0, E_1, \ldots, E_i$. Define a normalized function

$$\phi_0(\mathbf{r}) = \sum_{j=0}^{N} t_j \psi_j(\mathbf{r}) \tag{2}$$

for any $N \geqslant 1$. Let the eigenvalues of the secular equation of H *in the subspace orthogonal to* ϕ_0 be $\lambda_1, \lambda_2, \ldots, \lambda_n$.

Consider λ_i as a function of the $\{t_j\}$; then λ_i is an extremum, and $\lambda_i = E_i$ when $t_i = 0$. In fact, we can show that[15]

$$\lambda_i = E_i + t_i^2(E_0 - E_i) + \sum_{k=1}^{N}{}' \frac{(E_0 - E_i)^2}{E_i - E_k} t_i^2 t_k^2 + 0(t^6) \tag{3}$$

where $k = i$ is excluded in the summation. Notice that the dependence of λ_i on t_0 is eliminated by the normalization condition

$$t_0^2 = 1 - \sum_{j=1}^{N} t_j^2 \tag{4}$$

It is clear from Eq. (3) that as long as $t_i = 0$, $\lambda_i = E_i$ will be true regardless of the values of t_j for $j \neq i$. For the right-hand side of Eq. (3) to converge quickly, t_j should be small for all $j = 1, 2, \ldots, N$. If this is the case, then for $E_i > E_0$, λ_i will appear as a maximum, and, correspondingly, if $E_i < E_0$, it will appear as a minimum. In reality, it is the lower orbitals that we wish to project out. In this way, a vacancy is built into the Hilbert space. Therefore, Eq. (3) shows that λ_i appears as a maximum with respect to variations of the set $\{t_j\}$.

Since the eigenfunctions $\psi_j(\mathbf{r})$ may be unavailable, an expansion such as Eq. (2) cannot be actually carried out. In a variation calculation, we adopt a certain orbital wave function $\phi_0(\mathbf{r})$ with parameters q; the unprojected wave function contains linear parameters C and nonlinear parameters α. Thus, the energy expectation value becomes

$$E(C, \alpha, q) = \frac{\langle \Psi | H | \Psi \rangle}{\langle \Psi | \Psi \rangle} \tag{5}$$

If we minimize this E with respect to the linear parameters C, a secular equation is obtained.[16] The roots of this secular equation λ will be functions of α and q. For atomic systems, it is well known that the energy should be a minimum with respect to α. The theorem discussed in this section based on Eq. (3) shows that λ is a maximum with respect to variations of q. Hence, the energy appears as a saddle point. For this reason, this method is also called the *saddle-point* technique.

This theorem is rigorous for one-electron systems. In order to generalize it to many-electron systems, we must resort to the variational principle inherent in quantum mechanics. It is important to realize the searching for the stationary value of the energy is equivalent to solving the Schrödinger equation; that is

$$\delta E = \delta \frac{\langle \Psi | H | \Psi \rangle}{\langle \Psi | \Psi \rangle} = \frac{\langle \delta\Psi | H - E | \Psi \rangle + \langle \Psi | H - E | \delta\Psi \rangle}{\langle \Psi | \Psi \rangle} \tag{6}$$

with the trial function Ψ covering the proper Hilbert space. To require $\delta E = 0$ to be true for an arbitrary variation of any and all possible parameters in Ψ, the parameters must be at a value where

$$(H - E)\Psi = 0 \tag{7}$$

To calculate the inner-shell vacancy state of a many-particle system, we can parametrize the ϕ_0 and ψ in Eq. (1). Equations (6) snf (7) suggest that we should search for the stationary energy solution with respect to the variation of these parameters. Equation (3) suggests that the stationary solution should be a maximum with respect to the parameters in the vacancy orbital ϕ_0.

The basis of this method bears some interesting similarities to the mini-max principle in the literature.[17] We have also made the basic assumption that the excited states of a quantum system can be considered as states with inner-shell vacancies and that these vacancies can be built into the wave function with single-particle orbitals. For autoionizing states, this "inner-shell vacancy state picture" is only an approximation, because the full solution to the Schrödinger equation will contain some open-channel segment where no vacancies are present. However, if this inner-shell vacancy picture is a good description of these quantum states and the appropriate vacancies are built into the many-particle wave function, then the solution to Eq. (6) must be a good approximation to the true wave function.

Before we go on to specific calculations, let us consider the mechanics of building vacancies into a many-particle wave function. A typical configuration interaction wave function in the *LS* coupling scheme is expanded in terms of angular and spin partial waves

$$\Psi(1, 2, \ldots, n) = \mathscr{A} \sum_{(l), i} \psi_{(l), i} | LM l_1 l_2 \cdots l_n \rangle \chi_i \tag{8}$$

In this expression, χ_i is a spin function with good quantum numbers S and S_z; (l) represents the set $(l_1, l_2, \ldots, l_n)$ collectively. The angular part is an eigenfunction of L^2 and L_z with eigenvalues $L(L+1)\hbar^2$ and $M\hbar$, respectively, and $\psi_{(l),i}$ is the corresponding radial wave function.

If $1s$ and $2s$ vacancies for electron j and a $2p$ vacancy for electron k are present, a wave function with these vacancies is given by

$$\Psi = \mathscr{A}[1 - P_{1s}(\mathbf{r}_j) - P_{2s}(\mathbf{r}_j)]\left[1 - \sum_m P_{2pm}(\mathbf{r}_k)\right]\sum_{(l),i} \psi_{(l),i}|LM(l)\rangle\chi_i \tag{9}$$

Note that antisymmetrization is done *after* the vacancies are built into the wave function. For atomic systems, the single-particle projection operators P_{nlm} are constructed from single-particle orbitals ϕ_{nlm}, e.g.,

$$P_{nlm}(\mathbf{r}) = |\phi_{nlm}(\mathbf{r})\rangle\langle\phi_{nlm}(\mathbf{r})| \tag{10}$$

with

$$\phi_{nlm}(\mathbf{r}) = R_{nl}(q|r)Y_{lm}(\hat{r}) \tag{11}$$

where

$$R_{nl}(q|r) = e^{-qr/n}\left(\frac{2qr}{n}\right)^l F_{nl}(q|r) \tag{12}$$

is a normalized radial function and $F_{nl}(q|r)$ is a polynomial of r. An important constraint on the vacancy orbitals ϕ_{nlm} is that they be mutually orthogonal, which prevents a variational breakdown. We have found it adequate for two-electron systems to use Laguerre polynomials for the F_{nl} and to use the same q for all the vacancy orbitals for a particular resonance. The many calculations that we have done show that the optimized q can be interpreted as the effective nuclear charge seen by the vacancy orbital. While the Laguerre polynomials seem to be a good approximation for two-electron systems, care must be taken when fixing the functional form of F_{nl} for systems with three or more electrons. This can be seen by considering the $(1s3s3p)^2P^0$ resonance. We would expect the $1s$ vacancy orbital to be half shielded by the $1s$ electron (see Section 3) but the $2s$ and $2p$ vacancy orbitals to be fully screened by the $1s$ electron. In order to implement these ideas, different q's would be needed for different orbitals. If we choose F_{10} to be the proper Laguerre polynomial, then the coefficients in F_{20} must be chosen such that ϕ_{2s} is orthogonal to ϕ_{1s}. In this case, F_{20} can no longer be a Laguerre polynomial.

The projection operation in Eq. (9) effects only the radial function, $\psi_{(l),i}$; this is because

$$\sum_{m=-l}^{l} P_{nlm}(\mathbf{r}_k) = \sum_{m=-l}^{l} |R_{nl}(r_k)Y_{lm}(\Omega_k)\rangle\langle R_{nl}(r_k)Y_{lm}(\Omega_k)| \tag{13}$$

and

$$\sum_{m=-l}^{l} |Y_{lm}(\Omega_k)\rangle\langle Y_{lm}(\Omega_k)|LMl_1l_2\cdots l_n\rangle = \delta_{ll_k}|LMl_1l_2\cdots l_n\rangle \tag{14}$$

This greatly simplifies the computational efforts. Notice also that the vacancy orbital does not contain any spin coordinate. This is because resonances are formed by orbital excitations. The correct spin eigenfunctions χ_i are assumed at the outset of the calculation. These functions are not affected by the presence of the vacancies.

3. EXCITED BOUND STATES—A TEST CALCULATION

The most common procedure for verifying the effectiveness of a theoretical method is to apply it to a system for which experimental results are available for comparison. For the energy of a Feshbach resonance, caution should be exercised in this procedure. This is because the calculated energy is usually too high if the correlation effect is not fully accounted for, but it may be lowered if the continuum is not properly handled. Hence, a seemingly accurate energy may result from canceling errors. The resulting wave function would then be inaccurate.

Fortunately, it is possible to perform a more definitive test of the method proposed in Section 2, because, while the hole-projection method is a general approach to inner-shell vacancy states, not all inner-shell vacancy states are Feshbach resonances. The metastable states $(1s2s)^3S$ and $(1s2s)^1S$ are well-known examples. Unlike the 2^3S, the $1s$ hole in the 2^1S could be filled by the excited electron. Hence, the energy of this state can be calculated by the hole-projection technique. On the other hand, according to a theorem proven by MacDonald,[18] the energy of this bound state can also be calculated with the Rayleigh–Ritz variation method by minimizing the energy of the second lowest root. If the same basis functions are used for both calculations, then the correlation effects should be similar in both wave functions, and therefore a meaningful comparison can be made. The accuracy of this method can then be easily accessed.

For this test calculation, we use the nonrelativistic Hamiltonian

$$H = -\frac{1}{2}\nabla_1^2 - \frac{1}{2}\nabla_2^2 - \frac{z}{r_1} - \frac{z}{r_2} + \frac{1}{r_{12}} \tag{15}$$

where $z = 2$ for helium. If the LS coupling scheme is used, the spatial part of the basis function can be expanded as the product of

$$\phi_{ij}(r_1, r_2) = r_1^i r_2^j e^{-(\alpha r_1 + \beta r_2)} \tag{16a}$$

and

$$|LMl_1l_2\rangle = \sum_{m_1m_2} \langle l_1l_2m_1m_2|LM\rangle Y_{l_1m_1}(\hat{r}_1)Y_{l_2m_2}(\hat{r}_2) \tag{16b}$$

The spin function for the singlet is represented by $\chi(\boldsymbol{\sigma}_1, \boldsymbol{\sigma}_2)$. An unprojected wave function takes the form of

$$\Psi = \mathscr{A} \sum C_{ijl_1l_2} \phi_{ij}(r_1, r_2) | LMl_1 l_2 \rangle \chi(\boldsymbol{\sigma}_1, \boldsymbol{\sigma}_2) = \mathscr{A} \psi(1, 2) \tag{17}$$

where $\mathscr{A}$ is the antisymmetrization operator and $C_{ijl_1l_2}$ are the linear parameters. For the hole-projection technique, we assume the $1s$ vacancy takes the form

$$\phi_0 = N e^{-qr} \tag{18}$$

where N is a normalization factor and q is the parameter to be optimized. Hence, the wave function in the hole-projection method becomes

$$\Psi' = \mathscr{A}[1 - P_{1s}(\mathbf{r}_2)] \psi(1, 2) \tag{19}$$

where

$$P_{1s}(\mathbf{r}) = |\phi_0(\mathbf{r})\rangle \langle \phi_0(\mathbf{r})| \tag{20}$$

For the $(1s2s)^1S$, $L = 0$, $M = 0$, and Eq. (17) can be reduced to

$$\Psi = \mathscr{A} \sum_{i,j,l} C_{ijl} \phi_{ij}(r_1, r_2) | 00ll \rangle \chi(\boldsymbol{\sigma}_1, \boldsymbol{\sigma}_2) \tag{21}$$

In this calculation, we first take a $l = 0$, 15-term function; α is fixed to be 2.0. The saddle point for the lowest root of the secular equation occurs at $\beta = 0.86$ and $q = 1.52$. The saddle-point energy is $\lambda^{(1)} = -2.144095$ a.u. Here, the superscript gives the number of angular partial waves l used. If we use the same α and β for Eq. (21) and carry out a Rayleigh–Ritz variational calculation for the second lowest root, we find $E^{(1)} = -2.144089$ a.u. Next, we repeat these calculations by including a 10-term $l = 1$ basis; we find again $q = 1.52$, but $\beta = 0.90$. In this case, $\lambda^{(2)} = -2.145366$ a.u. and $E^{(2)} = -2.145334$ a.u. If a 10-term $l = 2$ basis is further included, we find $\lambda^{(3)} = -2.145406$ and $E^{(3)} = -2.145373$ a.u. with the same q and β as $\lambda^{(2)}$. Similar calculations are also made for Li^+ and Be^{++}; these results are shown in Table 1. It can be seen from this table that results from the two methods usually differ at the sixth or seventh digit. It should be emphasized that the purpose of this calculation is not to obtain high accuracy for the energy of 2^1S; rather, it is a relatively simple example for illustrating the soundness of the hole-projection technique. For this reason, only one nonlinear parameter β is varied in the basis functions. In this regard, it is interesting to note that accurate results for the 2^1S states of He, Li^+, and Be^{++} are -2.14597, -5.04088, and -9.18487 a.u., respectively.[20]

In this calculation, the value for q is found to be very stable against an increasing number of terms or number of partial waves. It is interesting to see whether the saddle-point energy for higher states of this system occur at

TABLE 1
Comparison of the Saddle-Point Energy and the Energy of an Unprojected Wave Function (in a.u.)

State[a]	$He(1s2s)^1S$	$Li^+(1s2s)^1S$	$Be^{++}(1s2s)^1S$
$\lambda^{(1)}$	-2.144095	-5.037560	-9.180608
$E^{(1)}$	-2.144089	-5.037553	-9.180601
$\lambda^{(1)} - E^{(1)}$	-0.6×10^{-5}	-0.7×10^{-5}	-0.7×10^{-5}
$\lambda^{(2)}$	-2.145366	-5.039907	-9.183645
$E^{(2)}$	-2.145334	-5.039884	-9.183624
$\lambda^{(2)} - E^{(2)}$	-3.2×10^{-5}	-2.3×10^{-5}	-2.1×10^{-5}
$\lambda^{(3)}$	-2.145406	-5.040020	-9.183821
$E^{(3)}$	-2.145373	-5.039997	-9.183800
$\lambda^{(3)} - E^{(3)}$	-3.3×10^{-5}	-2.3×10^{-5}	-2.1×10^{-5}

[a] For notation in the first column, see text.

the same q. The calculation for He $(1s3s)^1S$ again gives $q = 1.52$, suggesting that

$$\phi_0 = (1.52)^{3/2} e^{-1.52r}/(\pi)^{1/2} \tag{22}$$

is probably a good approximation to the vacant 1s orbital for this system. In this calculation, a 21-term $l = 0$ and 15-term $l = 1$ wave function is used. Here, $\lambda^{(1)} = 2.060712$ a.u., $\lambda^{(2)} = -2.060794$ a.u., $E^{(1)} = -2.060741$ a.u., and $E^{(2)} = -2.060824$ a.u. are obtained. The fact that $q = 1.52$ seems to suggest that the presence of the 1s electron half shields the nucleus, so that the effective charge seen by the vacancy is 1.52.

The results from this section seem to suggest that this inner-shell vacancy state picture offers a plausible description of an excited state. It is obvious that this hole-projection method is not needed for a bounded-excited-state calculation, because it is more cumbersome and it needs slightly more computer time as compared with the standard Rayleigh–Ritz variation method. This method becomes very useful for the autoionizing systems.

4. RESONANCES OF HELIUM BELOW THE $n = 2$, $n = 3$, AND $n = 4$ THRESHOLDS OF He^+

To study closed-channel resonances, the simplest system is a two-electron atomic system. Among these, the most extensively investigated is the helium atom, because many highly accurate experimental results are available. In this section, results from the hole-projection technique are compared with accurate theoretical and experimental works in the literature.

It is well known that the Feshbach formalism[5] provides an exact formulation for two-electron scattering systems. In this formulation, the Hilbert space is divided into open and closed channels with the corresponding projection operators P and Q. It has been shown by Feshbach that the closed-channel resonant energy associated with an eigenvalue of the QHQ operator is[5]

$$E_{\text{res}} = \varepsilon_n + \Delta_n(E_{\text{res}}) \tag{23}$$

where ε_n is obtained by solving

$$(QHQ - \varepsilon_n)Q\Psi_n = 0 \tag{24}$$

and Δ_n is a shift due to coupling $Q\Psi_n$ and the rest of the Hilbert space through the total Hamiltonian. It has been shown by Hahn *et al.*[21] that the Q operator for the two-electron system in the elastic-scattering energy region is given by

$$\begin{aligned} Q\Psi &= [1 - P_{1s}(\mathbf{r}_1)][1 - P_{1s}(\mathbf{r}_2)]\mathscr{A}\psi(1, 2) \\ &= \mathscr{A}[1 - P_{1s}(\mathbf{r}_1)][1 - P_{1s}(\mathbf{r}_2)]\psi(1, 2) \end{aligned} \tag{25}$$

Equation (25) has the same form as the wave function in the hole-projection method where two $1s$ holes are present. The difference is that in the Feshbach formalism, $P_{1s}(\mathbf{r})$ is strictly hydrogenic and contains no parameters. Since the energy is to be maximized with respect to the vacancy parameter, the saddle-point energy will always be higher than or equal to ε_n. This provides a particularly interesting comparison between the results of the two methods. This comparison for the $^1P^0$ and $^3P^0$ resonances below the $n = 2$ and $n = 3$ thresholds of He^+ has been presented in Ref. (19).*

In Tables 2–5, energies and wave function results for the ^1S, ^3S, ^1D, and ^3D helium resonances below the $n = 2$ threshold of He^+ are given together with the radial expectation values

$$\langle r^m \rangle = \tfrac{1}{2}\langle \Psi | r_1^m + r_2^m | \Psi \rangle \qquad \text{for } m = -1, 1, 2 \tag{26}$$

and the mass polarization effect operator's[22] expectation value $\langle \mathbf{P}_1 \cdot \mathbf{P}_2 \rangle$. The mass polarization correction to the energy is very small for these resonances. It is very interesting to note, however, that when the resonances are classified into series, $\langle \mathbf{P}_1 \cdot \mathbf{P}_2 \rangle$ always takes the same sign within each series. In these tables, we see the difference of the saddle-point energy from our

* Editorial comment: Since the hole-projection eigenvalue is always greater than the Feshbach $\mathscr{E} = \langle QHQ \rangle$, it is clear that for $\Delta < 0$ the hole-projection eigenvalue will be equal to or further from the resonant energy $E_r = \mathscr{E} + \Delta$ than is $\mathscr{E}$. This happens in at least one important case for precision calculations: the $\text{He}(^1P^0)$ autoionization state below the $\text{He}^+(n = 2)$ threshold; cf. Ref. (31) and the summary of this article.

TABLE 2

$^1S^e$ Resonances of Helium below the $n = 2$ Threshold of He^+ ($-E$ in a.u.)

Series	Saddle-point energy	QHQ eigen-values	Composition (%) ss	Composition (%) pp	N	L	q	$\langle r \rangle$	$\langle \mathbf{p}_1 \cdot \mathbf{p}_2 \rangle$	Other theoretical calculations	Experiment[o]	
A	0.777882	0.778761	72	28	78	8	1.919	3.2794	−0.0013	0.778037[i] 0.777870[j] 0.77786[k] 0.776682[m]	0.7789 (18)[a] 0.7793 (18)[c] 0.7789 (15)[e]	0.7760 (18)[b] 0.7778 (18)[d] 0.7804 (11)[f]
B	0.622356	0.622567	33	64	107	8	2.074	3.7429	0.0818	0.621942[i] 0.621928[j] 0.62013[k] 0.620232[m]	0.6198 (18)[a] 0.6201 (22)[d] 0.6216 (11)[f]	0.6253 (18)[b] 0.6231 (11)[e]
A	0.589972	0.590077	62	38	82	7	1.964	5.9795	−0.0134	0.589920[i] 0.589925[j] 0.58922[k] 0.588538[m]	0.5904 (18)[a] 0.5893 (11)[c]	0.5959 (18)[b] 0.5907 (11)[e]
B	0.548182	0.548187	50	50	96	8	2.009	8.3534	0.0121	0.548080[i] 0.548090[j] 0.544566[m] 0.544856[n]		
A	0.544901	0.544918	50	50	83	7	1.983	10.076	−0.0049	0.544865[l] 0.532986[m] 0.543606[n]	0.5437 (18)[a] 0.5426 (22)[h]	0.5437 (11)[g] 0.5452 (11)[e]
B	0.527736	0.527736	60	40	91	7	2.003	14.104	0.0051	0.527682[l] 0.526591[n]		
A	0.526675	0.526679	40	60	73	6	1.990	15.630	−0.0019	0.526392[l] 0.525914[n]	0.5257 (18)[a] 0.5272 (15)[e]	0.5261 (7)[g]

[a] Ref. (23). [b] Ref. (24). [c] Ref. (25). [d] Ref. (26). [e] Ref. (27). [f] Ref. (28). [g] Ref. (29). [h] Ref. (30). [i] Ref. (31). [j] Ref. (32). [k] Ref. (33). [l] Ref. (34). [m] Ref. (35). [n] Ref. (36).

[o] The numbers in parentheses give the experimental uncertainty in the last digits quoted.

TABLE 3

$^3S^e$ Resonances of Helium below the $n = 2$ Threshold of He^+

Series	Saddle-point energy	QHQ eigenvalue	Composition (%)		N	L	q	$\langle r \rangle$	$\langle r^2 \rangle$	$\langle 1/r \rangle$	$\langle \mathbf{p}_1 \cdot \mathbf{p}_2 \rangle$	Other theoretical calculations
			ss	pp								
A	−0.602586	−0.602600	65	35	34	4	1.987	5.5384	42.813	0.32849	0.0194	−0.602605[a] −0.60245[b]
B	−0.559739	−0.559747	38	62	53	4	2.013	6.7400	70.162	0.30959	0.0026	−0.559763[a] −0.558086[c]
A	−0.548844	−0.548846	60	40	42	3	1.994	9.5438	152.95	0.29070	0.0055	−0.548838[a] −0.548747[c]
B	−0.532500	−0.532501	42	58	43	4	2.005	11.993	259.73	0.28232	0.0010	−0.532507[a] −0.531941[c]
A	−0.528411	−0.528412	57	43	41	3	1.997	15.042	419.92	0.27476	0.0023	−0.528332[d] −0.528380[c]

[a] Ref. (31). [b] Ref. (33). [c] Ref. (36). [d] Ref. (34).

TABLE 4

$^1D^e$ Resonances of Helium below the $n = 2$ Threshold of He^+ ($-E$ in a.u.)

Series	Saddle-point energy	*QHQ* eigen-value	Composition (%) *pp*	*sd*	*pf*	*N*	*L*	*q*	$\langle r \rangle$	$\langle \mathbf{p}_1 \cdot \mathbf{p}_2 \rangle$	Other theoretical calculations	Experiment
A	0.702161	0.7021169	93	7	0	84	8	1.753	3.1900	0.0004	0.701924[g] 0.70182[h]	0.7039 (7)[a] 0.7025 (18)[b] 0.7036 (11)[c] 0.7025 (18)[d] 0.7028 (11)[e] 0.7036 (11)[f]
B	0.569324	0.569329	70	29	1	90	8	1.987	6.4353	−0.0087	0.569211[g]	0.5702 (7)[a] 0.5702 (11)[e]
C	0.556407	0.556407	31	59	10	86	8	1.999	6.6544	0.0100	0.556420[g]	
B	0.536650	0.536651	63	34	3	88	8	1.996	11.156	−0.0033	0.536559[g]	0.5378 (11)[a] 0.5375 (11)[e]
C	0.531474	0.531474	37	54	9	84	8	2.000	11.894	0.0048		

[a] Ref. (29). [b] Ref. (24). [c] Ref. (25). [d] Ref. (26). [e] Ref. (27). [f] Ref. (28). [g] Ref. (31). [h] Ref. (33).

TABLE 5

$^3D^e$ *Resonances of Helium below the* $n = 2$ *Threshold of* He^+

Series	Saddle-point energy	*QHQ* eigenvalue	Composition (%)										
			pp	*sd*	*pf*	*N*	*L*	*q*	$\langle r \rangle$	$\langle r^2 \rangle$	$\langle 1/r \rangle$	$\langle \mathbf{p}_1 \cdot \mathbf{p}_2 \rangle$	Other theoretical calculations
A	−0.583774	−0.583790	75	24	1	69	8	1.977	5.6173	45.973	0.32065	0.0288	−0.5837556[a]
B	−0.560663	−0.560663	25	67	8	73	6	2.00	6.0845	55.107	0.31038	−0.0179	−0.5606790[a]
A	−0.541654	−0.541656	70	28	2	65	6	1.993	10.072	176.55	0.28708	0.0099	−0.5416534[a]
B	−0.533431	−0.533431	30	58	12	78	6	2.000	11.102	220.54	0.28324	−0.0068	−0.5334135[a]
C	−0.529305	−0.529306	1	14	85	60	6	2.006	10.847	206.82	0.28008	0.0023	

[a] Ref. (31).

TABLE 6

Energies of the $^1P^0$ Resonances of Helium below the $n = 4$ Threshold of He^+ ($-E$ in a.u.)

State	Ho[a]	Herrick and Sinanoglu[b]	Oberoi[c] *QHQ*	Present *QHQ*	Present saddle point	Madden and Coding[d] $E_{max} = E_{res} + (\Gamma/2g)$	Woodruff and Samson[e] $E_{min} = E_{res} - (\Gamma g/2)$
1	0.1945	0.1946	0.19488	0.195032	0.193990	0.19271 (81)	0.1962 (11)
2	0.1788	0.1760	0.17610	0.177720	0.177024		0.1784 (15)
3		0.1688	0.16884	0.168923	0.168862		
4		0.1604	0.16065	0.161401	0.161022	0.16023 (83)	0.1630 (11)
5		0.1603	0.16044	0.160742	0.160630		
6		0.1517	0.15179	0.151804	0.151793		
7		0.1507	0.15079	0.151108	0.151091		
8		0.1488	0.14885	0.149671	0.149671		
9		0.1473	0.14747	0.148807	0.148807		
10		0.1468	0.14686	0.147365	0.147307	0.14712 (84)	0.1497 (11)
11		0.1465		0.146635	0.146618		

[a] Ref. (42). [b] Ref. (43). [c] Ref. (44). [d] Ref. (39); results were converted using $E = 455.6947/\lambda$ (Å) and $E_g = -2.903724$ a.u. [e] Ref. (45).

corresponding QHQ eigenvalue. The composition columns give the contribution (in percentages) to the normalization from the main angular partial waves. The N is the total number of linear parameters, L is the number of partial waves used, and q is the optimized parameter in the vacancy orbital. Some of the most accurate experimental[23–30] and theoretical[31–36] results in the literature are also quoted for comparison. Since this is one of the most extensively studied systems, it is impossible to quote all the relevant work in the literature, and references given here are representative rather than inclusive. It can be seen from these tables that results from this hole-projection technique compare favorably with those from accurate theoretical calculations and experimental measurements in the literature.

For the resonances below the $n=3\,He^+$ threshold, the hole-projection method has been carried out for the ^1P and ^3P helium resonances. The low-lying resonances compare well with the complex-rotation calculation by Ho[37] and the close coupling with correlation calculation by Burke and Taylor.[38] Results from the ^1P resonances also agree extremely well with experiments by Madden and Codling[39] and Dhez and Ederer[40] if the shift of the resonance position from the position of maximum absorption line intensity is considered.[4] These results are presented in Ref. (19).

Recently, the ^1P resonances below the $n=4$ threshold of He^+ have also been calculated with the hole-projection technique.[41] These energy levels are given in Table 6 together with the theoretical work of Ho,[42] Herrick and Sinanoglu,[43] and Oberoi.[44] The experimental results of Madden and Codling[39] and Woodruff and Samson[45] are also given in this table. The QHQ results of our work are lower than those of Oberoi[44] and Herrick and Sinanoglu,[43] which indicates that more correlation has been accounted for in our work. Our results lie lower than the maximum absorption line intensity positions of Madden and Codling. This position is related to the resonance energy by[4]

$$E_{\max}= E_{\text{res}} + \Gamma/2g \tag{27}$$

where Γ is the width of the resonance and g is its line profile parameter. The minimum position is given by

$$E_{\min} = E_{\text{res}} - \Gamma g/2 \tag{28}$$

Recently, Woodruff and Samson[45] have measured the photoionization cross section to the He^+ $(n=2)$ as a function of photon energy. They observe several minima associated with resonances below $n=4$. If we assume these to be the approximate positions of Eq. (28) and use the saddle-point energy as E_{res} and Madden and Codling's results for $E_{\max}$ we can then solve for Γ and g of these resonances. If we apply this procedure to the two lowest observed resonances, we find $g=1.37$, 1.58, and $\Gamma=0.0035$, 0.0025 a.u., respectively, for the first and fourth resonances. Due to the large uncertainties quoted in the experiments, these figures can be considered only as estimates. But it is of interest to note that for the lowest resonance, the width Γ agrees with the width 0.0034 a.u. given by the complex rotation calculation of Ho.[42] Our identification of the observed lines in Table 6 is different from that by Oberoi[44] and Herrick and Sinanoglu.[43] This is probably due to the fact that they may not have included a sufficient amount of correlation in their wave functions.

5. He^- RESONANCES WITH A SINGLY OR DOUBLY EXCITED CORE

5.1. *Singly Excited Core*

The electron–helium system is of special interest in the study of electron–atom scattering theory. The system is simple enough to be able to include most correlation effects in a numerical calculation. It is also sufficiently complicated so that the elegant Feshbach formalism can no longer be rigorously implemented. Many resonances of He^- have been observed in experiments and measured to good accuracy; the lowest of these is the $(1s2s2s)^2S$ Feshbach resonance. This state has been studied extensively both experimentally[46] and theoretically;[34,47] results of the theoretical calculations generally agree with those of experiment.

Above this ^2S resonance is a broad ^2P resonance lying approximately 0.5 eV above the helium $(1s2s)^3S$ threshold. Until very recently, it was designated as a shape resonance. Shape resonances differ from Feshbach resonances in that they are associated with an open channel and formed by the centrifugal potential barrier of the scattered electron. On the other hand, Feshbach resonances involve the excitation of the target electrons into a closed channel. The energy of the incoming electron becomes negative, so that the scattering system becomes quasi-bound. It then decays through an open channel due to the coupling of the Coulomb interaction between the electrons.

To study this ^2P resonance, the helium targets of interest are $(1s^2)^1S$, $(1s2s)^{3,1}S$, and $(1s2p)^{3,1}P$. The 2^3S and 2^1S are higher than the ground state by 19.824 and 20.620 eV, respectively. Experimentally, the position of the ^2P resonance is found to be at about 20.3–20.45 eV. Therefore, if the resonance is associated with the $(1s2s)^3S$ channel only, it must be a shape resonance. However, if it is formed by the virtual excitation to the other closed channels, it should be interpreted as a Feshbach resonance with the most likely configuration of $[(1s2s)^1S, 2p]^2P$. The wave function for this configuration can be constructed by building in a $1s$ vacancy through the hole-projection technique.

The nonrelativistic Hamiltonian is given by

$$H = -\sum_{i=1}^{3}\left(\frac{1}{2}\nabla_i^2 + \frac{2}{r_i}\right) + \sum_{i>j}^{3}\frac{1}{r_{ij}} \tag{29}$$

The basis functions for this three-electron system are chosen to be

$$\psi_{mnk}^{l_1 l_2 l_{12} l_3}(1,2,3) = \phi_{mnk}(r_1, r_2, r_3)\, Y_{l_1 l_2, l_{12}}^{l_3, LM}(\hat{r}_1, \hat{r}_2, \hat{r}_3)\; \chi(\boldsymbol{\sigma}_1, \boldsymbol{\sigma}_2, \boldsymbol{\sigma}_3) \tag{30}$$

where the spatial part is given by

$$\phi_{mnk}(r_1, r_2, r_3) = r_1^m r_2^n r_3^k e^{-(\alpha_j r_1 + \beta_j r_2 + \gamma_j r_3)} \tag{31a}$$

and

$$Y_{l_1 l_2, l_{12}}^{l_3, LM}(\hat{r}_1, \hat{r}_2, \hat{r}_3) = \sum_{m_1, m_2, m_3, \mu} Y_{l_1}^{m_1}(\hat{r}_1) Y_{l_2}^{m_2}(\hat{r}_2) Y_{l_3}^{m_3}(\hat{r}_3) \times \langle l_1 l_2 m_1 m_2 | l_{12} \mu \rangle \langle l_{12} l_3 \mu m_3 | LM \rangle \tag{31b}$$

and the spin part is given by

$$\chi(\boldsymbol{\sigma}_1, \boldsymbol{\sigma}_2, \boldsymbol{\sigma}_3) = [\alpha(1)\beta(2) \pm \alpha(2)\beta(1)]\alpha(3) - (1 \pm 1)\alpha(1)\alpha(2)\beta(3) \tag{31c}$$

The minus sign is chosen when electrons 1 and 2 form a singlet and plus is used for a triplet. The exponents α_j, β_j, and γ_j are the nonlinear parameters for the jth angular and spin partial wave. In the LS coupling scheme, L, M, S, and S_z are the good quantum numbers, and the energy is independent of M and S_z. To simplify the notation, we define

$$Y_{l_1 l_2, l_{12}}^{l_3, LM}\chi = [(l_1, l_2)^k l_{12}, l_3] \tag{32}$$

where k is the multiplicity of the (l_1, l_2) core. The total wave function then takes the form

$$\Psi = \mathscr{A} \sum C_{mnk}^{l_1 l_2 l_{12} l_3} [1 - P_{1s}(i)] \psi_{mnk}^{l_1 l_2 l_{12} l_3}(1,2,3) \tag{33}$$

where

$$P_{1s}(i) = |\phi_0(\mathbf{r}_i)\rangle\langle\phi_0(\mathbf{r}_i)| \tag{34}$$

and $\mathbf{r}_i$ is the coordinate of the $2s$ electron; ϕ_0 takes the same form as Eq. (18).

In order to see how this ^2P resonance is formed, a test calculation is carried out with only three partial waves: a 19-term $[(s,s)^1S, p]$, a 10-term $[(p,p)^1S, p]$, and a 10-term $[(s,p)^1P, d]$. The vacancy orbital parameter q is chosen to be the optimized value 1.5. The α_j and β_j in the $(s,s)^1S$, $(p,p)^1S$ and $(s,p)^1P$ two-electron cores of the partial waves are chosen such that a low energy is obtained as $\gamma_j \rightarrow 0$. With the number of terms in this wave function fixed, the energy of He^- is calculated as a function of γ from 0 to 1. This is equivalent to bringing the third electron in from infinity toward the two-electron target. The result of this calculation is shown in Fig. 1.

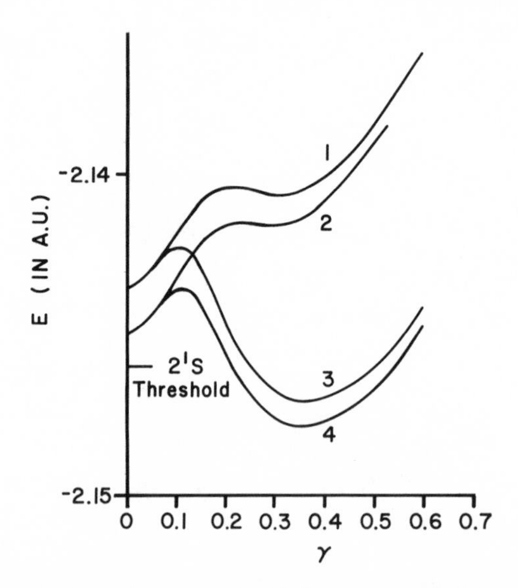

FIGURE 1. Energy calculation for the $(1s2s2p)^2P^0$ resonance of He^- as a function of the nonlinear parameter of the third electron γ. Curve 1 is a 19-term $[(s,s)^1S,p]$ calculation, $\alpha_1 = 2.05$, $\beta_1 = 0.415$. Curve 2 is a two-partial-wave calculation that includes a 10-term $[(p,p)^1S,p]$ and the calculation for curve 1, $\alpha_2 = 1.90$, $\beta_2 = 1.80$. Curve 3 is also a two-partial-wave calculation, where a 10-term $[(s,p)^1P,d]$ is used in addition to that for curve 1, $\alpha_2' = 2.0$, $\beta_2' = 0.49$. Curve 4 is a 39-term calculation obtained by combining all three partial waves.

In Fig. 1, four curves are plotted; curve 1 is the result of a 19-term $[(s,s)^1S,p]$ calculation. As $\gamma \to 0$, the energy is about -2.1435 a.u.; this energy rises as γ increases, showing the effect of the centrifugal potential as the electron approaches the target. There is a slight dip in energy when $\gamma \simeq 0.3$; however, it is well above the 2^1S threshold. This situation remains even if the 2^1S target wave function is improved, which can be seen from curve 2 where a 10-term $[(p,p)^1S,p]$ is added to curve 1. It gives an overall lowering of the total energy, but it does not lead to a resonance. The situation is drastically changed if a 10-term $[(s,p)^1P,d]$ partial wave is used instead of $[(p,p)^1S,p]$; this is illustrated in curve 3. As expected, this term does not contribute to the energy as $\gamma \to 0$. But as γ increases, it leads to a dramatic lowering of the total energy. The minimum at $\gamma = 0.37$ is about -2.1471 a.u., significantly below the 2^1S threshold. This shows that the ^2P resonance is a Feshbach resonance, arising from the strong dipole coupling of the $[(1s2s)^1S,2p]$ and $[(1s2p)^1P,3d]$ configurations. Curve 4 is the net result of all three partial waves. Compared with curve 3, the energy is generally lowered, but the results are qualitatively unchanged.

A similar calculation is performed for the triplet target states. In this case, a 19-term $[(s,s)^3S,p]$, 10-term $[(p,p)^3S,p]$, and a 10-term $[s,p)^3P,d]$ are

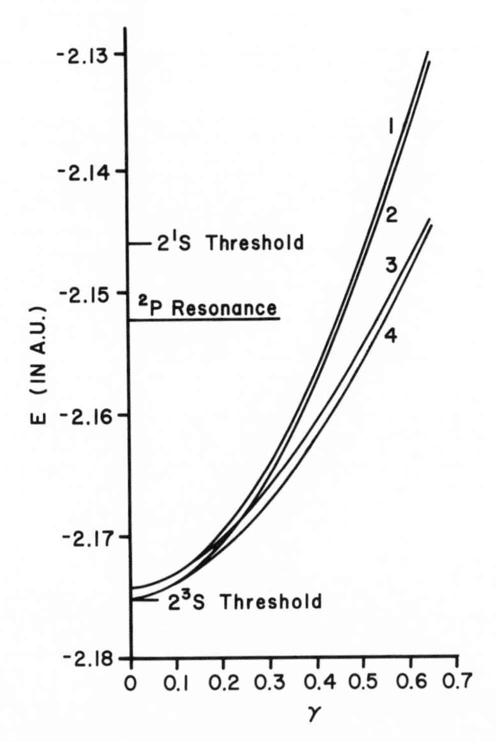

FIGURE 2. The corresponding results obtained as in the case of Fig. 1 if triplet target states are used rather than singlets. Here, the partial waves used are: a 19-term $[(s,s)^3S,p]$, a 10-term $[(p,p)^3S,p]$, and a 10-term $[(s,p)^3P,d]$. All four curves show that the energy of this system becomes monotonically higher as the third electron approaches the triplet targets. The third electron therefore sees a repulsive potential, and a shape resonance cannot be formed.

used. The results appear in Fig. 2, which shows that if the third electron is moved toward the triplet targets, the energy of the system will become monotonically higher. This electron sees a repulsive potential, so that a shape resonance cannot be formed. It has been argued in the past that this shape resonance is formed because of the attractive polarization potential of the $(1s2s)^3S$ target. However, our calculation shows (as indicated in Fig. 2) that the exchange energy is positive for this ^2P state, hence, canceling the effect due to polarization. The results from Figs. 1 and 2 show that it is a Feshbach resonance!

The energy of this ^2P resonance is calculated by using a seven partial wave 54-term wave function;[48] the result is -2.14905 a.u. (20.536 eV). The comparison of this result with other theoretical[49–51] and experimental[52,55] results is given in Table 7. It should be pointed out that the result of Bhatia and

TABLE 7
$(1s2s2p)^2P$ Resonance Energy for He^-

	Energy (eV)	Method
Theory	20.17	Matrix-variation method, Ref. (49)
	20.19	Close-coupling and R-matrix method, Ref. (50)
	20.536[a]	Hole-projection technique, Ref. (48)
	20.525, 20.560[b]	Quasi-projection-operator technique, Ref. (51)
Experiment	20.3 ± 0.3	Ref. (53)
	20.35 ± 0.3	Ref. (52)
	20.45 ± 0.05	Ref. (54)
	20.5	Ref. (55)

[a] Relative to the "exact" nonrelativistic ground-state energy of He at -2.903724 a.u. The conversion factor used is 1 a.u. = 27.211652 eV. The 2^3S and 2^1S threshold energies are at 19.824 V and 20.620 eV, respectively.
[b] The two results are from the closed-shell and open-shell target state wave functions, respectively.

TABLE 8
Energy of the $(2s2s2p)^2P$ Resonance of He^- (in a.u.)

Partial wave	No. of terms	α, β, γ[a]	$-\Delta E$[b]
$[(s,s)^1S, p]$	21	0.74, 0.74, 0.725	0.752566
$[(p,p)^1S, p]$	13	0.8, 0.8, 0.7	0.027756
$[(s,p)^1P, d]$	10	0.47, 0.71, 0.9	0.007683
$[(s,p)^1P, s]$	8	0.36, 0.72, 0.85	0.001838
$[(s,p)^3P, d]$	10	0.72, 0.81, 0.80	0.006789
$[(s,d)^1D, f]$	2	0.59, 0.85, 1.2	0.000411
$[(p,d)^1P, d]$	8	0.63, 0.99, 0.65	0.002637
$[(p,p)^1D, f]$	4	0.84, 0.84, 1.05	0.000572
$[(d,d)^1S, p]$	7	1.2, 1.2, 0.7	0.000158
$[(f,f)^1S, p]$	2	1.4, 1.4, 0.7	0.000031
$[(p,d)^1P, s]$	4	0.72, 0.72, 0.9	0.000077
$[(p,d)^3P, d]$	4	0.75, 0.75, 0.75	0.000089
$[(s,d)^3D, f]$	2	0.47, 1.05, 1.4	0.000012
$[(s,p)^3P, s]$	7	0.43, 0.9, 0.7	0.000146
Total	102		0.800765 (57.225 eV)
Expt.	Kuyatt *et al.* [Ref. (56)]		57.1 ± 0.1 eV
	Grissom *et al.* [Ref. (59)]		57.21 eV
	Hicks *et al.* [Ref. (60)]		57.22 ± 0.04 eV

[a] α, β, and γ are the nonlinear parameters used in the partial wave.
[b] $-\Delta E$ is the contribution to the binding energy due to the partial wave.

Temkin[51] also shows it to be a Feshbach resonance. Their calculated width is 0.44 eV, which agrees with that of the experiments.

5.2. *Doubly Excited Core*

In 1965, Kuyatt *et al.*[56] reported observing He^- resonant structures at 57.1 and 58.2 eV. These were subsequently confirmed by many later experiments. The lower resonance was analyzed by Fano and Cooper[57] to be the $(2s2s2p)^2P$ state. Even though the energy of this state has been calculated by many methods, the results are either slightly too high or too low. Recently, the energy and wave function of this state were recalculated with the hole-projection technique. In general, the wave function of a three-electron system with a doubly excited core can be written as

$$\Psi(1,2,3) = \mathscr{A}[1 - P_{1s}(\mathbf{r}_1)][1 - P_{1s}(\mathbf{r}_2)][1 - P_{1s}(\mathbf{r}_3)]\psi(1,2,3) \tag{35}$$

The three projections taken in this equation prevent any of the electrons from filling the $1s$ hole. In practice, the computation is greatly simplified because of Eq. (14). By using a 14 partial wave, 102-term wave function, the energy of this $^2P^0$ state was found to be -0.800765 a.u. (57.225 eV). This result is given in Table 8. For this ^2P state, the $2s2s2p$ configuration contributes about 80.2% to the normalization. The $2p2p2p$ configuration contributes 14.7% to the normalization and 0.755 eV to the binding energy. The other configurations contribute 0.502 eV to the binding energy but only 5.1% to the normalization; among them, 4.4% comes from $2s2p3d$ configurations. The result presented in Table 8 is a significant improvement over that of Chung.[58] This is because the computer code used was improved so that each angular partial wave is allowed to have a different set of optimized nonlinear parameters and a term selection option was incorporated into the code to eliminate the ineffective linear parameters. The new result 57.225 eV is still higher than that of Kuyatt *et al.*[56] but in excellent agreement with the result of 57.21 eV obtained by Grissom *et al.*[59] and the more recent result of 57.22 ± 0.04 eV from Hicks *et al.*[60]

The second structure at 58.2 ± 0.1 eV observed by Kuyatt *et al.*[56] is found at 58.31 eV by Grissom *et al.*[59] and 58.30 ± 0.04 eV by Hicks *et al.*[60] This structure has been interpreted to be a $(2s2p2p)^2D$ Feshbach resonance.[57] Many theoretical calculations have been carried out that seem to support this assignment. However, it is important to note that the position of this structure is very close to the $(2s2p)^3P$ resonance of helium, which has a measured position of 58.34 ± 0.05 eV[23] and a calculated position of 58.32 eV.[31] It is higher than the $(2s2s)^1S$ but lower than that of the second ^1S resonance by about 3.8 eV and lower than the $(2s2p)^1P$ by about 1.9 eV.

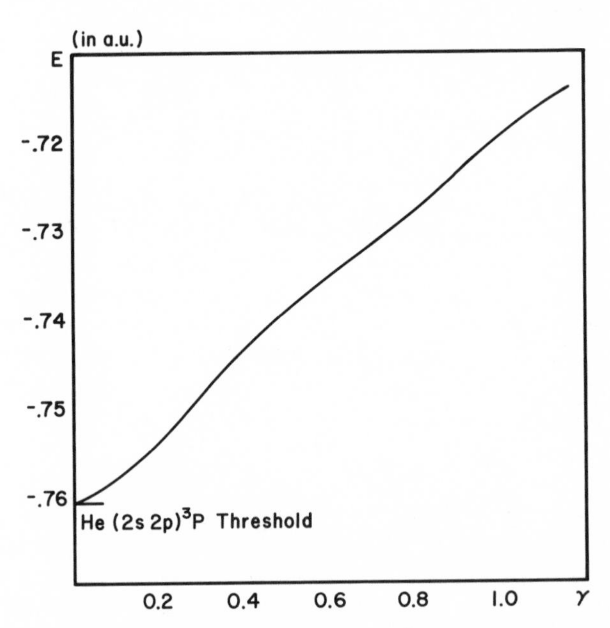

FIGURE 3. Energy calculation for the supposed $(2s2p2p)^2D$ resonance of He^- as a function of the nonlinear parameter of the third electron γ. The wave function used has 46 terms with three partial waves: $[(s,p)^3P,p]$, $[(s,p)^1P,p]$, and $[(p,d)^3P,p]$; $\alpha = 1.0$, $\beta = 0.96$, and $q = 1.92$. The fact that the energy monotonically decreases to the $(2s2p)^3P$ threshold as the third electron is moved out to infinity shows that the $(2s2p2p)^2D$ Feshbach resonance cannot be formed.

Hence, the $(2s2p)^3P$ is the single most important channel to consider. As in the case of the $(1s2s2p)^2P$ resonance, the exchange effect again tends to cancel the effect of polarization; hence, the $(2s2p2p)^2D$ Feshbach resonance may not be formed. A three partial wave, 46-term calculation supports this conclusion, which is shown in Fig. 3. The result does not corroborate the previous assignment by Fano and Cooper.[57] What has been seen in the experiment could be the result of a postcollision interaction effect.[61]

In direct contrast to this $(2s2p2p)^2D$ is the $(2s2p2p)^4P$ state of He^-, both polarization and exchange effects give rise to an attractive potential sufficient to sustain a Feshbach resonance. With a nine partial wave, 76-term wave function, the energy is found to be -0.79357 a.u., which is lower than the $(2s2p)^3P$ threshold by about 0.900 eV. The result of this calculation is shown in Table 9; it is an improvement over the result of Chung,[62] due to the improved computer code. This resonance does not show up in an electron-helium collision experiment, but it could be found in an ion-atom collision experiment. It should also be seen in a photoabsorption experiment from the $He^-(1s2s2p)^4P$ state.

TABLE 9
Energy and Wave Function of $2s2p2p\,^4P$ Resonance of He^-

Partial wave	No. of terms	α, β, γ	$-\Delta E$(a.u.)
$[(s,p)^3P, p]$	20	0.75, 0.66, 0.775	0.783953
$[(s,d)^3D, d]$	13	0.60, 1.05, 0.90	0.004233
$[(p,d)^3P, p]$	9	0.72, 0.91, 0.725	0.004321
$[(s,f)^3F, f]$	6	0.59, 1.35, 1.10	0.000207
$[(d,f)^3P, p]$	9	0.91, 1.10, 0.675	0.000553
$[p,f)^3D, d]$	5	0.75, 1.05, 0.80	0.000183
$[(f,g)^3P, p]$	4	1.20, 1.40, 0.90	0.000040
$[(p,p)^3P, d]$	6	0.78, 0.78, 0.75	0.000048
$[(s,g)^3G, g]$	4	0.35, 1.4. 1.90	0.000030
Total	76		0.793568

6. Li RESONANCES WITH A SINGLY OR DOUBLY EXCITED CORE

Although increasingly higher resolution has been achieved in recent years in electron–atom collision experiments, the resolution of the photoabsorption experiments remains best in measuring the position of a resonant state. In many cases, a standard deviation of 0.003 eV has been reported.[63,64] The dipole selection rule also makes identifying resonances seen in these experiments much simpler, especially for higher members of a Rydberg series.

The hole-projection technique has generated accurate results for the He^- resonances. However, they are the lowest resonances of each symmetry in the particular energy region of interest. It is not clear whether the method is also useful for higher members of a resonant series. In the case of lithium, there are many highly accurate experimental results available. For example, the $(1s2snp)^2P$ resonances from $n = 2$ to 10 have been measured accurately.[63] Other highly accurate results are also available for the ^2S and ^2D resonances in the elastic and inelastic scattering region.[65] Hence, lithium is an ideal system for testing the effectiveness of the hole-projection technique for higher resonances. Detailed calculations have been carried out for lithium ^2P and ^2D resonances, and these results agree excellently with those obtained from experiments.[66]

The lowest Feshbach resonance for lithium is $(1s2s2s)^2S$. By using a 10 partial wave, 79-term wave function, the energy from the hole-projection method is -5.40522 a.u. (56.404 eV). This is higher than the experimental result of Pegg *et al.* (56.31 ± 0.03 eV)[67] and the result of Zeim *et al.* (56.362 ±

TABLE 10

Energy and Wave Function of the $1s2s2s\,^2S$ Resonance of the Lithium Atom (in a.u.)

Partial wave	No. of terms	α, β, α	$-\Delta E$
$[(s,s)^1S, s]$	15	3.0, 1.85, 0.85	5.326666
$[(s,s)^3S, s]$	17	3.0, 1.1. 0.90	0.031081
$[(p,s)^1P, p]$	5	1.0, 3.0, 1.0	0.045122
$[(p,p)^3S, s]$	8	3.2, 1.84, 0.9	0.001012
$[(p,p)^1S, s]$	7	3.4, 2.24, 1.1	0.000784
$[(p,d)^1P, p]$	5	1.4, 2.24, 1.7	0.000219
$[(d,d)^1S, s]$	7	2.2, 1.28, 2.7	0.000159
$[(p,s)^3P, p]$	5	1.4, 3.0, 1.0	0.000063
$[(d,d)^3S, s]$	3	3.0, 2.8, 0.8	0.000067
$[(p,d)^3P, p]$	7	2.8, 2.24, 1.2	0.000046
Total	79		5.405219

0.01 eV).[68] It agrees, however, with more recent experiments by Rassi *et al.* (56.395 ± 0.015 eV)[69] and Røbdro *et al.* (56.37 ± 0.05 eV).[70] Our result is higher than the calculated energy of Bhatia and Temkin (56.3677 eV),[71] but lower than the result of Bhatia (56.424 eV)[72] and of Weiss (56.54 eV).[73]

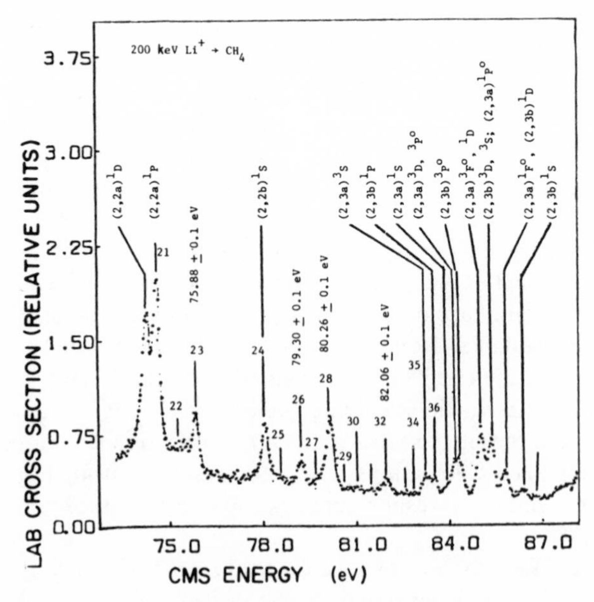

FIGURE 4. The Auger spectra of $Li^+ \rightarrow CH_4$. The Auger lines from Li^+ in this figure were identified in Ref. (70).

TABLE 11
Autoionizing Channel Energies for Some Triply Excited Resonances of the Lithium Atom (in eV)

Resonances	Energy[a]	Autoionizing channel energy[b]				Observed line,[c] (Fig. 4) number and position	
		$(1s2s)^3S$	$(1s2s)^1S$	$(1s2p)^3P$	$(1s2p)^1P$		
						21	74.68 ± 0.1
$(2s2s2p)^2P^0$	142.272	77.854	75.953	75.595	74.660	22	75.36 ± 0.1
$(2s2p2p)^4P^e$	142.547			75.870		23	75.88 ± 0.1
						24	78.13 ± 0.1
$(2s2p2p)^2D^e$	144.817	80.399	78.498	78.140	77.195	25	78.7 ± 0.2
$(2s2p2p)^2S^e$	146.517	82.099	80.198	79.840	78.905	26	79.30 ± 0.1
						27	79.9 ± 0.1
$(2p2p2p)^2D^0$	146.960			80.283	79.348	28	80.26 ± 0.1
$(2s2p2p)^2P^e$	146.990			80.313	79.378	29	80.52 ± 0.1
						30	81.2 ± 0.2
$(2p2p2p)^2P^0$	148.972	84.554	82.653	82.295	81.360	31	81.6 ± 0.2
$(2s2s3s)^2S^e$	149.344	84.926	83.025	82.667	81.732	32	82.06 ± 0.1
						35	83.36 ± 0.1
$^2P^e(2)$	149.765			83.088	82.153	36	83.60 ± 0.1
$^4P^e(2)$	150.125			83.448			
$^2D^e(2)$	150.251	85.833	83.932	83.574	82.639		

[a] Relative to the lithium ground state at −7.478025 a.u. See S. Larsson, *Phys. Rev.* **169**, 49 (1968).
[b] Threshold energies are form Y. Accad. C. L. Pekeris, and B. Schriff, *Phys. Rev. A* **4**, 516 (1971).
[c] Rødbro, Bruch, and Bisgaard [Ref. (70)] Lines 21, 24, 35, and 36 are identified with Li^+ Auger lines.

The energy and wave function of this calculation are given in Table 10.

The most important application of the hole-projection technique is probably to multiply core-excited (i.e., Feshbach) resonances. This is because for resonances in the elastic and low-lying inelastic region, it is possible to obtain good approximations from other methods, such as the close-coupling[(10)] or the quasi-projection-operator methods.[(14)] However for the doubly core-excited many-particle systems, there could be an infinite number of open channels. Although many theoretical methods are available,[(74–79)] the reliability of these methods is not clear. In addition to improved experimental techniques, many accurate doubly core-excited lithium results have become available. For example, by colliding Li^+ on CH_4, Rødbro *et al.*[(70)] observed many Auger lines that seem to be coming from the triply excited lithium atom. In this type of collision process, resonances of many assorted symmetries are formed, therefore, various Auger decays may result. None of these observed Auger lines have been positively identified.

Recently, a detailed calculation has been carried out for the 11 low-lying triply excited lithium resonances.[78] The symmetries considered are $^2P^0$, $^4P^e$, $^2D^e$, $^2S^e$, $^2P^e$, and $^2D^0$. The results give very interesting identifications for the experimental data[70] (see Fig. 4). For example, lines 23, 26, 28, and 32 have appreciable intensity, but they are not coming from the Auger lines of Li^+. The calculated results show that these lines are coming from upper states with only one or two decay channels. All the other resonances have five decay channels [i.e., $1s^2$, $(1s2s)^{3,1}S$, and $(1s2p)^{3,1}P$]; these calculated results are compared with experiment in Table 11.

Rødbro *et al.*[70] also reported Auger lines coming from the decay of triply excited Be^+; these lines have not been identifical in the literature. A detailed calculation[79] again shows that the most prominent lines are coming from the symmetry-preferred decays. The agreement with the experiment is excellent; the experimental spectrum is given in Fig. 5. The calculated results are given in Table 12. Details of this calculation are given elsewhere.[79]

TABLE 12

Autoionizing Channel Energies for Some Triply Excited Resonances of Be^+ (in eV)

Resonances	Energy[a]	Autoionizing channel energy[b]				Observed line,[c] (Fig. 5) number and position	
		$(1s2s)^1S$	$(1s2s)^1S$	$(1s2p)^3P$	$(1s2p)^1P$		
$(2s2s2p)^2P^0$	268.748	131.973	128.918	128.648	126.901	32	126.9 ± 0.2
						33	127.1 ± 0.3
						34	127.4 ± 0.2
$(2s2p2p)^4P^e$	269.063			128.963		35	128.3 ± 0.2
						36	128.9 ± 0.2
						37	129.6 ± 0.3
$(2s2p2p)^2D^e$	272.653	135.878	132.823	132.553	130.806	38	131.6 ± 0.3
						39	132.4 ± 0.2
$(2s2p2p)^2S^e$	275.330	138.555	135.499	135.230	133.483	40	132.8 ± 0.2
						41	134.0 ± 0.2
$(2s2p2p)^2P^e$	275.842			135.743	133.995	42	135.4 ± 0.3
						43	135.8 ± 0.2
						44	137.0 ± 0.3
$(2p2p2p)^2D^0$	275.998			135.898	134.151	45	138.5 ± 0.3
						46	138.8 ± 0.2
$(2p2p2p)^2P^0$	278.978	142.201	139.145	138.875	137.128	47	140.6 ± 0.3
						48	142.4 ± 0.3
						49	143.7 ± 0.3
$^2P^e(2)$	285.917			145.818	144.071	50	144.0 ± 0.3

[a] Energy is relative to the Be^+ ground state at -14.32350 a.u. See A. W. Weiss, *Phys. Rev.* **122**, 1826 (1961).
[b] Threshold energies are from Y. Accad, C. L. Perkeris, and B. Schiff, *Phys. Rev. A* **4**, 516 (1971).
[c] See Ref. (70).

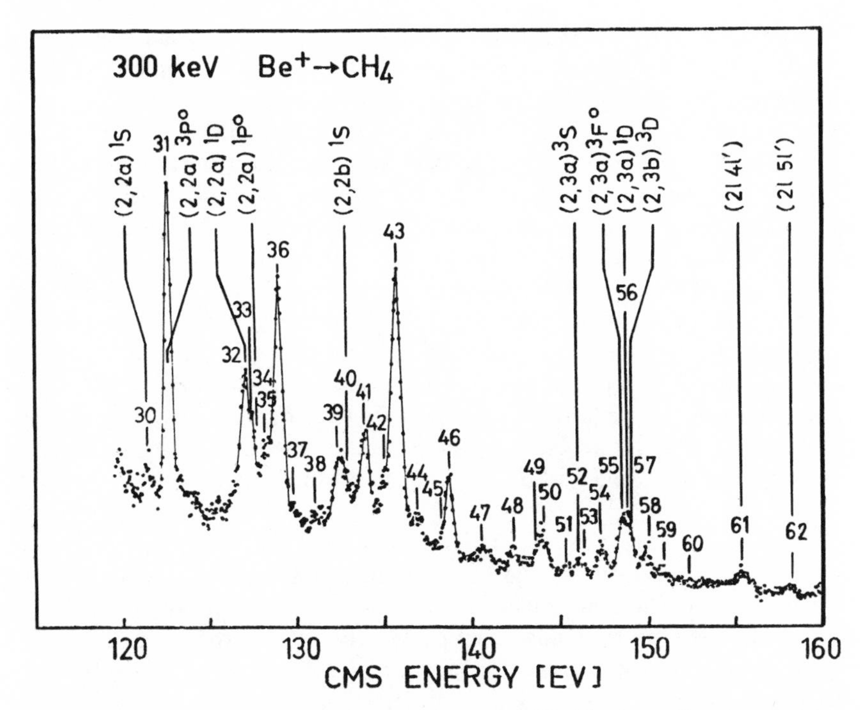

FIGURE 5. The Auger spectra of $Be^{+} \rightarrow CH_4$. From Ref. (70). The Auger lines from Be^{++} in this figure were identified in Ref. (70).

7. SUMMARY

In this chapter, some of the results based on the application of the hole-projection technique have been presented. The method is still relatively new, having been developed about two years ago. Nevertheless, many results have been generated.

The basic assumption in this method is to consider Feshbach resonances as quantum states with well-defined inner-shell vacancies. Each vacancy can be represented by a single-particle orbital function. These vacancies can then be built into the total trial wave function. It is well known that in the conventional Rayleigh–Ritz variation method, the energy is a minimum with respect to the parameters in the wave function. The theorem in Section 2 clearly shows that for inner-shell vacancy states, the energy will be a maximum with respect to the parameters in the vacancy orbital. This shows that the energy is a saddle point.

An advantage of this method is that it is very simple to apply: calculating a Feshbach resonance with this method is very similar to calculating a bound state. The MacDonald theorem can again be used to find the higher resonances. The maximization and minimization for each eigenvalue can easily be carried out. If suitable basis functions are chosen, convergence is usually rapid. There are no spurious solutions in the secular equation. Although we have applied the method to only two- and three-electron atomic systems, there should be no difficulty using if for more complex systems.

The results presented in Sections 4–6 seem to be highly accurate. It should be emphasized that these results have been obtained only after many radial and angular correlations have been included in the wave function. By maximizing the energy with respect to the parameters in the vacancy orbitals, the open-channel continua are excluded from the wave function. If some of the crucial correlations are not included in the wave function, the calculated result is likely to be poor. This is especially true for the triply excited three-electron systems. However, if all the necessary correlations are included, it is likely that we would find the result from this hole-projection method very rewarding.

The major weakness of this method is that the width of a resonance cannot be obtained unless the open channels are explicitly included in the wave function. There may also be some residual shift from the saddle-point energy to the true resonance energy. On the other hand, since the open-channel wave function is not square integrable, including these functions may increase the computational effort substantially.

One possibility of resolving this problem is to combine this hole-projection method with the complex-rotation method, where the width and energy of a resonance can be calculated with a square-integrable wave function. For the complex-rotation method, the major obstacle to extensive applications of larger systems has been poor convergence. Hence, it would be most desirable if we could combine the merits of both methods into a rapidly convergent theoretical method. Recently, such a method has been proposed. Calculations have been carried out for the He $2s2s\,^1S$ and $2s2p\,^1P$ resonances with very encouraging results.[80] The energy and width are both stable over a wide range of rotational angles. Currently, investigations are underway to extend this method to multiplichannel problems.

REFERENCES

1. For a review, see W. Bambynek, B. Crasemann, R. W. Fink, H. V. Freund, H. Mark, C. D. Swift, R. E. Price, P. V. Rao, *Rev. Mod. Phys.* **44**, 716 (1972).
2. T. Y. Wu, *Phys. Rev.* **46**, 239 (1934); **66**, 291 (1944).
3. P. G. Burke and H. M. Schey, *Phys. Rev.* **126**, 147 (1962).
4. U. Fano, *Phys. Rev.* **124**, 1866 (1961).

5. H. Feshbach, *Ann. Phys.* **5**, 357 (1958); **19**, 287 (1962).
6. G. K. Bhatia, A. Temkin, and I. J. Perkins, *Phys. Rev.* **153**, 177 (1967); and A. K. Bhatia and A. Temkin, *Phys. Rev.* **182**, 15 (1969).
7. K. T. Chung and J. C. Y. Chen, *Phys. Rev. Lett.* **27**, 1112 (1971); *Phys. Rev.* **A13**, 1655 (1976).
8. G. J. Seiler, R. S. Oberoi, and J. Callaway, *Phys. Rev. A* **3**, 2006 (1971).
9. A. L. Sinfailam and R. K. Nesbet, *Phys. Rev. A* **6**, 2118 (1972).
10. P. G. Burke, J. W. Cooper, and S. Ormonde, *Phys. Rev.* **183**, 245 (1969).
11. J. Macek, *J. Phys. B* **1**, 831 (1968).
12. E. Balslev and J. M. Combes, *Commun. Math. Phys.* **22**, 280 (1971).
13. A. U. Hazi and H. S. Taylor, *Phys. Rev. A* **1**, 1109 (1970).
14. A. Temkin, A. K. Bhatia, and J. N. Bardsley, *Phys. Rev. A* **5**, 1663 (1972).
15. K. T. Chung, *Phys. Rev. A* **20**, 1743 (1979).
16. E. Merzbacher, *Quantum Mechanics*, 2d ed., p. 315, Wiley, New York (1970).
17. See, for example, A. Weinsein and W. Stenger, *Methods of Intermediate Problems for Eigenvalues*, Academic, New York (1972).
18. J. K. L. MacDonald, *Phys. Rev.* **43**, 830 (1933).
19. B. F. Davis and K. T. Chung, *Phys. Rev. A* **22**, 835 (1980).
20. Y. Accad, C. L. Pekeris, and B. Schiff, *Phys. Rev. A* **4**, 516 (1971).
21. Y. Hahn, T. F. O'Malley, and L. Spruch, *Phys. Rev.* **128**, 932 (1962).
22. H. A. Bethe and E. E. Salpeter, *Quantum Mechanics of One and Two-Electron Atoms*, p. 168, Springer-Verlag, Berlin (1957).
23. M. E. Rudd, *Phys. Rev. Lett.* **15**, 580 (1965).
24. N. Oda, F. Nishimura, and S. Tahira, *Phys. Rev. Lett.* **24**, 42 (1970).
25. J. Comer and F. H. Reed, *J. Electron. Spectrosc. Relat. Phenom.* **1**, 3 (1972).
26. A. Bordenave-Montesquieu, A. Gleizes, M. Rodiere, and P. Benoit-Cattin, *J. Phys. B* **6**, 1997 (1973).
27. P. J. Hicks and J. Comer, *J. Phys. B* **8**, 1866 (1975).
28. F. Gelebart, R. J. Tweed, and J. Peresse, *J. Phys. B* **9**, 1739 (1976).
29. K. Siegbahn *ESCA Applied to Free Molecules*, North-Holland, Amsterdam (1969).
30. A. Bordenave-Montesquieu and P. Benoit-Cattin, C. R. H., *Acad. Sci. Ser. B.* **272**, 1473 (1971).
31. A. K. Bhatia and A. Temkin, *Phys. Rev. A* **11**, 2018 (1975).
32. Y. K. Ho, *Phys. Rev. A* **23**, 2137 (1981).
33. P. G. Burke, in *Advances in Atomic and Molecular Physics*, vol. 4 (D. R. Bates and I. Estermann, eds.) p. 173, Academic, New York (1968).
34. A. K. Bhatia, A. Temkin, and J. F. Perkins, Ref. 6
35. E. Holøien and J. Midtdal, *J. Phys. B* **3**, 592 (1970).
36. P. G. Burke and D. D. McVicar, *Proc. Phys. Soc. London* **86**, 989 (1965).
37. Y. K. Ho, *J. Phys. B* **12**, 387 (1979).
38. P. G. Burke and A. J. Taylor, *J. Phys. B* **2**, 44 (1969).
39. R. P. Madden and K. Codling, *Astrophys. J.* **141**, 364 (1965).
40. P. Dhez and D. L. Ederer, *J. Phys. B* **6**, L59 (1973).
41. B. F. Davis and K. T. Chung, *Phys. Rev. A* **26**, 2743 (1982).
42. Y. K. Ho, *J. Phys. B* **12**, L543 (1979); *Phys. Lett.* **79A**, 44 (1980).
43. D. R. Herrick and O. Sinanoglu, *Phys. Rev. A* **11**, 97 (1975).
44. R. S. Oberoi, *J. Phys. B* **5**, 1120 (1972).
45. P. R. Woodruff and J. A. R. Samson, *Phys. Rev. A* **25**, 848 (1982).
46. D. E. Golden, in. *Advances in Atomic and Molecular Physics*, vol. 14 (D. R. Bates and B. Bederson, eds.), p. 1, Academic, New York (1978), and the references therein.
47. B. R. Junker and C. L. Huang, *Phys. Rev. A* **18**, 313 (1978), and the references therein.

48. K. T. Chung, *Phys. Rev. A* **23**, 1079 (1981).
49. R. S. Oberoi and R. K. Nesbet, *Phys. Rev. A* **8**, 2969 (1973); R. K. Nesbet, *Phys. Rev. A* **12**, 444 (1975).
50. P. G. Burke, J. W. Cooper, and S. Ormonde, *Phys. Rev.* **183**, 245 (1969); W. C. Fon, *J. Phys. B* **11**, 325 (1978).
51. A. K. Bhatia and A. Temkin, *Phys. Rev. A* **23**, 3361 (1981).
52. F. M. J. Pichanick and J. A. Simpson, *Phys. Rev.* **168**, 64 (1968).
53. J. N. H. Brunt, G. C. King, and F. H. Read, *J. Phys. B* **10**, 433 (1977).
54. H. Ehrhardt and K. Willmann, *Z.* Phys. **203**, 1 (1967).
55. J. M. Phillips and S. F. Wong, *Phys. Rev. A* **23**, 3324 (1981).
56. C. E. Kuyatt, J. A. Simpson, and S. R. Mielczark, *Phys. Rev. A* **138**, 385 (1965).
57. U. Fano and J. W. Cooper, *Phys. Rev. A.* **138**, 400 (1965).
58. K. T. Chung, *Phys. Rev. A* **22**, 1341 (1980), and the references therein.
59. J. T. Grissom, R. N. Compton, and W. R. Garrett, *Phys. Lett.* **30**, 117 (1969).
60. P. J. Hicks, C. Cvejanovic, J. Comer, F. H. Read, and J. M. Sharp, *Vacuum* **24**, 573 (1974).
61. See, for example, F. H. Read, *J. Phys. B* **L10**, 207 (1977).
62. K. T. Chung, *Phys. Rev. A* **20**, 724 (1979).
63. D. L. Ederer, T. Lucatorto, and R. P. Madden, *Phys. Rev. Lett.* **25**, 1537 (1970).
64. A. M. Cantu *et al. J. Opt. Soc. Am.* **67**, 1030 (1977).
65 T. J. McIlrath and T. B. Lucatorto, *Phys. Rev. Lett.* **38**, 1390 (1977).
66. K. T. Chung, *Phys. Rev. A* **23**, 2957 (1981); *Phys. Rev. A* **24**, 1350 (1981).
67. D. Pegg, *Phys. Rev. A* **12**, 1330 (1975).
68. P. Ziem, R. Bruch, and N. Stolterfoht, *J. Phys. B* **8**, L480 (1975).
69. D. Rassi, V. Pejcev, and K. J. Ross, *J. Phys. B* **10**, 3535 (1977).
70. M. Rødbro, R. Bruch, and P. Bisgaard, *J. Phys. B* **12**, 2413 (1979).
71. A. K. Bhatia and A. Temkin, *Phys. Rev. A* **13**, 2322 (1976).
72. A. K. Bhatia, *Phys. Rev. A* **18**, 2523 (1978).
73. A. W. Weiss, quoted in Ref. 65.
74. U. I. Safronova and V. S. Senashenko, *J. Phys. B* **11**, 2623 (1978).
75. R. L. Simons, H. P. Kelly, and R. Bruch, *Phys. Rev. A* **19**, 682 (1979).
76. M. Ahmed and L. Lipsky, *Phys. Rev. A* **12**, 1176 (1975).
77. C. A. Nicolaides and D. R. Beck, *J. Chem. Phys.* **66**, 1982 (1977).
78. K. T. Chung, *Phys. Rev. A* **25**, 1596 (1982).
79. B. F. Davis and K. T. Chung, *J. Phys. B* **15**, 3113 (1982).
80. K. T. Chung and B. F. Davis, *Phys. Rev. A* **26**, 3278 (1982).

CHAPTER FOUR

COMPLEX STABILIZATION METHOD

B. R. JUNKER

1. INTRODUCTION

Many phenomena, such as field ionization, electron resonances, absorption and emission of radiation by composite systems, etc., which arise when photons and electrons interact with atomic and molecular systems or when atomic and molecular systems interact have a resonance structure. This observed structure (i.e., its position and width) can be described in terms of the real and imaginary parts, respectively, of the complex energy E^{res} of a pole of the scattering matrix.* In general, a number of closely spaced poles may occur. Theoretically, a number of techniques have been developed to determine these poles. The various methods have been classified in one of two categories—either a complete scattering calculation is performed at a number of energies in the vicinity of the resonance, and then the resonance parameters (position and width) are extracted by fitting the calculated cross section to a resonance expression, such as a Breit–Wigner form, or the complex energy where the S matrix has a pole is computed directly.

Until recently, all of these methods required that the wave function used or determined in the calculation have some explicit functional form asymptotically or some explicit boundary conditions to be satisfied at some boundary in order to extract the resonance parameters. Such requirements make these methods difficult to apply to many resonant phenomena, such as field ionization or electron resonances in molecules. For example, in the case of the former, the asymptotic wave function[2] for a simple hydrogenic system in a static electric field functions as an exponential containing a linear combination of products of $r^{3/2}$, $r^{1/2}$, $\cos(\chi/2)$, $\sin(\chi/2)$, $\cos^3(\chi/2)$, and $\sin^3(\chi/2)$, where χ is the angle between the electron and the electric field.

On the other hand, the Balslev–Combes theorem[3–5] stated in part,

*See, for example, Taylor,[1] Chap. 12.

B. R. JUNKER ■ Office of Naval Research, Code 412, 800 N. Quincy St., Arlington, Va 22217.

that under the coordinate transformation

$$r \to r \exp(i\theta) \tag{1}$$

the resonant wave function associated with the complex pole of the resolvent* is square-integrable for

$$\theta > \tfrac{1}{2}|\arg(E^{\text{res}})| \equiv \beta \tag{2}$$

This being the case, it is hoped that the exact asymptotic form of the wave function is not overly important, as is the case in bound-state calculations. The complex-coordinate method,† based on the results of the Balslev–Combes theorem, has been applied to numerous resonant phenomena including electron resonances, field ionizations, photoionization, etc. Essentially, all of the applications have been confined to atomic systems, although progress has recently been made with regard to the application of this method to molecular systems.[7–9]

In its present form the complex-coordinate method is directly applicable to molecular resonant phenomena (e.g., resonances, associative detachment, Penning ionization, etc.) if both electronic and nuclear coordinates are rotated. Within this framework, we could still perform a Born–Oppenheimer separation. The fact that the internuclear separation is complex is not a theoretical problem, since it is not an observable. That is, the observable quantity is either an electronic matrix element multiplied by a nuclear matrix element or averaged over the nuclear wave function. In either case, procedure can be performed with respect to complex as well as real internuclear coordinates. There are, however, computational problems that could arise in identifying particular resonances.[9] In addition, the pictorial concepts offered by real internuclear separations are appealing.

The power of the complex-coordinate method is considered to be the square-integrability of the resonant wave function. However, we will later present the argument that the square-integrability of the exact complex-coordinate resonant wave function and the analyticity of the Hamiltonian and exact resonant wave function imply that the complex poles of the resolvent corresponding to resonances can be computed directly *using the unrotated real Hamiltonian with a square-integrable basis without explicitly imposing any boundary condition*, such as a Siegert boundary condition. Thus, this technique is directly applicable to molecular systems without the questions of complex internuclear separations, distorted integration contours, or explicit

* In potential scattering, poles of the resolvent can be shown to correspond to poles of the S matrix. This is assumed to also be the case in N-particle scattering. Resonances are assumed to be associated with complex poles of the resolvent in the lower right-hand quadrant of the second or higher sheets in the complex energy plane.

† The entire special issue of *Int. J. of Quantum Chem.* **14** (1978)[6] is devoted to complex scaling.

Siegert boundary conditions ever arising. This technique, which is based on a number of observations regarding previous calculations and the analytic nature of the Hamiltonian and resonance wave functions, is actually also applicable to the rotated Hamiltonian.

In Section 2, we discuss the Balslev–Combes theorem and some of its implications, while in Section 3, we discuss a number of analytic and numerical calculations. Here, we present an interpretation of the θ trajectories of Doolen.[10] The proposed complex stabilization technique itself is discussed in Section 4 and the variational principle in Section 5. Possible choices of basis function and structures of variational wave functions are considered for various resonant phenomena in Section 6. These topics should provide examples of how we might study different phenomena. Then, in Section 7, a model potential problem is studied. Concluding remarks are given in Section 8.

We will refer to our previous work as I,[11] IA[12] II,[13] and III[14]. There are several notational changes that we will adopt that differ from our previous papers. First, we will adopt the prevalent use of θ for the rotation angle despite its possible confusion with the polar coordinate. In previous papers, we used the symbols r and ρ as radial coordinates in the sense of the transformation

$$r \rightarrow \rho \exp(i\theta) \tag{3}$$

although we noted that there is, in fact, only one radial coordinate. Here, we will drop the use of ρ and use only r. Atomic units are assumed except where explicitly noted otherwise.

2. THE BALSLEV–COMBES THEOREM

For simplicity, we will consider a one-particle system described by a time-independent Schrödinger equation for which the Hamiltonian possesses spherical symmetry. Then the wave function factors into the product of a radial function and a spherical harmonic. The extension to the general case is straightforward and will be noted at the end of this section.

2.1. *Spectrum of the Rotated Hamiltonian*

The rotated Hamiltonian $H[r \exp(i\theta)] \equiv H(\theta)$ is obtained from the real Hamiltonian $H(r)$ by means of the transformation in Eq. (1). Its spectrum contains the following classes.[3,4]

1. Real discrete eigenvalues corresponding to bound states that are identical to those of $H(r)$.

2. Continuum energies, associated with the scattering states, given by $E\exp(-2i\theta)$, where E is the corresponding energy of $H(r)$.
3. Isolated complex energies that are poles of the resolvent and can be shown to be poles of the S matrix for potential scattering.
4. Other elements that will not concern us here.

Balslev and Combes[3] proved that the domain of the bound states of the rotated Hamiltonian is obtained from the domain of the bound states of $H(r)$ by analytically continuing the latter according to Eq. (1). Simon[4] has stated that the resonant wave functions are analytic. In IA, we gave an explicit proof of the functional dependence of the bound, resonant, and scattering wave functions on r and θ. We will repeat it here, since it forms the basis for later arguments.

Theorem. Apart from a possible multiplicative phase factor depending on θ, the eigenfunctions for the bound states and isolated complex poles of the resolvent depend functionally on r and θ in only the combination $r\exp(i\theta)$. On the other hand, the wave functions corresponding to the scattering states are not a function of $r\exp(i\theta)$ for $\theta > 0$ and $E > 0$.

Proof. The rotated Hamiltonian $H(\theta)$ depends on r and θ only in $r\exp(i\theta)$. In the rotated Hamiltonian, write the product $r\exp(i\theta)$ as η. Assume that wave functions depend on r and θ other than in the product $r\exp(i\theta)$. Then ψ_n is a function of, say, η and θ, i.e., $\psi_n(\eta, \theta)$. For bound states and resonances, the Schrödinger equation then becomes

$$[H(\eta) - E_n]\psi_n(\eta, \theta) = 0 \tag{4}$$

In order to specify ψ_n, we must not only specify $H(\eta)$, but also an appropriate boundary condition, or equivalently, the corresponding energy E_n. For bound states and isolated complex energies, E_n is independent of η and θ. Therefore, the operator $[H(\eta) - E_n]$ for these states depends only on η and, consequently, so must the wave function ψ_n. Thus, for these states, ψ_n depends on $r\exp(i\theta)$, i.e., ψ_n is an analytical continuation of some function $\psi_n(r)$ associated with $H(r)$. Since the energies for the scattering states are of the form $E\exp(-2i\theta)$ for $E > 0$ and $\theta > 0$, the operator $[H(\eta) - E\exp(-2i\theta)]$ depends on both η and θ. Thus, the scattering wave functions depend on η and θ independently.

2.2. Gamow–Siegert Boundary Conditions

In the preceding, we have used the spectrum of $H(\theta)$ as given by the Balslev–Combes theorem to obtain information about the functional dependences of the wave function on r and θ. Additional insight can be obtained if we seek boundary conditions that must be imposed on various wave functions of $H(\theta)$ to obtain the spectrum given by the Balslev–Combes theorem.

First, we find the poles of the resolvent operator associated with $H(r)$. For the potential scattering case that we are considering here, they also correspond to the poles of the S matrix[1] and, consequently, the bound states and resonances. Gamow[15] and Siegert[16] showed that the radial part of the eigenfunctions associated with these poles obeyed the following boundary conditions:

$$\left\{\frac{d[r\psi_n(r)]}{dr}\right\} \sim i[k_n]r\psi_n = i[|k_n|\exp(-i\beta n)] \times \exp\{i[|k_n|\exp(-i\beta n)]r\} \tag{5a}$$

$$\lim_{r\to 0} r\psi_n(r) = 0 \tag{5b}$$

where

$$E_n = \tfrac{1}{2}k_n^2 = \tfrac{1}{2}|k_n|^2 \exp(-2i\beta n) \tag{6}$$

Such functions are pure outgoing waves with a complex wave number k_n. From standard scattering theory,[1] the values of k_n are restricted to lie on the positive imaginary k axis and the lower half of the complex k plane. The former set corresponds to bound states and have $\beta_n = -\pi/2$, while the latter set, which corresponds to $0 < \beta_n < \pi$, contains the resonances. While the resonances are sometimes defined as that subset where $0 < \beta_n < \pi/4$, this restriction is not significant in the present discussion. Since the wave functions behave asymptotically as

$$\psi_n \sim \exp(ik_n r) = \exp\{i[|k_n|\exp(-i\beta n)]r\} \tag{7}$$

we see that bound-state wave functions are square integrable (proper eigenfunctions), but resonant-state wave functions diverge exponentially asymptotically (improper eigenfunctions).

We now consider solutions for the analytically continued Hamiltonian and boundary conditions in Eq. (5). That is, we want solutions for

$$[H(\theta) - E_n]\psi_n(r,\theta) = 0 \tag{8}$$

satisfying the boundary conditions

$$\exp(-i\theta)\{d[r\exp(i\theta)]\psi_n(r,\theta)/dr\} \sim ik_n[r\exp(i\theta)]\psi_n(r,\theta) \tag{9a}$$

$$\lim_{r\to 0}[r\exp(i\theta)]\psi_n(r,\theta) = 0 \tag{9b}$$

Since the factor multiplying ψ_n on the right-hand side of Eq. (9a) is independent of θ, the same argument implies that

$$[r\exp(i\theta)]\psi_n(r,\theta) = [r\exp(i\theta)]\psi_n[r\exp(i\theta)] \sim \exp\{i|k_n|\exp[i(\theta-\beta_n)]r\} \tag{10}$$

and E_n is independent of θ. Equation (10) shows that ψ_n is square integrable when $0 \leqslant \theta \leqslant \pi/2$ for bound states and $\beta_n < \theta < \beta_n + \pi$ for resonances in agreement with the Balslev–Combes theorem. The fact that $\psi_n[r\exp(i\theta)]$ is a solution of a differential operator that is simply the analytical continuation of the differential operator for $\psi_n(r)$ and that $\psi_n[r\exp(i\theta)]$ satisfies boundary conditions that are simply analytical continuations of those satisfied by $\psi_n(r)$ implies that $\psi_n[r\exp(i\theta)]$ is the analytical continuation of $\psi_n(r)$. That is, the bound- and resonant-state wave functions are the analytical continuations of the bound and resonant Siegert state functions of $H(r)$. This result was first suggested in I.

The dependence of these eigenfunctions on r and θ in only the combination $r\exp(i\theta)$ is a constraint similar to other symmetry constraints. The fact that they are functions of $r\exp(i\theta)$ results in the independence of their energies on variations in θ. The complex resonant energies result from the fact that the corresponding resonant wave functions are explicitly *complex* functions of $r\exp(i\theta)$ and not because the resonant wave functions depend on r and θ in some manner* other than $r\exp(i\theta)$. The bound states, on the other hand, are *real* functions of $r\exp(i\theta)$.

Care must be exercised in discussing solutions for the scattering states. This is because they are, in general, formulated in terms of functions that asymptotically behave as eigenfunctions of the free particle or momentum operators. The transformation, in Eq. (1) for r implies the transformation

$$p \rightarrow p\exp(-i\theta) \tag{11}$$

for the momentum. Note that this preserves commutivity relations for conjugate variables as well as the uncertainty principle. Thus, eigenvalues for the momentum operator are the original eigenvalues multplied by $e^{-i\theta}$. As a result, the asymptotic boundary condition remains invariant in form, i.e., the scattering states move asymptotically as

$$\psi_k \sim A(\theta)\exp(ikr) + B(\theta)\exp(-ikr) \tag{12}$$

Consequently, the associated energy becomes

$$E = \tfrac{1}{2}kk\exp(-2i\theta) = |E|\exp(-2i\theta) \tag{13}$$

where k is real.

This result corresponds to the "natural" choice for the cut as discussed by Newton.[17] The dependence of scattering states with energy $E > 0$ on r and θ in a manner other than $r\exp(i\theta)$ is also consistent with the boundary condition at $r = 0$.

The reason some eigenvalues of $H(\theta)$ are invariant with respect to θ

*Note that this is contrary to the assumptions of Poole *et al.*, 1974.

while others are not is clear. Those eigenvalues associated with the bound states and resonances are associated with eigenvectors of an analytically continued Hamiltonian and satisfy analytically continued boundary conditions. Consequently, the eigenvalue is invariant. On the other hand, eigenvectors associated with the cuts are required to be eigenvectors corresponding to an analytically continued Hamiltonian but an invariant boundary condition. Thus, the eigenvalues depend on θ.

Unfortunately, the potential scattering results previously quoted have not been proven for the general N-particle problem.* We will, however, assume they are valid, so that the exact *unrotated* resonance wave function is assumed to behave asymptotically as

$$\Psi_n(\mathbf{r}_0, \mathbf{r}_1, \ldots) \sim \mathscr{A} \sum_{\nu} c_\nu \phi_T^\nu(\mathbf{r}_1, \mathbf{r}_2, \ldots) r_0^{-1} \exp[i|k_n^\nu| r_0 \exp(-i\beta_n^\nu)] Y_l^m \tag{14}$$

where $\mathscr{A}$ is the antisymmetrizer and ϕ_T is the target function for the νth channel. In light of the results in this section, we will now discuss some analytical models and previous numerical calculations.

3. PREVIOUS CALCULATIONS

3.1. Analytical Models and Analyticity of the S Matrix

Wave functions for the bound states of hydrogenic atoms have been discussed in I and IA and later by Nicolaides and Beck,[18] who also discuss the bound states of the harmonic oscillator. In these examples, the wave functions of $H(\theta)$ were obtained by assuming they were square-integrable just as is done in elementary quantum mechanics texts. The wave functions obtained are the normal hydrogenic and harmonic oscillator wave functions with r replaced by $r \exp(i\theta)$. This is in agreement with results from the previous section.

3.1.1. Hydrogenic Atoms. Since the bound-state wave functions are square-integrable for $0 \leqslant \theta < \frac{1}{2}\pi$. We might ask what happens for the interval $\frac{1}{2}\pi \leqslant \theta \leqslant \frac{3}{2}\pi$ or, equivalently, $\frac{1}{2}\pi \leqslant \theta \leqslant \pi$ and $-\pi \leqslant \theta \leqslant -\frac{1}{2}\pi$. If the preceding derivation of these wave functions is repeated with the divergence of $\exp[-\alpha\gamma \exp(i\theta)]$ simply ignored and requiring a truncation of the power series, the same results would be obtained. Another way of approaching the problem is by defining the bound states in terms of the poles of the S matrix. Consider the Schrödinger equation for a hydrogenic system.

$$0 = \left\{ \left[\frac{-\exp(-2i\theta)}{2} \right] \nabla^2 + \frac{zz' \exp(-i\theta)}{r} - \frac{1}{2} k^2 \right\} \psi[\mathbf{r} \exp(i\theta)] \tag{15}$$

* However, for advances in special cases see, for example, Hagedorn[17] and Sigal.[17]

or

$$0=\left\{-\frac{1}{2}\nabla^2+\frac{zz'\exp(i\theta)}{r}-\frac{1}{2}[k\exp(i\theta)]^2\right\}\psi[\mathbf{r}\exp(i\theta)] \tag{16}$$

where $z=1$ and $z'=-1$ for the hydrogen atom.

If we define

$$\begin{aligned}\psi[\mathbf{r}\exp(i\theta)] &\equiv F_l(k',r)Y_l^m(\chi,\phi)\\ &\equiv r^{-1}\exp(ik'r)(k'r)^{l+1}V_l(k',r)Y_l^m(\chi,\phi)\end{aligned} \tag{17}$$

with

$$k'\equiv k\exp(i\theta) \tag{18a}$$

$$\zeta\equiv -2ik'r \tag{18b}$$

and

$$\beta\equiv zz'\frac{\exp(i\theta)}{k'} \tag{18c}$$

we obtain

$$\{\zeta(d^2/d\zeta^2)+[2(l+1)-\zeta](d/d\zeta)-[(l+1)+i\beta]\}V_l(k',\zeta)=0 \tag{19}$$

For $F_l(k',r)$, this yields

$$\begin{aligned}F_l(k',r) &= c_l\exp[ikr\exp(i\theta)][kr\exp(i\theta)]^l\\ &\qquad\times {}_1F_1[l+1+i\beta;2l+2;-2ikr\exp(i\theta)]\\ &= \exp(-i\sigma_l)[(u_l^+S_l-u_l^-)/2i]\end{aligned} \tag{20}$$

where u_1^+ and u_1^- are the outgoing and incoming solutions, respectively, and

$$S_l=\exp(2_i\sigma_l)=\Gamma(l+1+i\beta)/\Gamma(l+1-i\beta) \tag{21}$$

The poles of S_l correspond to the poles of $\Gamma(l+1+i\beta)$, since $\Gamma(l+1-i\beta)$ has no zeros. The poles of $\Gamma(l+1+i\beta)$ occur for $(zz'=-z)$

$$l+1+i\beta=-n,\qquad n=0,1,2,\ldots \tag{22}$$

or, using Eq. (18a), we obtain

$$k=-izz'/(n+l+1)=iz/(n+l+1) \tag{23}$$

and

$$E=\tfrac{1}{2}k^2=-\tfrac{1}{2}[z^2/(n+l+1)^2] \tag{24}$$

If we now let $N=n+1+1$, $\psi[\mathbf{r}\exp(i\theta)]$ becomes

$$\begin{aligned}\psi[\mathbf{r}\exp(i\theta)] &\propto [r\exp(i\theta)]^l\exp[-zr\exp(i\theta)/N]\\ &\quad\times L_{N-l-1}^{2l+1}[2zr\exp(i\theta)/N]Y_l^m(\chi,\phi)\end{aligned} \tag{25}$$

where $L_{N-l-1}^{2l+1}(\gamma)$ is an associate Laguerre polynomial.

The scattering states for a hydrogenic system are chosen to satisfy the equation

$$[\eta(d^2/d\eta^2) + (2l + 2 - \eta)(d/d\eta) - (l + 1 + i\gamma)]V_l(k, \eta) = 0 \tag{26}$$

where

$$\eta = -2ikr \tag{27a}$$

and

$$\gamma = zz' \exp(i\theta)/k \tag{27b}$$

We now obtain the regular solution

$$\psi(\mathbf{r}, \theta) \propto r^l \exp(ikr)\,{}_1F_1(l + 1 + i\gamma; 2l + 2; -2ikr)\, Y_l^m(\chi, \phi) \tag{28}$$

Note here that the only dependence on the rotation angle θ is in γ.

3.1.2. Doolen Model Potential. Recently, Doolen[19] has presented a one-dimensional model potential that contains resonances and can be solved analytically. The Hamiltonian has the form

$$H = -\tfrac{1}{2}r^{-2}(d/dr)r^2(d/dr) + r^{-1} - \gamma r^{-2} \tag{29}$$

For the purpose of this discussion, we will consider a slight generalization of the Hamiltonian in Eq. (29), i.e.,

$$H = -\tfrac{1}{2}r^{-2}(d/dr)r^2(d/dr) + r^{-1} + \tfrac{1}{2}[l(l + 1) - 2\gamma]r^{-2} \tag{30}$$

First, we define a new parameter l' by

$$l'(l' + 1) = l(l + 1) - 2\gamma \tag{31}$$

This implies

$$l' = -\tfrac{1}{2}\{1 \mp [(2l + 1)^2 - 8\gamma]^{1/2}\} \tag{32}$$

In order for l' to go to l as γ goes to zero, the upper sign must be chosen. The $H(\theta)$ is then identical to the Hamiltonian in Eq. (18) with l replaced by l'. Thus, the solutions are

$$\begin{aligned} F_l(k', r) = C_{l'} \exp[ikr \exp(i\theta)][kr \exp(i\theta)]^{l'} \\ \times {}_1F_1[l' + 1 + i\beta; 2l' + 2; -2ikr \exp(i\theta)] \end{aligned} \tag{33}$$

with

$$S_l(k) = \exp(i\sigma_l) = \frac{\Gamma(\frac{1}{2}\{1 + [(2l + 1)^2 - 8\gamma]^{1/2}\} + i\beta)}{\Gamma(\frac{1}{2}\{1 + [(2l + 1)^2 - 8\gamma]^{1/2}\} - i\beta)} \tag{34}$$

At this point, two cases have to be distinguished. Recall that $\gamma > 0$.

Case A: $8\gamma < (2l + 1)^2$. The poles of $S_l(k)$ are such that

$$\tfrac{1}{2}\{1 + [(2l + 1)^2 - 8\gamma]^{1/2}\} + (i/k) = -n, \qquad n = 0, 1, 2, \ldots \tag{35}$$

or

$$k = -i/(n + \tfrac{1}{2}\{1 + [(2l+1)^2 - 8\gamma]^{1/2}\}) \tag{36}$$

Since the denominator is always real and positive, the only poles lie on the negative imgainary exis.

Case B: $8\gamma > (2l+1)^2$. The poles of $S_l(k)$ now satisfy

$$\tfrac{1}{2}\{1 \pm i[8\gamma - (2l+1)^2]^{1/2}\} + (i/k) = -n \qquad n = 0, 1, 2, \ldots \tag{37}$$

$$\therefore k = 2\{\mp [8\gamma - (2l+1)^2]^{1/2} - i(2n+1)\}/\{(2n+1)^2 + [8\gamma - (2l+1)^2]\} \tag{38}$$

Since $[8\gamma - (2l+1)^2]$, $(2n+1)$, and the denominator are all positive, Eq. (38) shows there are no bound states.[20] In addition, it shows the complementary nature of the poles in the lower half of the k plane. The energies associated with the poles defining the resonances, that is, the values of k having a plus sign in Eq. (38), are

$$E_{nl}^{\text{res}} = \frac{1}{2}k^2 = \frac{2\{[8\gamma - (2l+1)^2] - (2n+1)^2 - 2i(2n+1)[8\gamma - (2l+1)^2]^{1/2}\}}{\{(2n+1)^2 + [8\gamma - (2l+1)^2]\}^2} \tag{39}$$

For the resonances,

$$l' = -\tfrac{1}{2}\{1 + i[8\gamma - (2l+1)^2]^{1/2}\} \tag{40}$$

so that the corresponding resonant wave functions from Eq. (33) become

$$\begin{aligned} F_{nl}(k', r) = {} & C_l' \exp\{i|k|r\exp[i(\theta - \beta)]\} \\ & \times \{|k|r\exp[i(\theta - \beta)]\}^{-(1/2)\{1 + i[8\gamma - (2l+1)^2]^{1/2}\}} \\ & \times {}_1F_1\{-n; 1 - i[8\gamma - (2l+1)^2]^{1/2}; -2i|k|r\exp[i(\theta - \beta)]\} \end{aligned} \tag{41}$$

where

$$\alpha \equiv -\tan^{-1}\left\{\frac{-(2n+1)}{[8\gamma - (2l+1)^2]^{1/2}}\right\} \tag{42}$$

From Eq. (41), the square integrability of the resonant wave function is assured for

$$\theta > \beta = \tfrac{1}{2}|\arg(E_{nl}^{\text{res}})| \tag{43}$$

in agreement with the Balslev–Combes theorem.

For $l = 0$, Eq. (39) reduces to Doolen's result

$$E_{n0}^{\text{res}} = \frac{2[(8\gamma - 1) - (2n+1)^2 - 2i(2n+1)(8\gamma - 1)^{1/2}]}{[(2n+1)^2 + (8\gamma - 1)^2]} \tag{44}$$

For $n=0$ and $\gamma=0.5$, the resonant energy is

$$E_{00}^{\text{res}}=[1-i(3)^{1/2}]/4 \tag{45}$$

with the wave function

$$\begin{aligned} F_{00}[kr\exp(i\theta)] = C_{l'}\{r\exp[i(\theta-\beta)]\}^{-(1/2)(1+i\surd 3)} \\ \times \exp[ir\cos(\theta-\beta)]\exp[-r\sin(\theta-\beta)] \end{aligned} \tag{46}$$

Doolen gives the energy as $[1-i(3)^{1/2}]/8$, although in a preprint,[21] it is given correctly. The exponent of r in Doolen's result is given as $-\frac{1}{2}+\frac{1}{2}i\ (3)^{1/2}$, as opposed to $-\frac{1}{2}-\frac{1}{2}i\ (3)^{1/2}$. Although he properly chose $-i$ as the square root of -1 to obtain the correct expression for k for the resonant energy, he evidently incorrectly chose $+1$ in the wave function.

3.1.3. Separable Potential Model Problem. As a final example, we will consider a model potential problem given by Yaris *et al.*,[22] who considered a separable potential of the form

$$V(\mathbf{r},\mathbf{r}')\equiv\lambda u(|\mathbf{r}|)u(|\mathbf{r}'|) \tag{47}$$

with the specific form

$$u(r)=e^{-br}/r \tag{48}$$

If we make the change of variables given by Eq. (1) and define

$$k'=k\exp(-i\theta) \tag{49}$$

their Eq. (Y23) does not dependence on θ as it should if $\psi(r)$ is only a function of $r\exp(i\theta)$. In fact, substituting their Eqs. (Y19) and (Y25) into Eq. (Y20) yields

$$\begin{aligned} f(r) = A\exp[ik'r\exp(i\theta)] + B\exp[-ik'r\exp(i\theta)] + (8\pi\lambda m/k'\hbar^2) \\ \times\int_0^\infty d[y\exp(i\theta)][y\exp(i\theta)]u[y\exp(i\theta)]f[y\exp(i\theta)] \\ \times\int_0^r d[x\exp(i\theta)][x\exp(i\theta)]u[x\exp(i\theta)] \\ \times\sin k'[r\exp(i\theta)-x\exp(i\theta)] \end{aligned} \tag{50}$$

where the only dependence on r and θ is in the form $r\exp(i\theta)$.

Equations (25), (28), (41), and (50) clearly illustrate the theorem in Section 2. That is, the bound and resonant states are functions of $r\exp(i\theta)$, while the scattering states are not. Finally, we note that if Eqs. (25), (41), and (50) are analytically continued back to $\theta=0$, the resulting wave functions satisfy a Siegert boundary condition.

3.2. Numerical Complex-Coordinate Calculations

Doolen,[10] Doolen *et al.*,[23] and Ho[24] have employed the complex rotated Hamiltonian with a basis set containing only real functions of r. This forces the required θ dependence of the resonant- and bound-state wave functions previously discussed into the linear coefficients. We would thus not expect this to be a good basis for these states, which, in fact, require extremely large basis sets simply for two-particle systems and do not yield eigenvalues that are independent of variations in θ. Instead, the authors compute the resonant eigenvalue of $H(\theta)$ for a number of values of θ (this set of eigenvalues is called a theta trajectory) and search for regions of quasistability of the eigenvalue with respect to variations of θ over some region. It is assumed that these are the regions where the variational estimate best approximates the exact resonant energy.

One way of viewing this calculation follows. The authors use a basis that is independent of θ and spans certain θ-independent vector space. On the other hand, the exact resonant wave function of the complex Hamiltonian $H(\theta)$ is a function of $r\exp(i\theta)$. At each value of θ, their basis thus describes some θ-independent component of the exact resonant wave function, and the quasi-stationary behaviour probably arises when that component is a maximum. In effect, θ is merely playing the role of a nonlinear variational parameter, and the behavior of the theta trajectory results from using a variational principle to locate an isolated complex pole of the resolvent.

In I, IA, and II, we noted that the complex-coordinate resonant wave function could be written as

$$\Psi = P\Psi + Q\Psi \tag{51}$$

where $P\Psi$ represents the open channels and $Q\Psi$ represents the closed channels. We do not imply here a rigorous Feshbach-like decomposition of the total wave function, since we do not require the two parts to be orthogonal. In addition, such a decomposition does not restrict[12] us to Feshbach resonances. The basis functions used for $Q\Psi$ and the target functions properly incorporated the $r\exp(i\theta)$ constraint on the resonant wave function. The part of the basis describing the scattering function in $P\Psi$, however, was chosen to be complex functions of just r. In this case, we are forcing only the θ dependence into the linear coefficients for one particle in the P-space part of the wave function. As a result, the real part of the resonant eigenvalue was stable to six significant figures and the imaginary part to four significant figures for $0.02 \leqslant \theta \leqslant 1.0$ rad.

However, the preceding comments on the θ dependence and the role of θ in the calculation apply here. The advantage in this latter method, however, is that the basis functions for the bulk of the particles have the proper $r\exp(i\theta)$

dependence. As a result, essentially arbitrarily large systems of particles can, in principle, be successfully studied using this approach. In addition, as indicated in I. IA, and II, the basis functions and wave-function structure also supply a reasonably good discrete representation of the states along the cuts. This is because the basis functions describing the bound particles in the target functions depend on $r\exp(i\theta)$, while those describing the unbound particle do not depend on r and θ in only the combination $r\exp(i\theta)$, which is approximately the correct behavior.

To understand the behavior of the imaginary part of the resonant eigenvalue for very small and very large θ in these calculations, we can compare the bases used in I, IA, and II to the analytical results in the first part of this section. Due to the nature of the basis functions and the structure of the trial wave function, the imaginary part of the complex energy results from the form of $P\Psi$. That is, $Q\Psi$ consists of real functions of $r\exp(i\theta)$ and thus alone could yield only real eigenvalues. At theta equal to zero, the Hamiltonian matrix is complex symmetric. There exist, however, due to the form of the basis, similarity transformations that can transform it into a real symmetric or complex Hermitian matrix, depending on the particular transformation used. In effect, these similarity transformations transform the set of complex basis functions used to describe the scattering function to a set of real basis functions. Thus, the eigenvalues are all real. As theta increases from zero, these similarity transforms are less and less effective in transforming the matrix to a Hermitian structure. On the other hand, as theta becomes very large the exact resonant function becomes highly oscillatory due to the $r\exp(i\theta)$ dependence, as is obvious from Eqs. (19), (25), (41), and (50). Thus, an extremely larger and larger number of basis functions would be required as theta increases.

Rescigno *et al.*[25] have described a computational technique that they applied to shape resonances in which they first construct an effective one-electron Hamiltonian for the $(N + 1)$-electron system from an N-electron reference function and then apply the complex-rotation transformation to the effective one-particle Hamiltonian. McCurdy[26] has noted that this method is equivalent to the technique they described in I and IA if we were to use a multiconfigurational reference wave function. This is certainly true if the linear coefficients in the multiconfiguration N-particle wave function are determined in the solution of the effective one-electron Hamiltonian resonance problem and *not* in the N-particle bound problem. The electron correlation is different in these two cases. In the former case, we are treating correlations in the total $(N + 1)$-particle problem, as in I IA, while in the latter case, we are using a sort of N-particle correlated frozen core for the $(N + 1)$-particle problem. Again, here the $r\exp(i\theta)$ constraint is effectively incorporated into the functions describing the bound particle.[5] Thus, θ is again playing the role of a nonlinear variational parameter, and the preceding comments made with regard to the behavior of the theta trajectories apply.

The calculations by McCurdy and Rescigno[7] and by Moiseyev and Corcoran[8] are based on the equivalence between matrix elements of the rotated Hamiltonian $H(\theta)$ with unrotated basis functions, i.e., basis functions that depend on r alone and matrix elements of the unrotated Hamiltonian with basis functions containing $\exp(i\theta)$. McCurdy and Rescigno consider scaling the nonlinear parameters in their basis functions by $\exp(-2i\theta)$ and argue that this is justified, since the basis functions that determine the asymptotic form of the wave function are those with small nonlinear parameters, and, in this case, the two matrix elements above become equivalent for their choice of basis functions. Moiseyev and Corcoran, on the other hand, consider scaling the electronic radial coordinate in only a Born–Oppenheimer molecular Hamiltonian. They use a general complex scaling of the form $\lambda \exp(i\theta)$ and require the complex virial theorem[27] to be satisfied. The authors note that scaling only the electronic coordinates effectively moves the basis functions off the nuclear centers. They also note that while this is no problem for a complete basis, it may affect the quality of the results for a finite basis if λ differs much from unity. We will return to disscussion of each of these calculations at the end of the next section.

4. THE COMPLEX-STABILIZATION METHOD

The dependence of the bound and resonant states on $r\exp(i\theta)$ is a symmetry constraint imposed on the wave function by the nature of the Hamiltonian and boundary conditions just as other symmetry constraints, such as permutational and point or continuous-group symmetries. It is, in fact, this constraint that results in the independence of the complex eigenvalue, approximating the resonant energy, on variations in θ. Thus, the crudest wave function that incorporates this constraint would give eigenvalues independent of θ.

The computation of θ trajectories and stabilization of the resonant eigenvalue with respect to a variation in θ by means of a variational principle simply involves determining the value of θ for which the θ-independent basis best approximates the exact θ-dependent wave function. That is, θ is, in effect, playing the role of a nonlinear variational parameter. This is clearly illustrated in a recent calculation by Donnelly and Simmons[28] of the position and width of the $(1s)^2\ (2s)^2\ (kp)^2$ P-shape resonance in Be^-. The value of θ where the complex eigenvalue $\tilde{E}^{\mathrm{res}}$ stabilized is *less than** $\frac{1}{2}|\arg(\tilde{E}^{\mathrm{res}})|$. Thus, although

*This was first noted by J. N. Bardsley at the Fourteenth Quantum Chemistry Symposium.

we use the symbol θ in the following calculations, $\exp(i\theta)$ will be used in the sense of a complex-scale factor for the nonlinear parameters in the basis set, and the stabilization of E^{res} with respect to variations in θ will determine the value of θ.

We will now suggest a technique for computing the complex resonant eigenvalues directly *with* the unrotated Hamiltonian *with* a square-integrable basis and *without* explicitly imposing any boundary condition. We will then give two arguments to justify the technique. One is based on the analyticity of the resonant wave functions and previous calculations, and the other is based on the external complex-scaling theorem of Simon.[19]

Basically, the method[14] consists of using the unrotated Hamiltonian with a square-integrable basis to compute nonlinear parameter trajectories instead of theta trajectories and stabilizing $\tilde{E}^{\text{res}}$ with respect to variations in the various nonlinear parameters γ_i in the basis set. This stabilization could be performed with each parameter individually, with groups of parameters, or globally with all parameters at one time. Scaling the parameters could be a real scaling or a complex scaling. In effect, complex-coordinate calculations, such as those in Refs. (23) and (24) correspond to a global complex scaling, while those in Refs. (11), (13), and (25) correspond to a complex scaling of just those basis functions representing the unbound particle.

One way of understanding why this method could work is by analyzing previous calculations. Consider, for example, the calculation by Doolen[10] for the $(2s)^2\ {}^1S\ H^-$ resonance. In that calculation, a number of wave functions constructed from Hylleraas basis functions were used. That is, the variational wave function had the form

$$\Psi^\theta = \sum_{l,m,n} C_{nlm}(\theta)(r_1^l r_2^m + r_1^m r_2^l) r_{12}^n \exp[-a(r_1 + r_2)] \tag{52}$$

where the C_{nlm} were determined variationally. For a given basis set and a nonlinear parameter a, an optimum value of θ, say, θ_1, was found for which $\tilde{E}^{\text{res}}$ was most stable. On the other hand, the preceding theorem states that the resonant wave function is a function of $r\exp(i\theta)$. Consequently, a wave function Ψ^{θ_1}, determined at $\theta = \theta_1$, has an exactly equivalent wave function Ψ^{θ_2}, evaluated at $\theta = \theta_2$, and Ψ^{θ_2} is obtained from Ψ^{θ_1} by the transformation

$$r \to r\exp[i(\theta_2 - \theta_1)] \tag{53}$$

Ψ^{θ_2} has the same eigenvalues as Ψ^{θ_1}. Now, consider the wave function Ψ^0 for $\theta = 0$, which is equivalent to Ψ^{θ_1} in Eq. (52). Employing Eq. (53), we obtain

$$\Psi^0 = \sum_{n,l,m} C_{nlm}(\theta_1)\exp[-i\theta_1(l+m+n)](r_1^l r_2^m + r_1^m r_2^l) r_{12}^n$$
$$\times \exp[ia\sin\theta_1(r_1 + r_2)]\exp[-a\cos\theta_1(r_1 + r_2)] \tag{54}$$

That is, Ψ^0 is square-integrable even though the exact resonant wave function diverges exponentially. The same result is found for all previous complex-coordinate calculations. This implies that the exact asymptotic boundary condition need not be imposed on the variational wave function. Note that this is a result of the analyticity of the Hamiltonian and resonant wave function and the fact that there exists a region in the complex-coordinate plane where the resonant wave function is square-integrable. Such a situation does not exist for the functions representing the cut. Also note that Siegert basis-set calculations[29] are a special case of complex-coordinate calculations where this exact boundary condition is explicitly incorporated into the wave function.

A second way* of understanding the validity of this method is by way of Simon's[9] external complex scaling. Simon has shown that for a transformation T, of the form

$$T(r) = r \qquad 0 \leqslant r \leqslant R \tag{55a}$$

$$= R_0 + \exp(i\theta)(r - R_0) \qquad R_0 \leqslant r \tag{55b}$$

with boundary conditions

$$\Psi(R_0 - 0) = \Psi(R_0 + 0) \tag{56a}$$

and

$$\Psi'(R_0 - 0) = \exp(-i\theta)\Psi'(R_0 + 0) \tag{56b}$$

the spectrum is identical to that given by the transformation in Eq. (1). Thus, for dilation analytic potentials, we still have the spectrum discussed in Section 2 for $H(\theta)$.

To illustrate complex external scaling, consider the solution of Eq. (30), equivalent to Eq. (46) but using the transformation in Eq. (55) instead of Eq. (1). The solution is

$$F_{00}(kr) = C_0[r\exp(-i\alpha)]^{-(1/2)(1+i\sqrt{3})}\exp[ir\exp(-i\alpha)] \qquad 0 \leqslant r \leqslant R_0 \tag{57a}$$

and

$$F_{00}(k\{R_0[1 - \exp(i\theta)] + r\exp(i\theta)\}) = C_0'[e^{-i\alpha}R_0(1 - e^{i\theta}) + re^{i(\theta-\alpha)}]^{-(1/2)(1+i\sqrt{3})}\exp[ie^{-i\alpha}R_0(1 - e^{i\theta})]\exp[ire^{i(\theta-\alpha)}] \qquad r \geqslant R_0 \tag{57b}$$

For $r = R_0$, Eq. (56a) is directly satisfied for C_0 equal to C_0'. Similarly, we

* In a private communication,[34] Barry Simon suggested that this method could be viewed as a realization of external complex scaling.

have

$$F'_{00}(kr) = -\tfrac{1}{2}[1 + i(3)^{1/2}]C_0 e^{-i\alpha}(re^{-i\alpha})^{-(1/2)(3+i\sqrt{3})} \exp(ire^{-i\alpha}) + iC_0 e^{-i\alpha}(re^{-i\alpha})^{-(1/2)(1+i\sqrt{3})} \exp(ire^{-i\alpha}) \qquad 0 \leqslant r \leqslant R_0 \tag{58a}$$

and

$$e^{-i\theta}F'_{00}\{k[R_0(1-e^{i\theta}) + re^{i\theta}]\} = -\tfrac{1}{2}(1 + i\sqrt{3})C_0 e^{-i\alpha} \times [e^{-i\alpha}R_0(1-e^{i\theta}) + re^{i(\theta-\alpha)}]^{-(1/2)(3+i\sqrt{3})} \exp[ie^{-i\alpha}R_0(1-e^{i\theta})] \times \exp(ire^{i(\theta-\alpha)}) + iC_0 e^{-i\alpha}[e^{-i\alpha}R_0(1-e^{i\theta}) + re^{i(\theta-\alpha)}]^{-(1/2)(1+i\sqrt{3})} \times \exp[ie^{-i\alpha}R_0(1-e^{i\theta})] \exp[ire^{i(\theta-\alpha)}] r \geqslant R_0 \tag{58b}$$

Again, Eq. (56b) is trivially satisfied. In Eqs. (57b) and (58b), we see that the factor $\exp[ie^{-i\alpha}R_0(1-e^{i\theta})]$ guarantees that Eqs. (56a) and (56b) are satisfied at $r = R_0$, but the factor $\exp[-r\sin(\theta - \alpha)]$ in Eq. (57b) guarantees the square-integrability of Ψ^{res}.

In general, we can consider Ψ^{res} as being of the form

$$\Psi^{\text{res}} = \Psi^{\text{res}}(r) \qquad 0 \leqslant r \leqslant R_0 \tag{59a}$$

$$= \Psi^{\text{res}}[R_0(1-e^{i\theta}) + re^{i\theta}] \qquad r \geqslant R_0 \tag{59b}$$

Since the bound and resonant eigenvalues are independent of R_0 and θ, we can, in principle, make the contribution from Eq. (59b) arbitrarily small (and thereby negligible) by choosing R_0 and θ to be large within their permissible ranges. In effect, in the complex stabilization method, R_0 is implicitly set by the extent of the basis-set functions, although no explicit constant "effective R_0" is defined, since the extent of the basis set varies as the nonlinear parameters are varied. In a way, the calculation is a sort of calculation within a spherical box, although here the boundary is not specifically defined or constant and is, in fact, rather diffuse. From this point of view, we might expect convergence difficulties in the case of the long-range Coulomb interaction that can arise in He resonances, for example. However, Ho[31] has used the complex-coordinate method to study several autoionizing states of He, and from the preceding discussion, this method should converge faster than the complex-coordinate type of calculation performed by Ho.

We would now like to return to a discussion of the calculations of McCurdy and Rescigno[7] and Moiseyev and Corcoran.[8] Both of these calculations can be viewed as complex-stabilization calculations using the unrotated Hamiltonian and a global complex scaling. Viewed in this manner, the arguments in Ref. (7) attempting to justify the method by the fact that certain integrals become equal in the limit that certain nonlinear parameters become small and that they are the ones that determine the width are unnecessary. In the same manner, the concern in Ref. (8) about the basis functions being moved to a complex center is not necessary. Instead, we merely have to stabilize $\tilde{E}^{\text{res}}$ with respect to variations in the parameters, which in these cases corresponds to a global complex scaling. In Section 6, however, we discuss the construction of variational wave functions that should converge much more rapidly and produce more stable bound and resonant eigenvalues.

5. THE VARIATIONAL PRINCIPLE

As previously discussed, the complex-stabilization method, or complex-coordinate method, is based on the variational principle. Here, we will discuss the variational principle from the standpoint of the rotated Hamiltonian and use the analyticity of the resonant wave function to argue that the same technique can be used for the unrotated Hamiltonian, i.e., we can use the complex-coordinate theorems to derive a variational principle for the unrotated Hamiltonian. In a later article, we will discuss the connection between this approach and variational principles that explicitly impose Kapur–Peierls[32,33] and Siegert[15,16,29] boundary conditions.

Consider the functional

$$F(\chi) \equiv \frac{\int \chi^{+} H \chi d\tau}{\int \chi^{+} \chi d\tau} \tag{60}$$

where the meaning of χ^{+} will become evident. For an eigenfunction Φ of H. $F(\Phi)$ is just the energy E. If Φ is changed by an infinitesimal amount $\delta\Phi$, Eq. (60) becomes

$$\begin{aligned} &\delta F(\Phi) \int \Phi^{+} \Phi d\tau + F(\Phi)\left(\int \delta\Phi^{+} \Phi d\tau + \int \Phi^{+} \delta\Phi d\tau\right) \\ &\quad = \int \delta\Phi^{+} H \Phi d\tau + \int \Phi^{+} H \delta\Phi d\tau \end{aligned} \tag{61}$$

where we have neglected second-order terms and used

$$F(\Phi + \delta\Phi) \equiv F(\Phi) + \delta F(\Phi) \tag{62}$$

Applying Green's theorem to the last term on the righthand side of Eq. (61) yields

$$\begin{aligned} \delta F(\Phi) \int \Phi^{+} \Phi d\tau &= \int \delta\Phi^{+} (H - E) \Phi d\tau + \int \delta\Phi (H - E) \Phi^{+} d\tau \\ &\quad + \int_{s} \left(\Phi^{+} \frac{\partial}{\partial n} \delta\Phi - \delta\Phi \frac{\partial}{\partial n} \Phi^{+} \right) dS \end{aligned} \tag{63}$$

In order for $\delta F(\Phi)$ to vanish, Φ and Φ^{+} must satisfy the same Schrödinger equation, and the surface term must vanish. The first condition is satisfied if the radial function in Φ^{+} are not complex conjugated, while the second condition is satisfied, since the rotated wave function is square-integrable for $\theta > \theta_c$.

Finally, using the facts that the exact resonant wave function is analytic in θ and a square-integrable basis can be used in the application of the variational principle, we can argue that a square-integrable basis can be used for the unrotated Hamiltonian. The convergence of a basis-set expansion to the exact resonant wave function is assumed for the variational principle, since this has not been proven.

6. STRUCTURE OF VARIATIONAL WAVE FUNCTIONS AND BASIS SETS

The structure of the wave function will depend on the phenomenon being studied. In this section, we discuss possible choices of the wave function for potential scattering; electron-atom, or molecule Feshbach, or shape resonances; and field ionization. These examples should suffice to illustrate how to construct variational wave functions for other phenomena.

6.1. *Basis Functions*

As previously noted, the fundamental difference between bound- and resonant-state wave functions is that the former are real functions of r, while the latter are explicitly complex functions of r. Thus, an expansion of the radial part of the resonant wave function χ^{res} must be explicitly complex. That is, there must not exist a similarity transformation that can transform the complex basis functions into a real basis. If such a transformation existed, the eigenvalues would all be real since the eigenvalues of a secular equation are invariant with respect to a similarity transformation. The other property required of basis functions is that they be square integrable.

Slater-type orbitals (STOs), hydrogenic orbitals, or Gaussian orbitals with complex nonlinear parameters could be used. In fact, such calculations as Doolen's[10] correspond to complex scaling all nonlinear parameters by the same scale factor, while those of Junker and Huang[11] and Rescigno *et al.*[25] correspond to complex scaling a portion of the nonlinear parameters, which, in these cases, correspond to the basis functions used to represent the unbound particle. It should be noted, however, that complex STOs and hydrogenic orbitals have a periodic oscillatory behavior, while Gaussian orbitals have an oscillatory behavior in which the nodes become closer in a regular fashion as r increases. On the other hand, the exact resonant wave function will, in general, have nodes located at irregular positions determined by the energy and potential function. Thus, a basis set with only complex nonlinear parameters may have to be quite large so that the unnecessary nodes in the basis functions can be removed or shifted. As a result, a basis where all orbitals except for one or two have real nonlinear parameters may have better convergence properties. Nonetheless, either basis should be capable of representing the resonant wave function.

6.2. *Potential Scattering*

Since bases of the former type have generally been used in previous calculations [e.g., see Refs. (10–13)], we will confine our discussion to the latter. For example, for potential scattering, we might consider variational functions of the form

$$\chi^{\text{res}} = \sum_{i=1}^{m} c_i\phi_i + r^n \exp(ikr)\{c_{m+1}N_{m+1}\exp(-\alpha r) + c_{m+2}N_{m+2}\exp[-(\alpha+\varepsilon)r]\}Y_l^m(\theta,\phi) \tag{64}$$

or

$$\chi^{\text{res}} = \sum_{i=1}^{m} c_i\phi_i + c_{m+1}N_{m+1}r^n\exp(ikr)\exp(-\alpha r)y_l^m(\theta,\phi), \qquad n \geqslant 1 \tag{65}$$

where the ϕ_i are real basis functions, the c_i (including c_{m+1} and c_{m+2}) are linear variational coefficients, and N_{m+1} and N_{m+2} are normalization constants. The $\tilde{E}^{\text{res}}$ would then be stabilized with respect to variations in nonlinear parameters in the ϕ_i, k, α, and ε. The first function, Eq. (64), has regular nodal structure in only the regions n/α and $n/(\alpha+\varepsilon)$ for $n \geqslant 1$ or in the region from $1/\alpha$ to $1/(\alpha+\varepsilon)$ for $n = 0$, while Eq. (65) has regular nodal structure in only the region n/α. If ε is chosen to be small in Eq. (64), this region can be made arbitrarily small subject to only numerical stability. This regular nodal structure best represents the exact resonant function for large values of r. In a way, we can envision these complex basis functions as *implicitly* imposing a complex boundary condition on the variational wave function through the variational calculation. This does not imply that the complex functions in Eqs. (64) and (65) correspond to Eq. (59b), since Eq. (59a) is also an explicitly complex function, and the complex functions in Eqs. (64) and (65) *do not* have to be located at the furtherest extent of the basis functions. A study of different types of basis sets including all complex functions for potential scattering will be discussed in the following section.

6.3. *Electron–Atom and Electron–Molecule Resonances*

For electron–atom and electron–molecule resonances, we discussed in IA how the wave function could be considered to be partitioned into two parts—Ψ_P for the open channels or scattering part and Ψ_Q for the closed channel or boundlike part; i.e.,

$$\Psi = \Psi_P + \Psi_Q \tag{66}$$

We need not rigorously require Ψ_P and Ψ_Q to be orthogonal, in which case this partitioning does not correspond to the rigorous projection in Feshbach theory. We do not have to require this orthogonality, since we are using the total wave function in the variational calculation. This partitioning is also not unique, but merely represents a convenient way of viewing the variational wave function so that physical insight can be incorporated into its structure. For shape resonances, this partitioning may, in fact, be less useful, although it can be used to partition different classes of configurations that we might include in a variational wave function for a shape resonance, as discussed later.

For Feshbach resonances, Ψ_Q contains closed-channel configurations as well as polarization and correlation configurations. For the same reasons just stated, these resonances can best be represented in terms of basis functions with real nonlinear parameters. On the other hand, Ψ_P represents the open-channel part, and as such, it will consist of an antisymmetrized product of target function(s) and functions of the form in Eq. (64) or Eq. (65): i.e.,

$$\Psi_P = \mathscr{P}\mathscr{A}\sum_{\nu} c_\nu \phi_\nu^T \chi_\nu^{\text{res}} \tag{67}$$

where $\mathscr{P}$ represents the projection of the appropriate spin and space group symmetries, $\mathscr{A}$ is the antisymmetrizer, and ϕ_ν^T is the target wave function for the νth open channel. The real basis functions in ψ_ν^{res} for different channels may be chosen to be the same or different functions. However, the complex functions should certainly be different for different channels.

In the case of shape resonances, Ψ_P is basically constructed in the same manner as previously discussed for Feshbach resonances. Correlation-type, including polarization, configurations could be grouped into Ψ_Q. For both Feshbach and shape resonances, correlation between all open and closed channels is automatically incorporated. Finally, the approximate resonant energy $\tilde{E}^{\text{res}}$ should be stabilized with respect to the parameters in ϕ_ν^T, since the presence of the additional electron will distort the target functions.

6.4. *Field Ionization*

Cerjan *et al.*[34] have used the numerical range* of

$$H^{\text{field}}(\theta) = -[\exp(-2i\theta)/2]\nabla^2 + \exp(i\theta)\mathbf{F}\cdot\mathbf{r} \tag{68}$$

and the fact that the Coulomb potential is compact relative to $H^{\text{field}}(\theta)$ to show that the transformation in Eq. (1) can be used to uncover the Stark broadened

* The numerical range $R(\theta)$[34] of an operator θ is defined as the set of all complex numbers that might be obtained as a diagonal matrix element with respect to a normalized function $|u\rangle$, i.e.,

$$R(\theta) \equiv \{z \equiv \langle u|\theta|u\rangle \in C : \langle u|u\rangle = 1\}$$

and shifted atomic states. The theorem stating that the resonant wave functions depend on r and θ in only the combination $r\exp(i\theta)$ applies here also. Thus, we should be able to compute the complex energies associated with the Stark broadened and shifted atomic states by using the unrotated Hamiltonian directly with a square-integrable basis. For the nonrelativistic case of a constant field directed along the z axis, for example, only M_L remains an exact quantum number that group theoretically characterizes the states. The hydrogenic case is an exception where the fortuitous form of the Coulomb potential permits complete separation of variables and, thus, the necessary three quantum numbers that completely characterize each state. For an N-electron atomic system, we could construct a variational wave function of the form

$$\Psi = \mathscr{A}\sum_L\sum_S \phi_S^L(N-1)\left\{\sum_{j=1}^{n_S} C_{Sj}^L\chi_j^L(N) + r^{L+l_j}\exp(ik_{sj}r)\right.$$
$$\left.\times\{C_{S(n_S+1)}^L r^{m_j}\exp(-\alpha_{Sj}r) + C_{S(n_S+2)}^L r^{n_j}\exp[-(\alpha_{Sj}+\varepsilon_{Sj}r)]\,Y_L^m\}\,Y_L^m\right\} \tag{69}$$

or

$$\Psi = \mathscr{A}\sum_L\sum_S \phi_S^L(N-1)\left[\sum_{j=1}^{n_S} C_{Sj}^L\chi_j^L(N) + C_{S(n_S+1)}^L r^{L+l_j}\right.$$
$$\left.\times\exp(ik_{Sj}r)\exp(-\alpha_{Sj}r)\,Y_L^M\right] \tag{70}$$

where $\mathscr{A}$ is as described above; C_{Sj}^L, $C_{S(n_S+1)}^L$, and $C_{S(n_S+2)}^L$ are linear variationally determined coefficients. The $\phi_S^L(N-1)$ represent wave functions for the $(N-1)$ electrons, which are *relatively* stable with respect to field ionization; the $\chi_j^L(N)$ are a set of basis functions describing the ionized electron; and l_j, m_j, and n_j are greater than or equal to zero. In writing Eqs. (69) and (70), the assumption that only one particle is unstable with respect to ionization is implicit, which is the case generally studied. The integrals required for the preceding wave function are straightforward, where as a Siegert-like wave function using basis functions representative of just the case of a hydrogenic system in a constant field would contain functions with exponentials of fractional powers of r, $\cos\chi$, and $\sin\chi$. Such integrals must be performed numerically. Finally, the effects of fields on resonances, such as those studied by Broad and Reinhardt[35] and Wendoloski and Reinhardt,[36] can be studied by a straightforward merger of Eqs. (66) and (69) or (70).

These examples should indicate the general approach that could be taken in constructing a variational wave function to study various resonant phenomena. The fact that boundary conditions can be implicitly incorporated into the variational calculation itself leads to a considerable potential simplification in studying any complex phenomena.

7. MODEL POTENTIAL CALCULATIONS

We will illustrate the above techniques by computing the position and width of a resonance in a model considered earlier[37]

$$V(r) = 7.5r^2 \exp(-r) \tag{71}$$

From direct numerical integration, a resonance is known to exist with position and width parameters of 3.42639 and 0.025549 a.u., respectively ($E_R - E_I =$ 3.42639–0.012775 i a.u.).

In these calculations, four types of variational wave functions are studied. The first one is of the form used by Doolen, the second is identical to the first except that a Siegert function is added, and the third is the same as the second except that the added Siegert function is altered to make it square integrable and eliminate its r^{-1} factor. The fourth variational function is of the form of Eq. (64).

$$\Psi_1(N) = \sum_{j=1}^{N} C_j[r\exp(-i\theta)]^{j-1} \exp[-\gamma r \exp(-i\theta)] \tag{72a}$$

$$\Psi_2(N) = \Psi_1(N) + C_{N+1} r^{-1}[1 - \exp(-\alpha r)] \exp(ik_R r) \exp(k_I r) \tag{72b}$$

$$\Psi_3(N) = \Psi_1(N) + C_{N+1} \exp(ik_R r) \exp(-k_I r) \tag{72c}$$

$$\Psi_4(N) = \sum_{j=1}^{N} C_j r^{n_j - 1} \exp(-\gamma_j r) + \exp(ikr) \times \{C_{N+1} \exp(-\alpha r) + C_{N+2} \exp[-(\alpha + \varepsilon)r]\} \tag{72d}$$

In the calculations with Eqs. (72a)–(72c), wave functions with N equal to 10 and 14 are used. The Siegert value for $k_R - ik_I$ is 2.6177891–0.00487989 i a.u. The calculations are performed at θ values of 0.0, 0.05, 0.1, and 0.5. At each value of $\theta, \tilde{E}_r$ is stabilized with respect to variations in γ. Figure 1 illustrates the variation of $\tilde{E}_r$ with respect to γ for $\Psi_1(10)$, $\Psi_2(10)$, and $\Psi_3(10)$ at several values of theta with α equal to one. The behavior of the curves in Fig. 1 is similar to that noted earlier by Moiseyev and Weinhold[38] except that the oscillations are better defined. Several points are obvious. First, including complex functions produced a significant improvement in the rate of convergence. Secondly, the non-Siegert function is as good as the Siegert function. This illustrates the comment in Section 4 that imposing the exact long-range behavior is unnecessary. Consequently, we do not have to in-

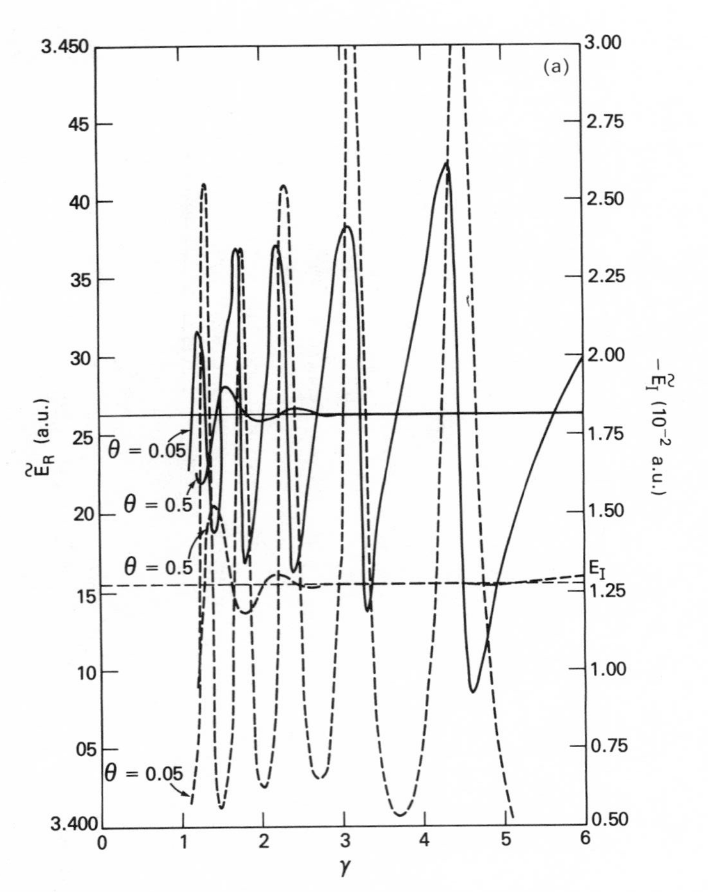

(a)
$\tilde{E}_R$ (a.u.)
$-\tilde{E}_I$ (10^{-2} a.u.)
γ
$\theta = 0.05$
$\theta = 0.5$
$\theta = 0.5$
$\theta = 0.05$
E_I
3.450
45
40
35
30
25
20
15
10
05
3.400
3.00
2.75
2.50
2.25
2.00
1.75
1.50
1.25
1.00
0.75
0.50
0
1
2
3
4
5
6

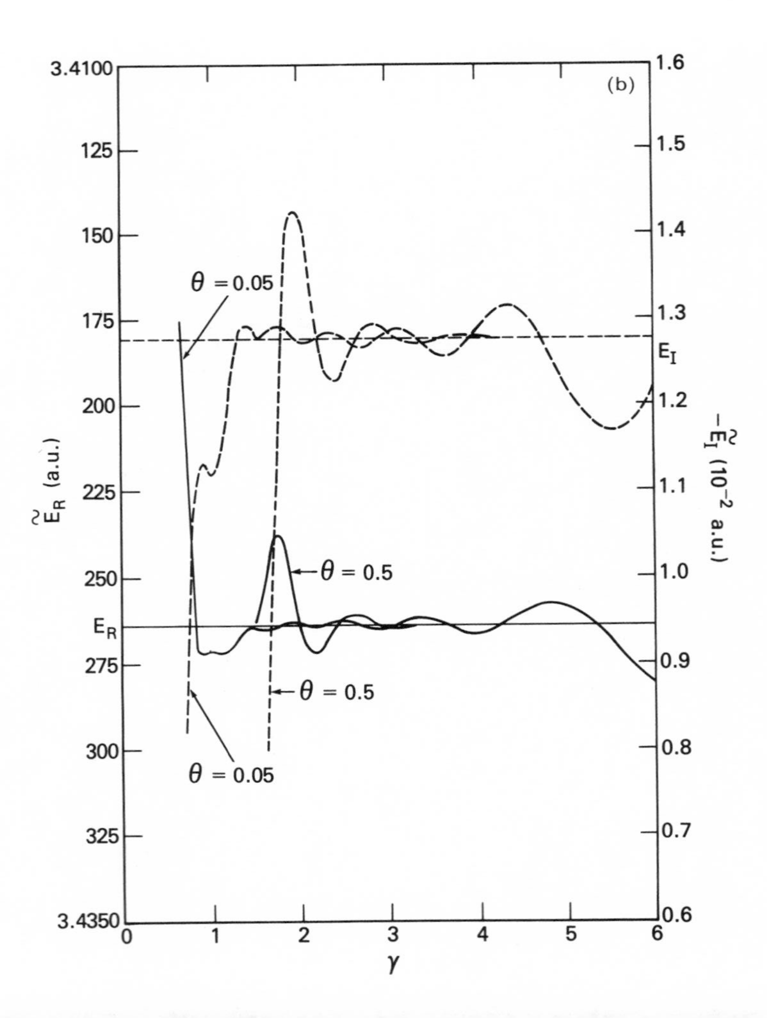

(b)
$\tilde{E}_R$ (a.u.)
$-\tilde{E}_I$ (10^{-2} a.u.)
γ
$\theta = 0.05$
$\theta = 0.5$
$\theta = 0.5$
$\theta = 0.05$
E_I
E_R
3.4100
125
150
175
200
225
250
275
300
325
3.4350
1.6
1.5
1.4
1.3
1.2
1.1
1.0
0.9
0.8
0.7
0.6
0
1
2
3
4
5
6

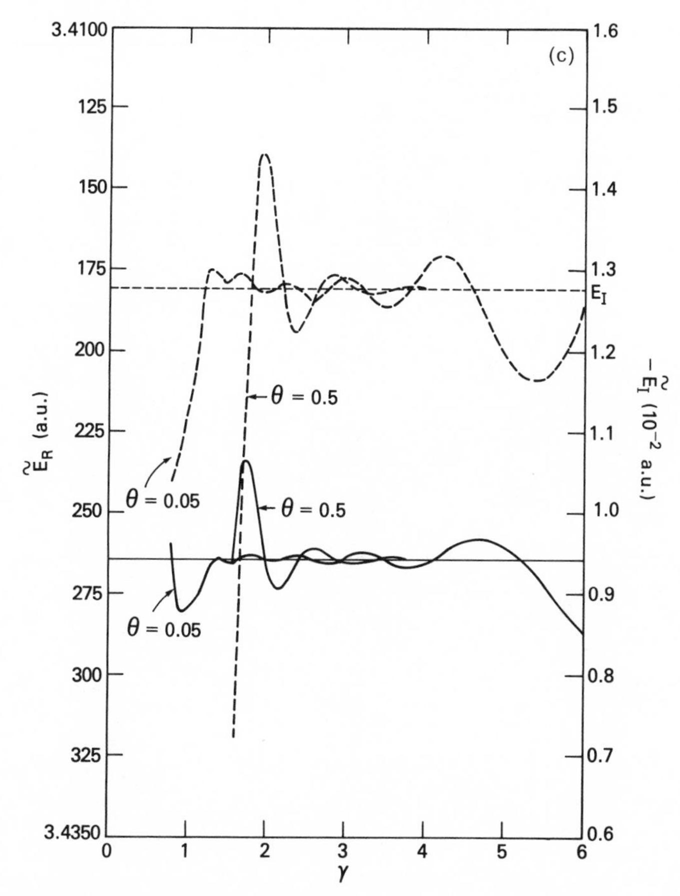

FIGURE 1. The $\tilde{E}_R(\gamma)$ and $\tilde{E}_I(\gamma)$ for $0.07 \leqslant \gamma \leqslant 6.0$ at θ equal to 0.05 and 0.5 for (a) $\Psi_1(10)$, (b) $\Psi_2(10)$, and (c) $\Psi_3(10)$.

corporate ill-behaved Siegert functions in resonance calculations. Thirdly, there is a region in the nonlinear parameter space over which both the real and imaginary parts of $\tilde{E}_r$ are most stable. For example, for both $\Psi_2(10)$ and $\Psi_3(10)$, $\tilde{E}_r$ is most stable in region $1.8 \leqslant \gamma \leqslant 2.5$. Fourthly, for a given value of θ, both the real and imaginary parts of $\tilde{E}_r$ will only accidently stabilize together at a given γ with respect to variations in γ. In fact, a more meaningful number is some sort of average of $\tilde{E}_r(\gamma)$ over the region of most stability of $\tilde{E}_r$. The oscillations are probably due to interference effects in the oscillating basis functions, $r^n \exp[-\gamma r \exp(-i\theta)]$.

Calculations in which γ was stepped in increments of 0.05 were then performed at the four angles noted for $\Psi_{1,2,3}{}^{(10)}$ and $\Psi_{1,2,3}{}^{(14)}$ The value

TABLE 1

Stabilized Value of γ and $\tilde{E}_r$ for $\Psi_1(N)$, $\Psi_2(N)$, and $\Psi_3(N)$

	0.00			0.05			0.10			0.50		
	γ	E_R	$-E_r$	γ	E_R	$-E_I$	γ	E_R	$-E_I$	γ	E_R	$-E_I$
$\Psi_1(10)$	—	—	—	—	—	—	—	—	—	4.70	3.42640	0.012739
$\Psi_2(10)$	2.70	3.42643	0.012481	2.25	3.42643	0.012844	2.75	3.42641	0.012749	5.60	3.42639	0.012776
$\Psi_3(10)$	2.60	3.42640	0.012448	2.20	3.42640	0.012858	2.70	3.42642	0.012752	5.55	3.42639	0.012776
$\Psi_1(14)$	—	—	—	—	—	—	—	—	—	4.45	3.42639	0.012774
$\Psi_2(14)$	2.80	3.42643	0.012820	2.70	3.42640	0.012773	2.65	3.42639	0.012773	5.20	3.42639	0.012774
$\Psi_3(14)$	2.90	3.42631	0.012771	2.65	3.42640	0.012782	2.50	3.42639	0.012772	5.15	3.42639	0.012774

of γ given in Table 1 is the value where $[(\Delta\tilde{E}_R/\Delta\gamma)^2 + (\Delta\tilde{E}_I/\Delta\gamma)^2]^{1/2}$ is a minimum subject to the constraint that both $(\Delta\tilde{E}_R/\Delta\gamma)$ and $(\Delta\tilde{E}_I/\Delta\gamma)$ also have minima at values of γ that differ by no more than a few tenths. For $\theta = 0$, $\Psi_2(10)$ and $\Psi_3(10)$ are most stable in the regions of γ between 1.9 and 2.1. The average of the maximum excursions of $\tilde{E}_R$ and $\tilde{E}_I$ here are 3.42635 and -0.012794 a.u., respectively. A similar comment can be made at the other angles.

Finally, calculations were performed with several wave functions of the form $\Psi_4(N)$. Initially, a wave function with N equal to ten was used with the (n_j, γ_j) pairs chosen such that the set of basis functions $r^{n_j-1}\exp(-\gamma_j r)$ formed a set of functions where each function had its maximum at successively larger values of r. The ε was chosen to be a small number for the reasons given in Section 4. Note that unlike in Ψ_2 and Ψ_3, here k it treated directly as a variational parameter.

The $\tilde{E}_r$ was stabilized with respect to real scalings of all γ_j, k, and the pair $[\alpha, (\alpha+\varepsilon)]$ individually. Since the nonlinear parameters k, α, and $(\alpha+\varepsilon)$ are the most significant parameters with respect to $\tilde{E}_I$, $\tilde{E}_r$ was stabilized with respect to them first. In effect, the last two terms are implicitly imposing a boundary condition on Ψ_4 through the stabilization procedure. The order

TABLE 2
$\tilde{E}_r(k)$

k	$\Delta\tilde{E}_R{}^a$	$\Delta\tilde{E}_I{}^a$
1.54000		
	-7.39	0.55
1.63625		
	-5.86	1.91
1.73250		
	-3.74	1.62
1.82875		
	-1.52	0.64
1.92500		
	0.69	-0.32
2.02125		
	2.93	-0.85
2.11750		
	5.32	-0.70
2.21375		
	8.00	0.19
2.31000		

[a] The units of $\Delta\tilde{E}_R$ and $\Delta\tilde{E}_I$ are 10^{-5} a.u. The value of $\tilde{E}_r(k=1.925)$ is 3.42639–0.012758i a.u.

TABLE 3
$\tilde{E}_r(\gamma_i)$ for γ_i with Largest P_i

γ_i	$\Delta\tilde{E}_R$ [a]	$\Delta\tilde{E}_I$ [a]
3.0912		
	2.07	−3.98
3.1584		
	1.75	−2.29
3.2256		
	1.38	−1.46
3.2928		
	1.11	−1.21
3.3600		
	1.02	−1.37
3.4272		
	1.19	−1.83
3.4944		
	1.70	−2.51
3.5616		
	2.60	−3.29
3.6288		

[a] The units of $\Delta\tilde{E}_R$ and $\Delta\tilde{E}_I$ are 10^{-5} a.u. $\tilde{E}_r(i = 3.36)$ is 3.42638–0.012757i a.u.

TABLE 4
Basis Set Parameters and Properties

n_i	γ_i [a]	k	r_{max} [b]	P_i
1	7.1436		0.140	0.0341
3	3.4944		0.572	0.0233
2	1.3104		0.763	0.0728
2	0.8580		1.166	0.0488
1	0.8398		1.191	0.0793
4	2.0748		1.446	0.0097
3	1.3104		1.526	0.0576
1	0.4940		2.024	0.0560
3	0.8450		2.367	0.0480
4	0.8398		3.572	0.0886
1	0.1700	2.00	5.882	0.2390
1	0.1500	2.00	6.667	0.2427

[a] The last two entries correspond to $(\alpha + \varepsilon)$ and α, respectively.
[b] r_{max} for the orbitals with n_i equal to one corresponds to the radial distance at which the orbital is at $\exp(-1)$ of its original value.

for stabilizing $\tilde{E}_r$ with respect to the γ_j was according to the size of the structure projection P_i for each term in the wave function. The P_i is defined as

$$P_i \equiv |\langle \chi_i | \Psi_4 \rangle| \tag{73}$$

where Ψ_4 is a variational wave function obtained after stabilizing $\tilde{E}_r$ with respect to k and $[\alpha, (\alpha + \varepsilon)]$. Ordering the stabilization in this way stabilized the procedure. The stabilization procedure was iterated several times. Tables 2 and 3 give the nonlinear parameter trajectories for $\tilde{E}_r(k)$ and $\tilde{E}_r(\gamma_j)$ for one of the γ_j's.

Since $\tilde{E}_r$ is a hypersurface in the nonlinear parameter space, all the nonuniqueness problems associated with normal bound-state variational calculations are, of course, present in these calculations. For some nonlinear parameters with small P_i's, $\tilde{E}_r$ does not stabilize well, since these basis functions are not significant and can generally be eliminated.

The value of the complex energy obtained with this function is 3.426395–0.012757 i a.u. A final global real scaling of all the γ_j yielded a complex energy of 3.426397–0.012764 i a.u. Table 4 gives the final stabilized values for the nonlinear parameters along with the location of the maximum for each function and its structure projection. When two more basis functions were added, with $\tilde{E}_r$ being stabilized with respect to them, k and $[\alpha, (\alpha + \varepsilon)]$ followed by a global scaling as before, a complex energy of 3.426395–0.012775 i a.u. was obtained.

8. DISCUSSION

We have suggested that theta in previous complex-coordinate calculations is simply playing the role of a nonlinear parameter. This along with the analyticity of the Hamiltonian and exact resonant wave functions implies that we can use the unrotated Hamiltonian with a square-integrable basis and without explicitly imposing a Siegert boundary condition to compute the complex poles of the resolvent.

Since earlier complex–coordinate calculations are only special cases of a more general complex-stabilization method, we might ask what advantage is obtained from this more general procedure. There are actually a number of advantages:[39] First, if θ is just a nonlinear scale factor, values of theta other than $\beta < \theta \leqslant \pi/4$, including negative values of theta, are meaningful. Second, by using $H(r)$, chemical and physical insights can be incorporated straightforwardly into the construction of the variational wave function [an approach using $H(\theta)$ was suggested earlier[11]]. Third, a number of different θ's can be incorporated into the variational calculation. This can be useful, for example, for different (s, p, d, etc.) polarization contributions. Another situation where more than one θ value is useful is when there are several thresholds below a given resonance, so that we must describe several continua. The value of θ that stabilized $\tilde{E}_r$ with respect to functions describing one continuum is generally quite differ from that describing another continuum. Fourth, molecules present no conceptual problem; that is, while the total molecular Hamiltonian is dilation analytic, the Born–Oppenheimer Hamiltonian is not. Since the Hamiltonian is not scaled in the complex-stabilization method, this method is directly applicable to molecular systems without the questions of complex internuclear coordinates, distorted integration contours, external complex scaling, or Siegert boundary conditions ever explicitly arising. Fifth, by using $H(r)$ instead of $H(\theta)$, we avoid the difficulty of describing the unnecessary kinematic θ-induced oscillations in the exact wave functions for bound and resonances states as well as for the bound particles in scattering states. Finally, wave functions of the form given by Eqs. (64), (65), (67), (69), and (70) can be used.

REFERENCES

1. J. Taylor, *Scattering Theory*, Wiley, New York (1972).
2. R. J. Damburg and V. V. Kolosov, *J. Phys. B.* **9**, 3149 (1976).
3. E. B. Balslev and J. M. Combes, *Commun. Math. Phys.* **22**, 280 (1971).
4. B. Simon, *Commun. Math. Phys.* **27**, 1 (1972); *Ann. Math.* **97**, 247 (1973).
5. C. Van Winter, *Math. Anal. Appl.* **49**, 88 (1975).
6. *Int. J. of Quantum Chem.* **14**, (1978).
7. C. W. McCurdy, Jr. and T. N. Rescigno, *Phys. Rev. Lett.* **41**, 1364 (1978).
8. N. Moiseyev and C. Corcoran, *Phys. Rev. A* **20**, 814 (1979).
9. B. Simon, *Phys. Lett.* **71A**, 211 (1979).
10. G. D. Doolen, *J. Phys. B* **8**, 525 (1975).
11. B. R. Junker and C. L. Huang, Abstracts of the Tenth International Conference on the Physics of Electronic and Atomic Collisions, Paris, 1977 (edited by M. Barat and J. Reinhardt), p. 614, Commissariat à l'Énergie Atomique, Paris 1977; *Phys. Rev. A* **18**, 313 (1978).
12. B. R. Junker, *Int. J. Quantum Chem.* **14**, 371 (1978).
13. B. R. Junker, *Phys. Rev. A* **18**, 2437 (1978).
14. B. R. Junker, *Phys. Rev. Lett.* **44**, 1487 (1980).
15. G. Gamow, *Z. J. Phys.* **51**, 204 (1928).
16. A. F. J. Siegert, *Phys. Rev.* **56**, 750 (1939).
17. R. G. Newton, *Scattering Theory of Waves and Particles*, pp. 261–263, McGraw-Hill, New York (1966); G. Hagedorn, *Bull. Am. Math. Soc.* **84**, 155 (1978); I. Sigal, *Bull. Am. Math. Soc.* **84**, 152 (1978).
18. C. A. Nicholaides and D. R. Beck, *Int. J. Quantum Chem.* **14**, 457 (1978).
19. G. D. Doolen, *Int. J. Quantum Chem.* **14**, 523 (1978).
20. L. D. Landau and E. M. Lifshitz, *Quantum Mechanics*, pp. 118–121, Pergamon, London (1958).
21. G. D. Doolen, private communication.

22. R. Yaris, J. Bendler, R. A. Lovett, C. M. Bender, and P. A. Fedders, *Phys. Rev. A* **18**, 1816 (1978).
23. G. D. Doolen, J. Nuttall, and R. W. Stagat, *Phys. Rev. A* **10**, 1612 (1974).
24. Y. K. Ho, *Phys. Rev. A* **5**, 1675 (1978).
25. T. N. Rescigno, C. W. McCurdy, Jr., and A. E. Orel, *Phys. Rev. A* **17**, 1931 (1978).
26. C. W. McCurdy, *Electron–Molecule and Photon–Molecule Collisions* (T. Rescigno, V. McKoy, and B. Schneider, eds.), p. 299, Plenum, New York 1979.
27. E. Brändas and P. Froelich, *Phys. Rev. A* **16**, 2207 (1977); P. Froelich, M. Hehenberger, and E. Brändas, *Int. J. Quantum Chem.* **11**, 295 (1977).
28. R. A. Donnelly and J. Simmons, *J. Chem. Phys.* **73**, 285 (1980).
29. J. N. Bardsley and B. R. Junker, *J. Phys. B.* **5**, L178 (1972); R. A. Bain, J. N. Bardsley, B. R. Junker, and C. V. Sukumar, *J. Phys. B.* **7**, 2189 (1974); A.D. Isaacson and W. H. Miller, *Chem. Phys.* **34**, 311 (1978); A. D. Isaacson and W. H. Miller, *Chem. Phys. Lett.* **62**, 374 (1979); and C. W. McCurdy and T. N. Rescigno, *Phys. Rev. A* **20**, 2346 (1979).
30. B. Simon, private communication.
31. Y. K. Ho, *J. Phys. B.* **12**, 387 (1979).
32. P. L. Kapur and R. E. Peierls, *Proc. R. Soc.* **A166**, 277 (1938).
33. A. Herzenberg and F. Mandl, *Proc. R. Soc.* **A274**, 256 (1963).
34. C. Cerjan, W. P. Reinhardt, and J. E. Avron, *J. Phys. B.* **11**, 1201 (1978).
35. J. T. Broad and W. P. Reinhard, *Phys. Rev. A* **14**, 2159 (1976).
36. J. Wendoloski and W. P. Reinhardt, *Phys. Rev. A* **17**, 195 (1978).
37. R. A. Bain, J. N. Bardsley, B. R. Junker, and C. V. Sukumar, *J. Phys. B* **7**, 2189 (1974).
38. N. Moiseyev. and F. Weinhold, *Int. J. of Quantum Chem.* **17**, 1201 (1980).
39. B. R. Junker, to appear in *Book of Invited Talks* (edited by S. Datz), Twelth International Conference on the Physics of Electronic and Atomic Collisions, Gatlinburg, TN, p. 491, North-Holland, Amsterdam (1981).

CHAPTER FIVE

MOLECULAR RESONANCE CALCULATIONS

APPLICATIONS OF COMPLEX-COORDINATE AND COMPLEX BASIS FUNCTION TECHNIQUES

C. WILLIAM McCURDY

1. INTRODUCTION

Resonances in low-energy (0–10 eV) electron scattering from molecules have been observed in a large number of cases[1,2] with target molecules ranging in size from H_2 to naphthalene ($C_{10}H_8$). In fact, almost every organic molecule containing a double bond displays at least one low-energy resonance.[2] The mechanism of formation for these low-energy resonances can be viewed in two superficially different, but essentially equivalent, ways, depending on whether we initially adopt the point of view of the atomic physicist or quantum chemist. From both points of view, most of the resonances below 5-eV incident energy in electron-molecule scattering can be thought of in the first approximation as "shape" resonances, that is, not involving electronic excitation of the target molecule.

To the atomic physicist, this picture calls to mind the situation in Fig. 1a. The potential experienced by the incident electron due to long-range polarization effects (proportional to $-r^{-4}$) is attractive, as is the screened Coulomb potential which dominates at short range. Combining the attractive long-range potential with the repulsive centrifugal potential produces a barrier, and the shape resonances are associated with tunneling through this barrier.

To the chemist, the picture is that the incident electron temporarily occupies the lowest unoccupied molecular orbital [LUMO in the parlance of

C. WILLIAM McCURDY ■ Department of Chemistry, 140 West 18th Avenue, Ohio State University, Columbus, OH 43210.

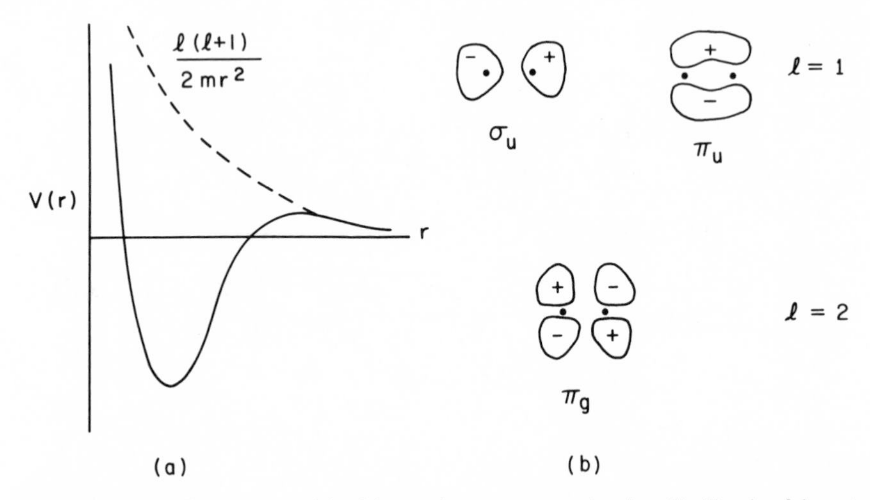

FIGURE 1. (a) Effective potential with angular momentum barrier. (b) Sketch of lowest unoccupied molecular orbitals (LUMOs) and most important l value in their single-center expansions.

"frontier molecular orbital theory".[2,3]] The LUMOs for resonances corresponding to orbital angular momenta $l = 1$ and $l = 2$ are indicated schematically in Fig. 1b. The l values associated with these orbitals are their principal components in an expansion in spherical harmonics around the center of the molecule. The chemist's picture associates an orbital of symmetry σ_u with the ${}^2\Sigma_u^+$ resonance in electron–H_2 scattering and a π_g orbital with the ${}^2\Pi_g$ resonance in electron–N_2 scattering. The utility of this approach is still more apparent if we consider the classification of resonances in molecules more complicated than diatomics, for example, in naphthalene, for which five resonances have been observed.[2,4]

A natural language for the qualitative description of resonances in electron–molecule scattering (and electron–atom scattering as well) is the language of bound-state electronic structure theory, because it reflects the correct physics of the temporarily bound states in question. Thus, an obvious and venerable problem in electron scattering is to devise a quantitative description of resonances by reducing the resonance problem to a form addressable by the arsenal of bound-state computational methods that have been developed for the electronic-structure problem.

The main subject of this chapter is a group of recent developments in that direction that make use of complex coordinates or complex basis functions. Specifically, we will describe the extension for resonance calculations of the principal tools of quantum chemistry, namely, the self-consistent-field (SCF) and configuration interaction (CI) methods. But first, we must specify precisely what it is that we want to calculate.

Resonances in electron–molecule scattering play an important role in vibrational excitation and dissociative attachment processes, sometimes greatly enhancing the cross section at energies near the resonance. The minimum information necessary for the approximate calculation of the resonance contribution to vibrational excitation or dissociative-attachment cross sections is the complex potential energy surface for nuclear motion in the resonance state. For example, for a diatomic molecule, the complex potential function $E_{\text{res}}(R)$ depends on the internuclear distance R, and it is written as

$$E_{\text{res}}(R) = E_R(R) - i\Gamma(R)/2 \tag{1}$$

where $E_R(R)$ is the real part of the potential function and $\Gamma(R)$ is the width function. The potential $E_{\text{res}}(R)$ together with the potential curve for the ground state of the neutral target molecule is the starting point for calculations using the "boomerang" model of Birtwistle and Herzenberg[5] and Dubé and Herzenberg[6] or the "energy-modified adiabatic approximation" of Nesbet.[7] Dubé and Herzenberg[6] point out that since the vibrational structure of the cross section for vibrational excitation is a quantum interference effect, it may be necessary to know the potential curves to extreme accuracy to reproduce that structure in a theoretical calculation. If the complex potential curves are to be obtained to the accuracy required from practical *ab initio* calculations on metastable negative ions, those calculations cannot be much more difficult than ordinary bound-state calculations.

Complex basis function approaches to the problem of calculating the complex energies corresponding to resonances in electron–molecule scattering promise to provide generally applicable methods that cast the calculation of the complex resonance energy essentially into the form of an ordinary self-consistent-field or configuration interaction calculation on a bound state. The complex basis function methods currently in use for resonances in molecules and many-electron atoms have evolved from the literal application of mathematical theorems on complex scaling of electronic coordinates to a point where they bear only distant resemblance to their antecedents. Fortunately, the complex basis function methods are simpler.

In this chapter, we will survey the state of the art in complex basis function calculations on molecular resonances as well as give some notion of the relationship of methods presently in use to the simple mathematical theorems that, in part, motivated their development. It is apparent that more efficient extensions of these methods will be developed in the future and that most molecular resonance calculations that have been performed to date are explorations of new approaches. Many unanswered questions remain, some of which we raise at the end of this chapter.

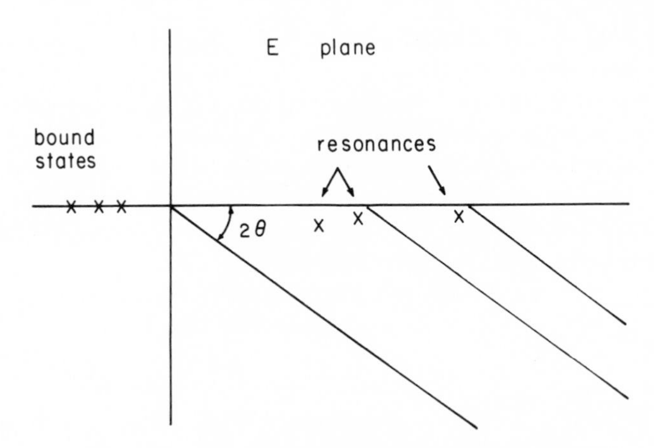

FIGURE 2. Spectrum of the complex Hamiltonian $\hat{H}_\theta$.

2. COMPLEX COORDINATES AND COMPLEX BASIS FUNCTIONS: THE ESSENTIAL IDEAS

2.1. Complex Coordinates

Let us begin by reviewing the basic "rotated-coordinates" theorem of Aguilar, Balslev, and Combes,[8] which we will state in a somewhat pedestrian fashion as follows: For a system of particles interacting through only pairwise Coulomb potentials (an atom, for example), if the coordinates $\{\mathbf{r}_j\}$ in the center-of-mass Hamiltonian $\hat{H}$ are scaled according to

$$\mathbf{r}_j \to \mathbf{r}_j e^{i\theta} \tag{2}$$

the resulting scaled Hamiltonian $\hat{H}_\theta$ has the spectrum shown schematically in Fig. 2.

1. The bound states of $\hat{H}_\theta$ have the same (real) energies as those of $\hat{H}$.
2. The continuum states of $\hat{H}_\theta$ lie on rays in the complex plane, each ray associated with a scattering threshold, making an angle of $-2\text{Re}(\theta)$ with the real energy axis. Each ray is a branch cut of the resolvent of $\hat{H}_\theta$.
3. If resonance states have been uncovered by the rotation of the branch cuts, they appear with discrete complex eigenvalues $E_R - i\Gamma/2$. They are the locations of poles of the resolvent of $\hat{H}_\theta$.

For a single-particle system, this theorem is actually not very surprising. In potential scattering, complex coordinates are a textbook device used in establishing the analytic properties of the S matrix as a function of momentum or energy.[9] What *is* remarkable about the rotated-coordinates theorem of Aguilar, Balslev, and Combes[8] is that it holds for *many*-electron systems.

A further property of the complex scaled Hamiltonian $\hat{H}_\theta$ (and the central idea behind the application of this formal mathematical idea in resonance calculations) is that the wave functions associated with its complex resonance eigenvalues are *square integrable*. To understand this important point better, consider a one-particle system for which $\hat{H}_\theta$ is just $\hat{H}(re^{i\theta})$. We are looking for solutions of

$$\hat{H}_\theta \Psi_{E_\theta} = E_\theta \Psi_{E_\theta} \tag{3}$$

for which the wave functions Ψ_{E_θ} satisfy the boundary conditions that either $\Psi_{E_\theta}(r)$ goes to zero as r goes to infinity, or is a bounded oscillatory function as r goes to infinity. For bound states, the eigenfunctions Ψ_E of the unscaled Hamiltonian

$$\hat{H}\Psi_E = E\Psi_E \tag{4}$$

satisfy

$$\Psi_E(r) \underset{r\to\infty}{\longrightarrow} ae^{-(-2mE^{1/2})r} \tag{5}$$

where a is a constant, and it can easily be verified by substitution that $\Psi_E(re^{i\theta})$ goes to zero as $r \to \infty$ for $\theta < \pi/2$. So, we have

$$\Psi_{E_\theta}(r) \equiv \Psi_E(re^{i\theta}) \qquad \text{(bound states)} \tag{6}$$

for the bound part of the spectrum. In other words, complex scaling amounts to only a change of variable in the Schrödinger equation in the case of a bound state. For resonance states (in s-wave scattering, for example), the solutions of the unrotated problem satisfy

$$\Psi_E(r) \underset{r\to\infty}{\longrightarrow} b\frac{e^{ikr}}{r} \to \infty \tag{7}$$

where b is a constant and k is the complex resonance momentum. Equation (7) is the condition that at the complex resonance energy, there are only outgoing waves, and since $\text{Im}(k) < 0$, the wave function increases exponentially with increasing r. But by substitution, we have

$$\Psi_E(re^{i\theta}) \underset{r\to\infty}{\longrightarrow} b\frac{e^{ikre^{i\theta}}}{re^{i\theta}} \to 0 \tag{8}$$

which goes to zero for $0 < [\theta + \arg(k)] < \pi$. So, we have for resonance states

$$\Psi_{E_\theta}(r) \equiv \Psi_E(re^{i\theta}) \underset{r\to\infty}{\longrightarrow} 0 \qquad \text{(resonance states)} \tag{9}$$

But for continuum states, complex scaling produces solutions that are unobtainable by a mere change of radial variable in the physical solutions of the unscaled problem, and by arguments similar to those previously given,[10] it can be seen that the energies of the continuum states of our one-particle example satisfy

$$E_\theta = Ee^{-2i\theta} \tag{10}$$

i.e., they are rotated down into the complex plane as shown in Fig. 2.

The first applications of the complex scaling to atomic resonances were literal computational translations of the Aguilar, Balslev, and Combes theorem.[11,12] These calculations used the additional property of complex scaling that it requires no computation of new matrix elements of the atomic Hamiltonian because $e^{-i\theta}$ simply factors out the Coulomb potentials. If T and V are the (real-valued) matrices of the kinetic and potential energies in an atomic CI calculation, the (complex symmetric) matrix of the rotated Hamiltonian H_θ is given by

$$\mathsf{H}_\theta = e^{-2i\theta}\mathsf{T} + e^{-i\theta}\mathsf{V} \tag{11}$$

The eigenvalues of H_θ then include the complex resonance eigenvalues that are independent of θ in the limit that the basis is complete. The idea expressed by Eq. (11) is instantly attractive to the computational atomic physicist. In essence, Eq. (11) says that an atomic CI calculation can be converted into a resonance calculation by trivial complex scaling of the kinetic and potential energy matrices.

This procedure was applied to resonances in two-electron systems by Doolen and co-workers[11], and it worked quite well, although large basis sets were required. As a further improvement of the method, Doolen[12] noted that in a finite basis, the energy is a function of θ but shows apparent stationary behavior near a particular value of θ in a given basis. It has developed since Doolen's calculations on the lowest ^1S autoionization state of He that rigorous stationary points where

$$\left(\frac{dE_{res}}{d\theta}\right)_{\theta_{\text{stationary}}} = 0 \tag{12}$$

do in fact exist for complex values of $\theta_{\text{stationary}}$. These stationary points have been analyzed in terms of a complex version of the virial theorem.[13-16] For two-electron atoms and the positronium negative ion, Ho[17] has made very successful applications of the simple scaling approach.

However, for systems with more than two electrons, the simple complex scaling idea runs into severe practical difficulties. The origin of these problems can be seen by asking what happens to the $1s$ orbital in a complex scaling calculation on the ^2P resonances states $(1s^2 2s^2 2p)$ of Be^-, for example. Restricting our attention to this single configuration for a moment, we see that the 1*s* orbital ϕ_{1s} behaves in a way quite similar to that described for a bound state. To a good approximation, we can represent this orbital by a single exponential, which, under the scaling transformation, becomes

$$\phi_{1s}(r) \simeq ae^{-(-2mE_{1s})^{1/2}re^{i\theta}} \tag{13}$$

This function is a complex oscillatory but decaying function of r that is difficult to represent as an expansion in real basis functions. Yet, in a complex scaling calculation, we are doing just that. The core electrons of the target atom or molecule have little to do with the formation of electron-scattering resonances, but in a complex scaling calculation, a great deal of effort must go into describing them.

2.2. Complex Basis Functions

The solution to this problem was suggested by Rescigno, McCurdy, and Orel[18] and by Junker and Huang[19] in superficially different but completely equivalent formulations. From our preceding discussion, it appears that we would like somehow to scale only the coordinates of the outgoing electron in the many-electron wave function $\Psi(\mathbf{r}_1, \mathbf{r}_2, \ldots, \mathbf{r}_{N+1})$, but if we are scaling the coordinates in the Hamiltonian, it is impossible to pick out one of the identical electrons. The basic device that allows us to apply complex scaling in a more specific and powerful way is a simple contour distortion in the matrix elements of the Hamiltonian. For a one-particle system and with square-integrable basis functions $\phi_i(r)$, for r times the radial function, it is easy to show by contour distortion that

$$\begin{aligned} H_{ij}^{\theta} &= \int_0^{\infty} \phi_i(r)\hat{H}(re^{i\theta})\phi_j(r)dr \\ &= \int_0^{\infty} e^{-i\theta/2}\phi_i(re^{-i\theta})\hat{H}(r)e^{-i\theta/2}\phi_j(re^{-i\theta})dr \end{aligned} \tag{14}$$

In other words, the matrix element of $\hat{H}_\theta$ in a real basis is the same as those of $\hat{H}$ in the basis $e^{-i\theta/2}\phi_i(re^{-i\theta})$. To arrive at Eq. (14), we first make the change of variable $r' = re^{i\theta}$, which yields an integral with the same integrand as that in Eq. (14) but with limits 0 to $\infty e^{-i\theta}$. Then, we make use of the fact that the basis functions we employ (Slaters or Gaussians) go to zero as their arguments goes to infinity along this ray as well as along the real axis and any ray in between. Thus, we can deform the contour 0 to $\infty e^{-i\theta}$ onto the real line 0 to ∞. An important point is that in practical calculations using complex coordinates, we generally deal with only the matrix elements of the Hamiltonian between square-integrable basis functions and never with the Hamiltonian itself or with continuum functions.

Equation (14) suggests that in the many-electron problem, we use *complex basis functions* of the form

$$\Phi(\mathbf{r}_1 \cdots \mathbf{r}_{N+1}) = A[\phi_i(\mathbf{r}_1)\phi_j(\mathbf{r}_2)\cdots\phi_k(\mathbf{r}_N)\phi_l(e^{-i\theta}\mathbf{r}_{N+1})] \tag{15}$$

—where A is the antisymmetrizer—to diagonalize the *real Hamiltonian* $\hat{H}(r_1 \cdots r_{N+1})$. In this prescription, the one-electron basis functions that appear in our calculation are meant to describe the outgoing electron and, hence, will generally include the most diffuse functions in the basis. Note that the reason complex eigenvalues are obtained by this method is that the matrix elements of $\hat{H}$ are defined *without complex conjugation*

$$H^{\theta}_{ab} = \int \Phi_a(\mathbf{r}_1 \cdots \mathbf{r}_{N+1})\hat{H}\Phi_b(\mathbf{r}_1 \cdots \mathbf{r}_{N+1})d^3r_1 \cdots d^3r_N \tag{16}$$

so that the matrix H^{θ}_{ab} is complex symmetric. We have written Eq. (15) with the coordinate vector $\mathbf{r}_{N+1}$ scaled by $e^{-i\theta}$. If our basis had radial and angular functions factored apart, the complex scaling would be applied to only the radial coordinate r_{N+1} and the angular functions (only) would be complex conjugated in Eq. (16).

Equations (15) and (16) express the basic idea of the method of complex basis functions, but more general prescriptions have been developed that are even more successful. For example, we can choose to use complex basis functions for only the most diffuse members of the basis, so that only some of the many-electron basis functions $\Phi_a(\mathbf{r}_1 \cdots \mathbf{r}_{N+1})$ have the form shown in Eq. (15), while the remainder are entirely real. In any case, we still employ θ as a nonlinear variational parameter to locate the stationary point of $E_{\text{res}}(\theta)$. In Section 3, we will see that such an approach is related to Simon's formal theory of exterior scaling.[20]

2.3. *The Complex Variational Principle*

A central theme of the present chapter is that by using complex basis functions we may take advantage of a complex variational principle for resonances which is far more powerful than the theorems on complex scaling (rotated coordinates) described in Section 2.1. If we define the energy functional without complex conjugation and with the real Hamiltonian (no rotated coordinates) as

$$I = \frac{\int \Psi \hat{H} \Psi \, d\tau}{\int \Psi\Psi \, d\tau} \tag{17}$$

then I is stationary

$$\frac{\delta I}{\delta \Psi} = 0 \tag{18}$$

at the solutions for both resonances and bound states if we allow arbitrary complex variations in the wave function Ψ. This variational principle is only a stationary expression and does not provide variational bounds on complex resonance energies. As we will see, the solution to the difficulties peculiar to molecular resonance calculations also lies in Eqs. (16)–(18). The fact that we do not have variational bounds is not a barrier to practical calculations.

In another chapter in this volume, Junker discusses application of Eqs. (17) and (18) in atomic resonance calculations. It was Junker who first articulated the idea of dropping the connection with complex scaling altogether in a complex basis function calculation in his method of "complex stabilization."[21] Although the choice of the complex basis function in the calculations described in the following section of this chapter is guided to some extent by the notion of exterior complex scaling, because that approach currently seems most successful for molecular applications, the essential concepts and spirit of our approach are the same that Junker describes. It should be mentioned that as with amost all good ideas, there is still earlier work that anticipates it. Herrick and co-workers implicitly used a complex variational principle to locate resonance energies as early as 1975.[22] However, there is still some questions about the precise mathematical footing of the complex variational principle[23] and to our knowledge, no rigorous discussion of its limits of applicability (to the Stark effect and noncompact potentials, for example) has been given. Regardless of the outstanding mathematical questions, numerous numerical applications attest to its utility.

3. FEATURES PECULIAR TO THE MOLECULAR PROBLEM

The complex scaling theorems of Aguilar, Balslev, and Combes apply to a collection of particles interacting through Coulomb potentials. A molecule is just such a system, so at first glance, there seems to be no problem in the molecular case that does exist for atomic resonance calculations—if complex scaling is applied to the coordinates of all the particles, both electrons *and* nuclei. To apply the complex scaling theorem literally in a molecular problem, we would either have to forego using the Born–Oppenheimer approximation or else compute electronic energies for complex positions of the nuclei. Neither of these alternatives is attractive, so, barring situations where we might actually need electronic energies for complex nuclear positions, it seems obvious that we should attempt to scale only the electronic coordinates in a molecular calculation.

3.1. Branch Points in the Nuclear Attraction Potentials

If we look at the part of the Hamiltonian containing the electron–nucleus interaction (one electron in this example)

$$V_{\text{nuc}} = -\sum_{\alpha} \frac{Z_\alpha}{[(\mathbf{r} - \mathbf{R}_\alpha)^2]^{1/2}} \tag{19}$$

where Z_α is the atomic number of nucleus α and $\mathbf{R}_\alpha$ denotes its coordinates and scale just the electronic coordinates, we obtain

$$V^{\theta}_{\text{nuc}} = -\sum_{\alpha} \frac{Z_\alpha}{[(\mathbf{r}e^{i\theta} - \mathbf{R}_\alpha)^2]^{1/2}} \tag{20}$$

Each term in Eq. (20) now has a circle of branch points induced by the square root we must use to define the distance between an electron at $\mathbf{r}e^{i\theta}$ and the nucleus at $\mathbf{R}_\alpha$. These square root branch points lie on the circle with $\mathbf{r}$ satisfying

$$\mathbf{r} = \mathbf{R}_\alpha, \qquad \frac{\mathbf{r}\cdot\mathbf{R}_\alpha}{|\mathbf{r}|\,|\mathbf{R}_\alpha|} = \cos\theta \tag{21}$$

Since the Hamiltonian is no longer an analytic function of $\mathbf{r}$, it would seem that the idea of complex scaling only the electronic coordinates fails completely.

This difficulty appears to be insurmountable, but, in fact, the branch point question has turned out to be something of a red herring in molecular resonance calculations. Simon[20] noted that it is only the transformation to complex coordinates in the asymptotic region that determines the spectrum of the complex Hamiltonian. What happens to the coordinates under complex scaling close to the target does not affect the locations of the exact eigenvalues of the complex Hamiltonian. To remedy the nonanalytic behavior of the Hamiltonian at small values of $\mathbf{r}$ in Eq. (20), Simon proposed to transform the coordinates into complex values only outside of some radius R_0. Simon's

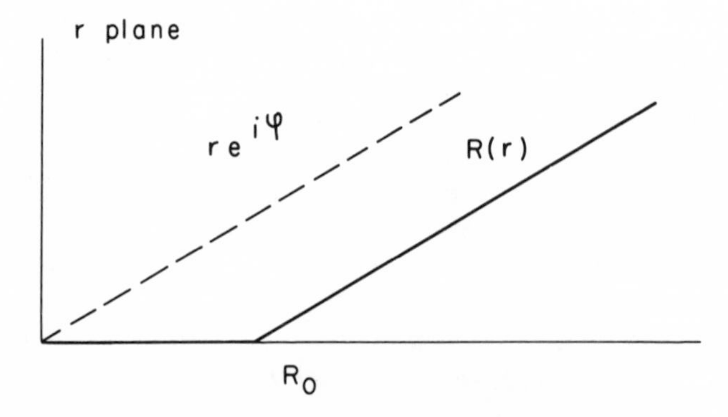

FIGURE 3. Simon's exterior scaling contour.

exterior scaling of the radial coordinate r of the electron (s) is $r \to R(r)$ where

$$R(r) = \begin{cases} r & 0 < r \leqslant R_0 \\ R_0 + e^{i\theta}(r - R_0) & R_0 \leqslant r \end{cases} \tag{22}$$

The exterior scaling contour $R(r)$ is shown in Fig. 3. Although Simon's formal theory would be difficult to implement literally for an electronic resonance in a real molecule, its essential idea can be demonstrated in the case of a simple potential scattering problem. Turner and McCurdy[24] have applied exterior scaling to finding rotational predissociation resonances in H_2 using a piecewise continuous representation of the potential as a function of the distance between the two H atoms. A resonance wave function from an exterior scaling calculation is shown in Fig. 4 along with a plot of the effective potential in Fig. 5. The features of the wave function that are apparent and illuminate the exterior scaling idea are the exponential oscillatory increase of the pure outgoing wave resonance wave function as long as the coordinates remain real and the sudden decrease to zero with large r along the complex portion of the exterior scaling contour. Clearly, it is what happens at large r

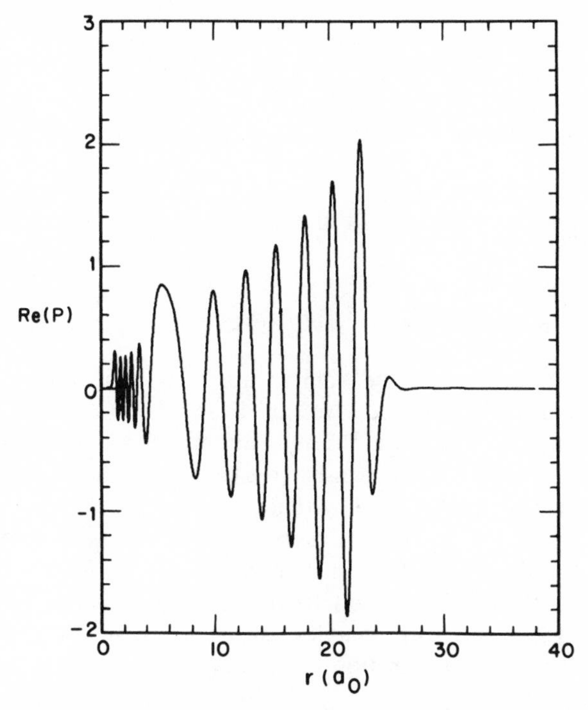

FIGURE 4. The wave function for the $v = 10, j = 17$ rotational predissociation resonance of H_2 from an exterior complex scaling calculation; see Ref. (24). The R_0 is $22.75a_0$.

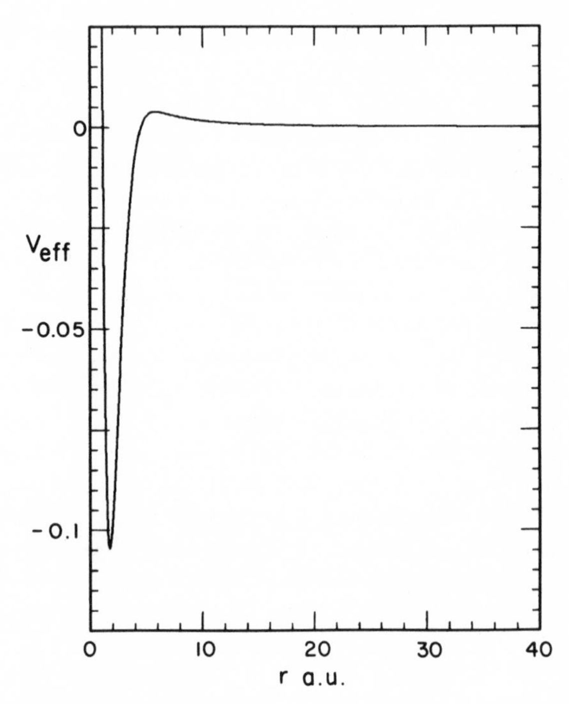

FIGURE 5. The effective potential $V(r) + j(j+1)\hbar^2/2\mu r^2$, which yields the wave function in Fig. 4.

that determines whether or not the resonance wave function is square-integrable.

3.2. *Practical Basis Set Methods and the Details of Why They Work*

Although we may not be able to apply exterior scaling directly to electronic molecular resonances, we can do several things that are at least indirectly related to that idea. Prior to Simon's work on exterior scaling, McCurdy and Rescigno[25] made use of the fact that, as previously mentioned in a practical calculation on an electronic resonance, we do not work with the Hamiltonian—we work with its matrix elements in a basis—to extend the method of complex basis functions to molecular resonance calculations. McCurdy and Rescigno[25] diagonalized a *real-valued* moleclar Hamiltonian in a basis of Gaussians with complex exponents chosen so that their basis functions had the form (s function, for example)

$$\phi(r) = \left(\frac{2\alpha e^{-2i\theta}}{\pi}\right)^{3/4} e^{-\alpha e^{-2i\theta}(\mathbf{r}-\mathbf{A})^2} \tag{23}$$

where the function is centered at **A** and its exponent α has been scaled to $\alpha e^{-2i\theta}$. This procedure leads to numerically stable results, even though it is not simply related to a particular contour distortion in r. Later, Moiseyev and Corcoran[26] presented calculations on molecular resonances using a method that at least superficially amounted to simply scaling electronic coordinates to $re^{i\theta}$ and ignoring the branch point problem entirely. The Moiseyev and Corcoran[26] calculations show less stability of the resonance energy with variations in θ, but they do not exhibit the gross failure we would expect from our preceding discussion of V_{nuc}.

Why do these procedures work? How do they avoid the branch point difficulty associated with the nuclear attraction potentials? Consider the matrix element I of one term in V_{nuc} between two normalized S-type Gaussians

$$I = \int \left(\frac{2\alpha}{\pi}\right)^{3/4} e^{-\alpha(\mathbf{r}-\mathbf{A})^2} \frac{1}{[(\mathbf{r}-\mathbf{R})^2]^{1/2}} \left(\frac{2\beta}{\pi}\right)^{3/4} e^{-\beta(\mathbf{r}-\mathbf{B})^2} d^3r \tag{24a}$$

$$= \left(\frac{32}{\pi}\right)^{1/2} \frac{(\alpha\beta)^{3/4}}{\alpha+\beta} F_0[(\alpha+\beta)(\mathbf{R}-\mathbf{P})^2] \exp\left[-\frac{\alpha\beta}{\alpha+\beta}(\mathbf{A}-\mathbf{B})^2\right] \tag{24b}$$

where

$$\mathbf{P} = \frac{\alpha\mathbf{A}+\beta\mathbf{B}}{\alpha+\beta} \tag{25}$$

and $F_0(z)$ is the entire function of z given by

$$F_0(z) = \tfrac{1}{2}(\pi/z)^{1/2}\operatorname{erf}(z^{1/2}) \tag{26}$$

In a complex basis function calculation or in a complex-coordinate calculation using a basis, it is matrix elements of the Hamiltonian we analytically continue to complex values of some parameter(s) and not the operator itself. In the procedure proposed by McCurdy and Rescigno,[25] the exponents α and β in Eq. (24) are scaled by $e^{-2i\theta}$ so that Eq. (24b) becomes

$$I_{\text{MR}} = e^{-i\theta}\left(\frac{32}{\pi}\right)^{1/2} \frac{(\alpha\beta)^{3/4}}{\alpha+\beta} F_0[e^{-2i\theta}(\alpha+\beta)(\mathbf{R}-\mathbf{P})^2] \times \exp\left[-e^{-2i\theta}\frac{\alpha\beta}{\alpha+\beta}(\mathbf{A}-\mathbf{B})^2\right] \tag{27}$$

and since Eq. (24b) gives the well-defined analytic continuation of I to that expressed by Eq. (27), no branch point ambiguities arise. Similarly, if in Eq. (24a) we scale **r** to $\mathbf{r}e^{i\theta}$, we know the integrand has branch points that would be encountered or encircled if the integration were performed numerically.[27] But in the Moiseyev and Corcoran[26] calculation, the formula in Eq. (24b) was used to continue analytically the matrix element. Factoring $e^{-i\theta}$ out of the integrand in Eq. (24a) with $\mathbf{r} = \mathbf{r}e^{i\theta}$ produces an integral with $\mathbf{R}e^{-i\theta}$ in the integrand for which Eq. (24b) becomes

$$I_{\mathrm{MC}} = e^{-i\theta}\left(\frac{32}{\pi}\right)^{1/2}\frac{(\alpha\beta)^{3/4}}{\alpha+\beta}F_0[(\alpha+\beta)(\mathbf{R}e^{-i\theta}-\mathbf{P})^2] \times \exp\left[-\frac{\alpha\beta}{\alpha+\beta}(\mathbf{A}-\mathbf{B})^2\right] \tag{28}$$

so the McCurdy–Rescigno *and* Moiseyev–Corcoran procedures use the analytic formulas formatrix elements to continue analytically the matrix elements even in the (Moiseyev–Corcoran) case where the operator may not be an analytic function of the complex parameter.

The reason both of these procedures work is that the analytic continuation they use is equivalent to choosing a particular contour along which to perform the integration in Eq. (24a), which avoids the branch points in case they might give trouble and produces a smooth analytic continuation of the integral. McCurdy[27] has discussed the fact that the implied integration contours are not unique but that a connection can be found between both procedures previously discussed and Simon's theory of exterior scaling.[20] Morgan and Simon have also discussed a connection.[28]

3.3. *Numerical Stability and the Calculation of Molecular Photoionization Cross Sections*

A note should be made at this point about numerical stability. It is apparent from examining Eqs. (27) and (28) that I_{MR} can be obtained from I_{MC} if the centers of the basis functions are scaled according to $\mathbf{A} \to \mathbf{A}e^{-i\theta}$ and $\mathbf{B} \to \mathbf{B}e^{-i\theta}$. In other words, the McCurdy–Rescigno procedure is the same as the Moiseyev–Corcoran procedure if the basis functions are located at particular complex positions. McCurdy and Rescigno[29] have argued that the enhanced numerical stability of their technique is associated with the fact that complex cusp conditions on the wave function are important in calculating molecular resonances. We will not reproduce this argument in detail here. However, dramatic numerical evidence of the superior numerical stability of the McCurdy–Rescigno procedure is obtained by applying these ideas when calculating photoionization cross sections.

The photoionization cross section $\sigma(\omega)$ as a function of frequency ω can be written n terms of a matrix element of the molecular resolvent[29]

$$\sigma(\omega) = \lim_{\varepsilon \to 0}\frac{4\pi\omega}{c}\,\mathrm{Im}\left\langle \Psi_0 \middle| \mu \frac{1}{H - E_0 - \omega - i\varepsilon} \mu \middle| \Psi_0 \right\rangle \tag{29a}$$

where E_0 and ψ_0 are the energy and wave function of the target molecule, respectively, and μ is the dipole operator. Using complex coordinates and/or complex basis functions allow us to evaluate the resolvent matrix element in Eq. (29a) using a spectral resolution in terms of the complex eigenfunctions and eigenvalues of the complex scaled Hamiltonian (complex coordinates following Moiseyev and Corcoran) *or* of the real-valued Hamiltonian in a complex basis (complex basis functions following McCurdy and Rescigno).

$$\sigma(\omega) \simeq \frac{4\pi\omega}{c} \operatorname{Im} \sum_n \frac{\langle \Psi_0^{(\theta)} | \mu^{(\theta)} | \Psi_n^{(\theta)} \rangle \langle \Psi_n^{(\theta)} | \mu^{(\theta)} | \Psi_0^{(\theta)} \rangle}{E_n^{(\theta)} - E_0 - \omega} \tag{29b}$$

The definitions of the terms in the numerator of this equation vary according to which method is used, and they are discussed in Refs. 27 and 29.

The important point is that in either case, calculating the photoionization cross section is reduced to a discrete sum over complex eigenfunctions and eigenvalues from a finite-basis calculation. Because the eigenvalues are complex, the limit $\varepsilon \to 0$ is unnecessary in Eq. (29b). The calculation of resolvent matrix elements is potentially a very useful application of complex-coordinate

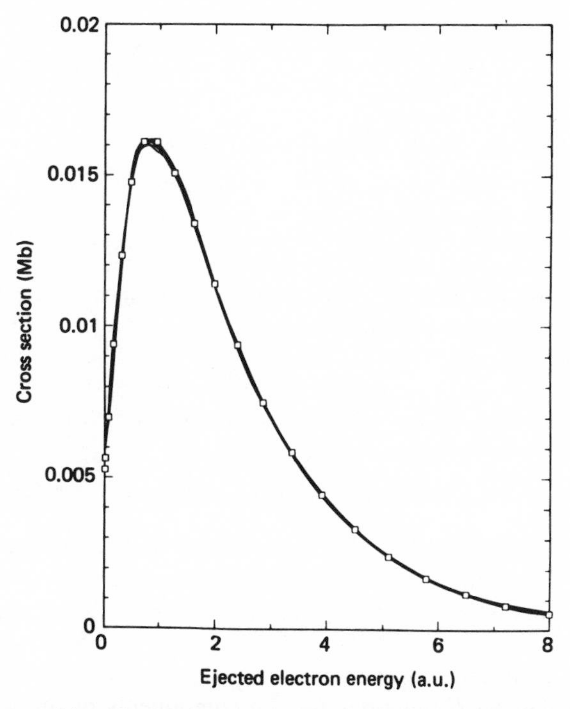

FIGURE 6. Superposition of results for the parallel component of the photoionization of the cross section of H_2^+ using the McCurdy–Rescigno method; see Ref. (29). The superimposed results are for θ over the interval 15–25° in 2.5° increments. □ denotes the exact values.

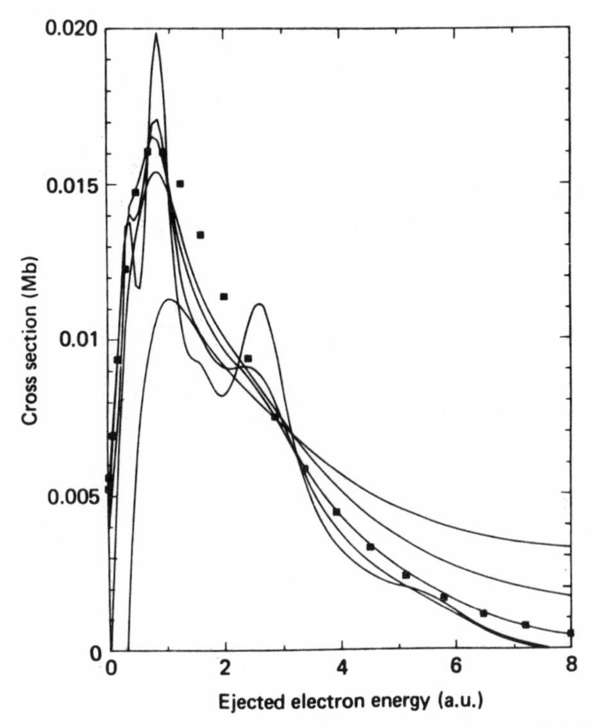

FIGURE 7. Same as Fig. 6 but with the Moiseyev–Corcoran procedure; see Ref. (26). The θ varies from 5 to 15° in 2.5° increments.

and basis function techniques. The question here concerns the stability with varying θ for a fixed number of terms in the sum in Eq. (29b). Figures 6 and 7 compare the stability with varying θ of the McCurdy–Rescigno and Moiseyev–Corcoran approaches to evaluating the cross section in Eq. (29) for molecule. Twenty-four terms were included in the sum in Eq. (29b) in both cases.

3.4. *Summary of the Complex Basis Function Approach for Molecules*

We can summarize the results of this section and the general approach to be taken to the molecular resonance problem in the remainder of this chapter as follows. Our methods consist of two steps:

1. Choose an appropriate trial function constructed from (a) real Gaussian basis functions for the components of the wave function localized in the region close to the target molecule and (b) complex Gaussian basis functions of diffuse and intermediate character of the

form

$$\chi^{\alpha}_{nlm} = N_{nlm}(\alpha e^{-2i\theta})(x - A_x)^n(y - A_y)^l(z - A_z)^m e^{-\alpha e^{-2i\theta}(r-A)^2} \tag{30}$$

where $N_{nlm}(\alpha e^{-2i\theta})$ is the normalization constant.

2. Make use of the complex variational principle for discrete states by setting

$$\delta \frac{\int \psi_t \hat{H} \psi_t d\tau}{\int \psi_t \psi_t d\tau} = 0 \tag{31}$$

for complex variations in the trial function ψ_t and using the real-valued Hamiltonian $\hat{H}$.

In this way, we can not only formulate ordinary configuration interaction calculations with both linear and nonlinear complex parameters but also SCF theory for resonances.

4. THE COMPLEX SELF-CONSISTENT-FIELD (CSCF) METHOD

4.1. *The Working Equations*

To obtain the self-consistent-field (SCF) equations in the complex case, we need the expectation value of the energy [without complex conjugation as in Eq. (17)] with respect to our trial function. Most of our applications are to open-shell systems, and our trial function is a single-configuration wave function, which is a spin eigenfunction and therefore a sum of several antisymmetric products.

We can obtain a fairly general formalism by beginning with an open-shell energy expresson like that originally suggested by Roothaan.[30] We write this expression here in terms of the spatial orbitals $\phi_{i\lambda\alpha}$, which will be the solutions of the CSCF equations for an atomic or molecular resonance. The subscript i is the shell index, which distinguishes principal quantum numbers in the atomic case; λ is the symmetry index distinguishing irreducible representations or orbital angular momentum quantum numbers; and α is the subspecies index distinguishing the members of the irreducible representation λ. Our energy expression is

$$E_{\text{CSCF}} = \sum_{i\lambda} N_{i\lambda} h^{\lambda}_{ii} + \tfrac{1}{2} \sum_{i\lambda j\mu} N_{i\lambda} N_{j\mu} [a_{i\lambda j\mu} J_{i\lambda j\mu} - \tfrac{1}{2}(b_{i\lambda j\mu} K_{i\lambda j\mu})] \tag{32}$$

where $N_{i\lambda}$ denotes the occupation number of the ith shell of λ symmetry and $\alpha_{i\lambda j\mu}$ and $b_{i\lambda j\mu}$ are coupling coefficients that vary for different open-shell cases. For interactions between two closed shells and between open and closed shells,

$\alpha_{i\lambda j\mu} = b_{i\lambda j\mu} = 1$. The energy expression in Eq. (32) differs from the ordinary SCF expression in only that the one-electron Coulomb and exchange matrix elements are defined without complex conjugation of the orbitals. With d_λ denoting the degeneracy of symmetry λ, those matrix elements are

$$h_{ii}^{\lambda} = d_{\lambda}^{-1} \sum_{\alpha} \int \phi_{i\lambda\alpha}(\mathbf{r}) \hat{h} \phi_{i\lambda\alpha}(\mathbf{r})\, d^3 r \tag{33a}$$

$$J_{i\lambda j\mu} = (d_\lambda d_\mu)^{-1} \sum_{\alpha,\beta} \int\int \phi_{i\lambda\alpha}(\mathbf{r}_1)\phi_{j\mu\beta}(\mathbf{r}_2) r_{12}^{-1} \times \phi_{i\lambda\alpha}(\mathbf{r}_1)\phi_{j\mu\beta}(\mathbf{r}_2) d^3 r_2 \tag{33b}$$

$$K_{i\lambda j\mu} = (d_\lambda d_\mu)^{-1} \sum_{\alpha,\beta} \int\int \phi_{i\lambda\alpha}(\mathbf{r}_1)\phi_{j\mu\beta}(\mathbf{r}_2) r_{12}^{-1} \times \phi_{j\mu\beta}(\mathbf{r}_1)\phi_{i\lambda\alpha}(\mathbf{r}_2)\, d^3 r_1\, d^3 r_2 \tag{33c}$$

Now, by adding the energy expression in Eq. (32) to a set of Lagrange multiplier terms constraining the orbitals to remain orthogonal to one another, we obtain an appropriate variational function I from which the open-shell CSCF equations can be derived.

$$I = E_{\text{CSCF}} - \sum_{ij\lambda\alpha} \varepsilon_{ij}^{\lambda} d_{\lambda}^{-1} \int \phi_{i\lambda\alpha}(\mathbf{r})\phi_{j\lambda\alpha}(\mathbf{r})\, d^3 r \tag{34}$$

In Eq. (34), $\varepsilon_{ij}^{\lambda}$ is the complex symmetric matrix of Lagrange multipliers.

From this point onward, the derivation proceeds in exactly the same fashion as the derivation of ordinary SCF equations. Expanding the orbitals $\phi_{i\lambda\alpha}$ in a basis of real Gaussian functions and complex ones of the form given in Eq. (30) leads to a formulation where the Fock matrices are complex symmetric instead of Hermitian. With the variational functional chosen as in Eq. (34), the matrix form of the CSCF equations is exactly the same as that of the ordinary SCF equations except that the one-electron, Coulomb, and exchange integrals are defined as in Eqs. (33a–c).

Iterating the matrix form of the CSCF equations to self-consistency has not proved to be significantly more difficult than in a real-valued SCF calculation. We have avoided extrapolation schemes for the orbital coefficients and/or density matrices in the calculations presented in Section 5, but we have no indication that such techniques for accelerating convergence cannot be made to work if suitably modified. Also, we have no clear indication from CSCF calculations on atoms and molecules that "one-Hamiltonian" methods, such as that of Davidson and Stenkamp,[31] are superior in CSCF calculations to "two-Hamiltonian" approaches, such as that of Roothaan and Bagus.[32]

However, a critical point is that the open-shell coupling coefficients $\alpha_{i\lambda i\lambda}$

and $b_{i\lambda i\lambda}$ for the open shell with indices $i\lambda$ are not unique, as is well known.[32] In resonance calculation (where there is, in general, only one open shell), we have found that it is absolutely essential to choose these coefficients so that the open shell does not experience a potential due to itself ($\alpha_{i\lambda i\lambda} = b_{i\lambda i\lambda} = 0$); otherwise, convergence of the SCF iterations is hard to obtain. Of course, this choice is for computational convenience only. Any choice satisfying the condition that

$$a_{i\lambda i\lambda} - \tfrac{1}{2} b_{i\lambda i\lambda} = 0 \tag{35}$$

gives the same value of E_{CSCF} in principal but may make convergence more difficult than the preceding choice.[33,54]

This completes the formulation of the CSCF method for atomic or molecular shape resonances. Unfortunately, when the CSCF method is applied to a Feshbach resonance, it yields a value of E_{CSCF} that is real and therefore provides no information about the width of the resonance. The reason for this failure is that the decay of a Feshbach resonance is a correlation effect, and it is not present in the Hartree–Fock description of the resonance state.[33] To illustrate this point, consider a CSCF calculation performed on the lowest ^3P autoionizing state of helium, with the trial function corresponding to just the $2s2p$ configuration. If A is the antisymmetrizer, the trial function is

$$\Psi_t = A(\phi_{2s}\phi_{2p}) \tag{36}$$

where ϕ_{2s} and ϕ_{2p} are the space orbitals (the triplet spin eigenfunction factors out). The CSCF equations then give ϕ_{2s} and ϕ_{2p}. This resonance decays to the $1s\ kp$ continuum, where kp denotes a continuum p orbital. The $1skp$ configuration is not in the variational space of the CSCF calculation. The open-shell version of Brillouin's theorem states,[34] in this case, that matrix elements of the Hamiltonian between the $2s2p$ Slater determinant and any determinant corresponding to a single excitation to an orthogonal orbital from $2s2p$ all vanish. The SCF calculation is equivalent to a CI calculation in this space of *single* excitations from $2s2p$. The configuration $1skp$ is a double excitation from $2s2p$ and does not appear in this space. The resonance, therefore, is prevented from decaying, and the energy of the $2s2p$ state computed in this manner is real valued.

Perhaps a convenient way of describing Feshbach resonances is to use a complex-coordinate version of multiconfiguration SCF theory (MCSCF)[35]. For example, we would treat the $2s2p$ triplet autoionizing the state of helium by choosing a two-configuration trial function of the form

$$\Psi_t = aA(\phi_{2s}\phi_{2p}) + bA(\phi_{1s}\phi_{kp}) \tag{37}$$

All four orbitals and the ratio of coefficients a/b would then be determined variationally. At the time of this writing, the CSCF procedure has been applied with only a single-configuration trial function, although there is no apparent practical or theoretical reason preventing the use of multiconfiguration trial functions.

4.2. *Continued Fraction Procedure for Finding the Stationary Point of E_{CSCF}*

In Section 2, we mentioned the fact that θ can (and, in general, should) be used as an additional nonlinear variational parameter, and this is also the case when θ appears as a parameter in the basis functions, as in Eq. (30). The stationary point satisfying Eq. (12) is the best estimate of the resonance energy from a given basis set. In this section, we will give a simple, practical method for finding that stationary point for E_{CSCF} or any basis set approximation to E_{res}.

First, we write the scale factor $e^{i\theta}$ as a single complex variable with respect to which we wish to find the stationary point.

$$\eta \equiv e^{i\theta} \tag{38a}$$

$$\left[\frac{dE_{\text{CSCF}}(\eta)}{d\eta}\right]_{\eta_{\text{stationary}}} = 0 \tag{38b}$$

We calculate $E_{\text{CSCF}}(\eta)$ for a few values of η. Usually, calculations at about five real values of η suffice. Then, we construct a rational-fraction representation of $E_{\text{CSCF}}(\eta)$

$$E_{\text{CSCF}}(\eta) \simeq \frac{P_n(\eta)}{Q_m(\eta)} \tag{39}$$

where P_n and Q_m are polynomials. A convenient way of determining such a rational fraction is to use the algorithm given by Schlessinger[36] to construct a continued-fraction representation of $E_{\text{CSCF}}(\eta)$. Once we have the continued fraction, the numerical evaluation of $\eta_{\text{stationary}}$ and E_{CSCF} $(\eta_{\text{stationary}})$ is a simple operation. In practice, we have found that $dE_{\text{CSCF}}/d\eta$ is very small over a substantial range of η in a calculation employing a large basis set and that a value of η for which Eq. (38b) is satisfied to within 10^{-8} a.u. is most easily found by a steepest descent numerical algorithm applied to the magnitude of the derivative.

This application of analytic continuation of $E_{\text{res}}(\eta)$ using continued fractions hints of far more general schemes for analytically continuing a resonance eigenvalue to complex values of the scale factor η starting from information calculated at only real values of η. We will return to this question in Section 7.2.

5. CSCF CALCULATIONS ON MOLECULAR SHAPE RESONANCES

5.1. *The Lowest* $^2\Sigma_u^+$ *Shape Resonance State of* H_2^-

The lowest $^2\Sigma_u^+$ shape resonance plays a role in determining the cross sections for vibrational excitation[1] of H_2^- by electron impact and for dissociative attachment.[37] In spite of the fact that this three-electon system should exhibit the simplest molecular resonance, the complex potential energy function $E_{\text{res}}(r)$ is not yet accurately known. The $^2\Sigma_u^+$ state of H_2^- dissociates to $H + H^-$, and, therefore, since H^{-1} is bound, the resonance state of H_2^- must become a bound state as R increases. The theoretical description of the conversion of resonance to bound states in molecules presents particular difficulties, as we will see.

McCurdy and Mowrey[38] performed CSCF calculations on the $^2\Sigma_u^+$ state for values of R between 1.4 and 4.0 a_0; the remainder of this section summarizes their results. The basis chosen for these calculations is described in detail in Ref. (38). Briefly, it consists of a $(7s, 2p)$ contracted basis of real Gaussian centered on the nuclei and 15 complex p_z functions centered in the

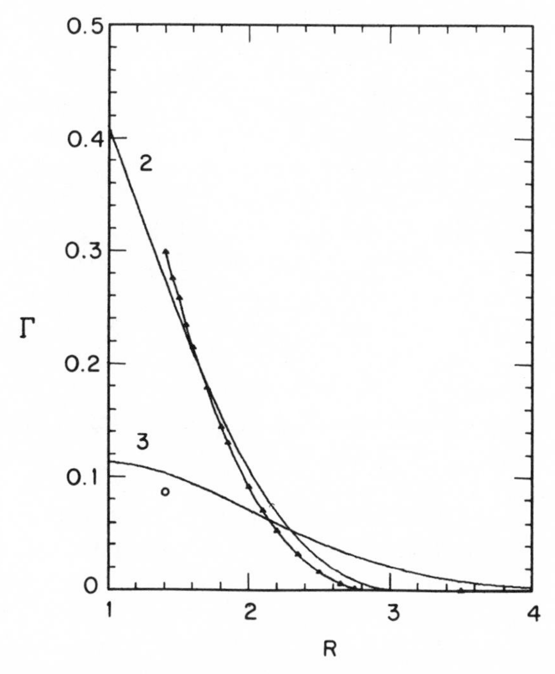

FIGURE 8. Width of $^2\Sigma_u^+$ resonance as a function of internuclear distance from CSCF calculation (▲), and from Ref. (39) (curve 2) and Ref. (40) (curve 3). The (○) is the theoretical result from Ref. (26).

middle of the molecule. The configuration for this resonance is $1\sigma_g^2 k\sigma_u$. There have been several empirical determinations of the $^2\Sigma_u^+$ complex potential[39–42] and one *ab inito* calculation of the real part of the potential.[43] Nesbet[42] has observed that there is remarkable disagreement among these calculations, particularly for the real part of the potential.

The CSCF results for the width function $\Gamma(R)$ are compared with the empirical determination of Wadehra and Bardsley[39] and Chen and Peacher[40] in Fig. 8. The CSCF results are in excellent agreement with results Wadehra and Bardsley[39] obtained for the width by fitting the dissociative attachment cross section. Wadehra and Bardsley constrained their width function to vanish at $3.0a_0$, and the CSCF width goes to zero between 2.80 and 2.81 in our calculations. On the other hand, the older Chen and Peacher[40] determination of the width from the isotope effect in dissociative attachment is in poor agreement with the Wadehra and Bardsley result and the CSCF values. Also shown in Fig. 8 is the result of Moiseyev and Corcoran,[26] which was obtained by a CI calculation in a very small $(3s, 2p, 1d)$ basis and using complex coordinates with the coordinates of all three electrons scaled. Scaling all the electronic coordinates leads to numerical instability in general, and the variation of the energy as a function of the scaling parameters in the Moiseyev and Corcoran calculation is far more severe than that in the CSCF calculations. Considering the accuracy with which the CSCF method predicts the width of the $^2\Pi_g$ resonance in N_2^- (see Section 5.2), it would be surprising if the effects of correlation affected the width as much as the Moiseyev and Corcoran calculation suggests. Extensive and accurate complex CI calculations will be necessary to completely resolve this discrepancy.

Figure 9a compares the real part of the CSCF energy for H_2^- with other empirical and theoretical results and also the CSF and exact values for the H_2 ground-state curve. We immediately observe that the CSF and CSCF results are shifted upward by the neglect of correlation energy. The empirical results of Chen and Peacher[40] and those of Wahedra and Bardsley[39] show remarkable disagreement with one another and with the early stabilization calculations of Taylor and Harris[43] and Eliezer, Taylor, and Williams.[43] In Fig. 9b, the H_2^- curves are shown shifted to coincide at $3.0\,a_0$. Although the CSCF curve agrees best in shape with that of Wahedra and Bardsley,[39] it is difficult to draw a definite conclusion on the basis of that comparison.

The Eliezer, Taylor, and Williams[43] stabilization calculation included mainly target correlation but did allow the σ_g orbital to respond somewhat to the presence of the continuum electron. It is interesting to ask if the sharper minimum in their results arises predominately from target correlation, since the CSCF calculations, which also include the response of the σ_g orbital but not the target correlation, do not show such a sharp minimum. Again, more work is necessary to resolve these discrepancies.

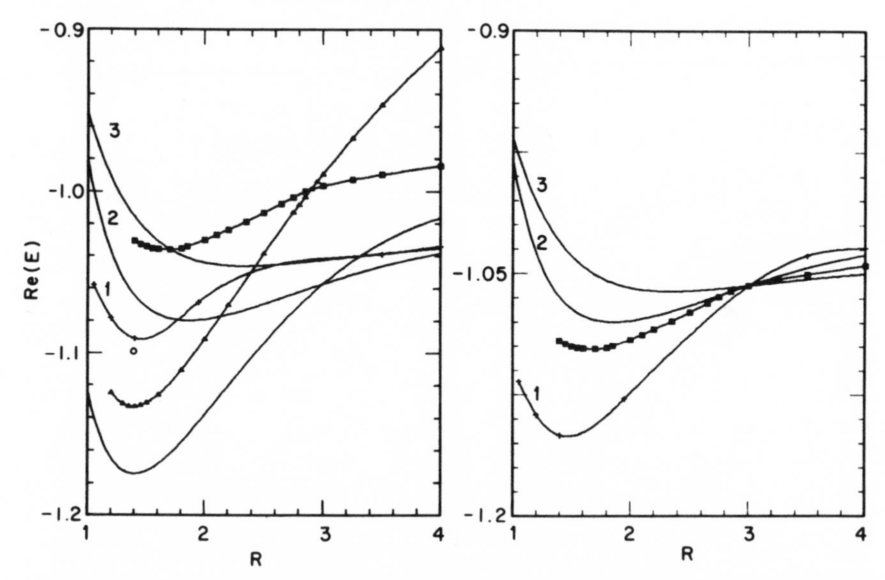

FIGURE 9. (a) The SCF energy for H_2 (▲) and the real part of the CSCF energy for H_2^- (■). Curves 1, 2, and 3 for H_2^- are from Refs. (43), (39), and (40), respectively, and (○) is from Ref. (26). The lowest curve is the exact ground-state potential for H_2 of Kolos and Wolniewicz [Ref. (45)]. (b) The H_2^- potentials shifted to coincide at $R = 3.0\, a_0$ for comparison.

Figures 10–12 demonstrate various aspects of the transition from the $^2\Sigma_u^+$ bound state of H_2^- at large distances to the resonance at a smaller internuclear distance. In Fig. 10, we show an enlargement of the crossing region, which reveals that as R decreases from infinity, the CSCF curve for H_2^- crosses the SCF curve for H_2 nearly 0.1 a_0 before the imaginary part of the CSCF energy becomes appreciably nonzero. Lauderdale *et al.*[44] have investigated this inconsistency in the SCF description of the crossing point, as discussed at the end of Section 5.2.

On the same graph is shown an ordinary real-valued SCF calculation on H_2^- performed with the large basis of diffuse functions (with $\theta = 0$) that is used in the CSCF calculation. Very near the point where the imaginary part of the energy in the CSCF calculation becomes nonzero, the σ_u orbital energy in the real-valued SCF calculation goes through zero. That orbital becomes very diffuse, and an attempt to follow the SCF calculation in this large basis to smaller internuclear distances resulted in an H_2^- solution that is essentially the H_2 SCF wave function with a nearly zero energy σ_u orbital. Real-valued SCF calculations can be made to stabilize (converge or almost converge) to shape resonances if small basis sets without diffuse functions are used. Our experience is that this sort of stabilization is very difficult, if not impossible, to achieve with large, diffuse basis sets.

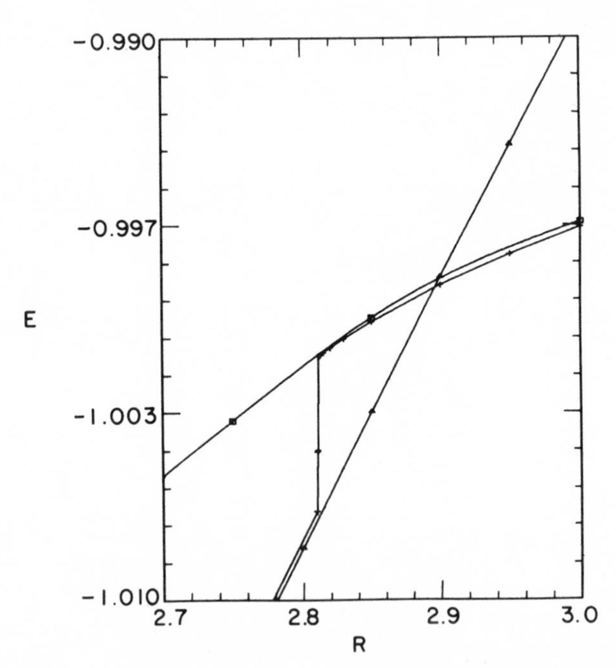

FIGURE 10. The SCF potential in the crossing region: (■) CSCF for H_2^-, (▲) SCF for H_2, (+) real-valued SCF for H_2^- as described in text.

In the CSCF calculation, the σ_u orbital energy never has to have a value of exactly zero, and it moves into the complex plane without reaching that point, as is shown in Fig. 11. The orbital remains quite well localized in the process. The σ_g orbital energy also becomes complex as the bound state becomes a resonance. Figure 12 shows how the modulus of the σ_u orbital, plotted along the internuclear axis, changes with internuclear separation R. At large R, the orbital is well localized and has as its only prominent feature the cusps at the nuclear positions. At $R = 2.8\,a_0$, it begins to become more diffuse and by $R = 1.7\,a_0$ has become exceedingly diffuse in accordance with the large width of the resonance (4.9 eV) at that point. All of the calculations that were done to form Fig. 12 were performed with the same value of the scaling parameter θ. Changing the scaling parameter can change the envelope of the wave function at larger values of the electronic coordinates but cannot change the qualitative trend shown in Fig. 12.

5.2. *The* $^2\Pi_g$ *Shape Resonance State of* N_2^-

The CSCF calculations have also been performed for the $N_2^-\,{}^2\Pi_g$ shape resonance over a range of internuclear distances.[44,46] This resonance is

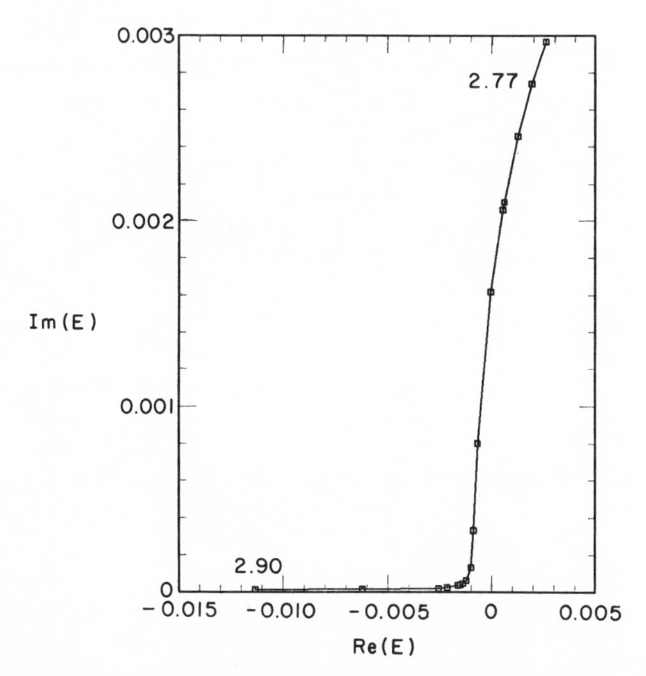

FIGURE 11. Trace in the complex plane of the σ_u orbital energy from the CSCF calculation as a function of the internuclear distance near the crossing point.

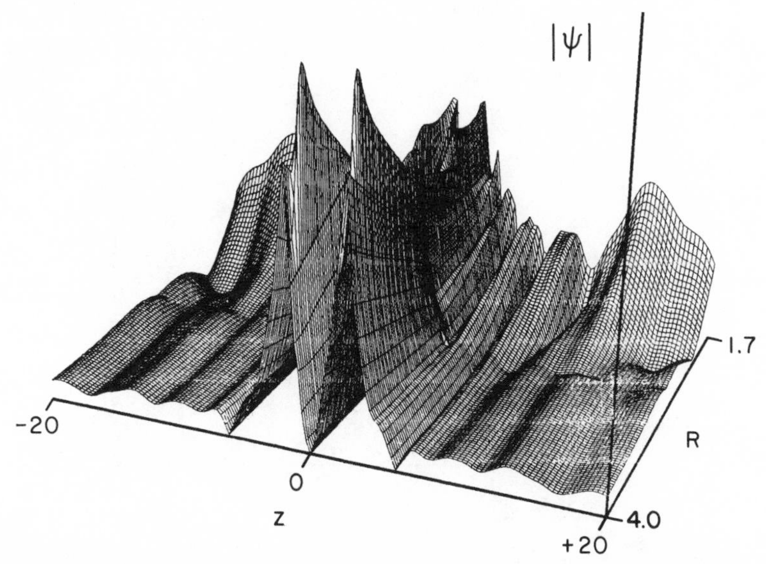

FIGURE 12. Modulus of the σ_u orbital plotted along the internuclear axis z as a function of the internuclear separation R.

among the best characterized electron-scattering resonances, both by experiment[1] and theory.[47,48] Consequently, it has become the prototype problem in resonant vibrational excitation. Details of the CSCF calculations summarized here are given in Ref. 44.

The complex basis set for these calculations consisted of a $(4s, 4p, 3d)$ set of real contracted Gaussian together with a $(2p, 2d)$ set of complex Gaussians in π_g symmetry on each nitrogen and 9d complex Gaussians at the middle of the molecule. The width function $\Gamma(R)$ from the CSCF calculations is shown in Fig. 13 compared with other determinations. A glance at these results shows that there is good agreement between the CSCF results for $\Gamma(R)$ and those of Hazi, Rescigno, and Kurilla;[49] Schneider, LeDourneuf, and Vo Ky Lan;[50] and Levin and McKoy.[51] The CSCF procedure is apparently capable of producing reliable values of the width function for a molecular shape resonance.

The real part of E_{CSCF} for N_2^- is plotted in Fig. 13; also shown in this figure

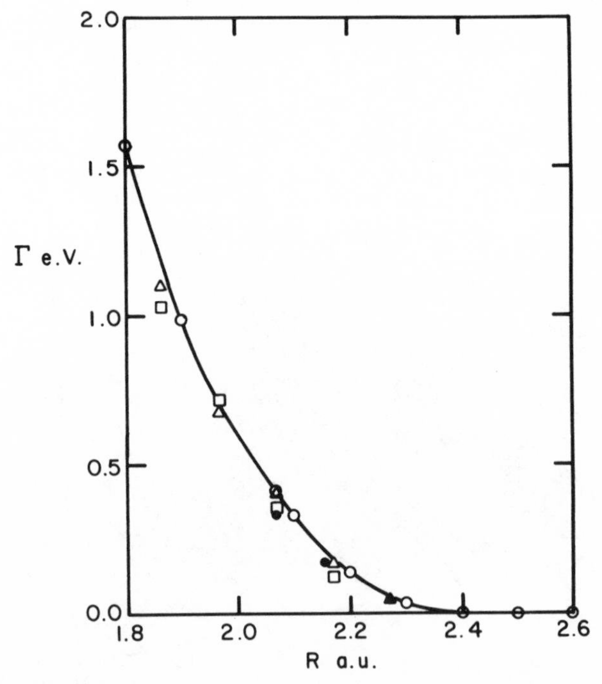

FIGURE 13. The Γ for the $^2\Pi_g$ resonance in N_2^- as a function of the internuclear distance: (○—○) CSCF, (△) Stieltjes imaging results from Ref. (49), (□) Ref. (51), (●) R-matrix results from Ref. (50).

is the SCF energy of N_2. If

$$E_R = \mathrm{Re}\,[E_{\mathrm{CSCF}}(N_2^-)] - E_{\mathrm{SCF}}(N_2) \tag{40}$$

is taken as the position of the resonance, the CSCF calculation predicts a result that is too large at $R = 2.069\,a_0$ by about 1 eV. On the other hand, if the real part of the π_g orbital energy is taken as the position

$$E_R = \mathrm{Re}\,(\varepsilon_{\pi_g}) \tag{41}$$

the CSCF essentially reproduces the position of the resonances quoted by Hazi, Rescigno, and Kurilla[49] exactly. The Hazi, Rescigno, and Kurilla results are presumably quite reliable, since they produce the experimental vibrational excitation cross sections in good agreement with experiment when used in the boomerang model for vibrational excitation.

Using the orbital energy ε_{π_g} to give the resonance energy is an application of Koopmans's theorem to the N_2^- CSCF calculation. According to bound-state quantum chemistry, this is the best approximation to the electron affinity or ionization potential we can obtain from an SCF calculation. Hence, it would appear that the way to obtain the resonance position and width from the CSCF procedure is simply to associate the real and imaginary parts of ε_{π_g} with the complex function $E_{\mathrm{res}}(R)$. However, Fig. 11 shows that the imaginary part of the CSCF orbital energy can have the wrong sign! In Fig. 11, the σ_u orbital energy moves into the *upper* half of the E plane as the $^2\Sigma_u$ bound state becomes a resonance. This occurs as the total SCF energy moves to the lower half of the E plane. Thus, while E_{CSCF} gives a positive Γ according to Eq. (1), Koopmans's theorem interpretation of the σ_u orbital energy gives a negative value. In Section 7, we come back to this question, which we refer to as the problem of constructing a physically meaningful version of Koopmans's theorem for the CSCF procedure for resonances.

Another point evident from Figs. 13 and 14 is that, as in the case of H_2 and H_2^-, the crossing of the SCF curve for the neutral N_2 target and the CSCF energy for N_2^- does not coincide with the point where Γ becomes nonzero. This phenomenon is an entirely general feature of the SCF description of resonance states and is due to the fact that the Hartree–Fock approximation neglects a different amount of correlation energy in the case of the N electron's neutral target than in the N + 1 electron's metastable anion. Lauderdale, McCurdy, and Hazi[44] have seen this behavior near the crossing point in CSCF calculations on F_2^-, and they have also noted this behavior in earlier SCF calculations by Hay[52] on SF_6^-. The different amount of correlation energy in N and N + 1 electron systems makes the difference in their total energies (whether real or complex) difficult to calculate. That problem raises the question of the role of correlation in determining the widths and positions of molecular shape resonances.

6. THE QUESTION OF CORRELATION IN SHAPE RESONANCES

The results of the previous section suggest that the effects of correlation on the width and position of a shape resonance are in a sense separate questions. Figure 13 demonstrates that the CSCF result for Γ can apparently be a good approximation even when the difference of SCF energies of the anion and neutral targets is a poor approximation (at least as an absolute value) to the position. In general, what we mean by the correlation energy is the difference between the exact nonrelativistic energy and the SCF approximation to it. The fact that the correlation energy of the anion is different from that of the neutral is the origin of a great deal of difficulty in predicting the resonance position.

Part of the reason for this difference in correlation energies can be explained by an example. In the H_2 molecule, the principal component of the wavefunction beyond the SCF σ_g^2 configuration is the σ_u^2 configuration necessary to allow dissociation to two H atoms. However, in the H_2^- ion, the $\sigma_g^2 k\sigma_u$ configuration does not admit the $\sigma_g^2 \rightarrow k\sigma_u^2$ double excitation because of the Pauli principle. Thus, correlation of the σ_g^2 shell in H_2^- arises from configurations different from those that add correlation to the H_2 neutral SCF wave function.

This problem might be addressed approximately for the position of the resonance by finding an appropriate version of Koopmans's theorem. Many-body theory methods for computing the electron affinity directly can also be pressed into service for this problem and may present a promising alternative to direct calculations on the neutral and anion resonance states separately.[53]

The effect of correlation on the width in a molecular shape resonance seems from the results in Fig. 13 to be a less pressing question *except near the crossing point* between the neutral and anion resonance potentials. The evidence is the fact that the phenomenon shown in Figs. 9, 13, and 14, where the point at which Γ becomes nonzero and the SCF crossing point do not coincide, is a general one. The difference[44] seems to be on the order of $0.1a_0$ in H_2^-, N_2^-, and F_2^-. Near this region, Γ can be quite small, and correlation effects may be very important in determining the width. If the nuclear dynamics for the process in question, e.g., dissociative attachment in $e + F_2$, are sensitive to the crossing region, then the question of the effects of correlation on the width becomes critical.

Before concluding this section, we should note the evidence that generalizing conclusions regarding the role of correlation in molecular shape resonances to atomic shape resonances is a dangerous business. McNutt and McCurdy[54] have performed CSCF and CI calculations on the lowest ^2P resonance state of Be^- for which the CSCF configuration is $1s^2 2s^2 kp$. Their

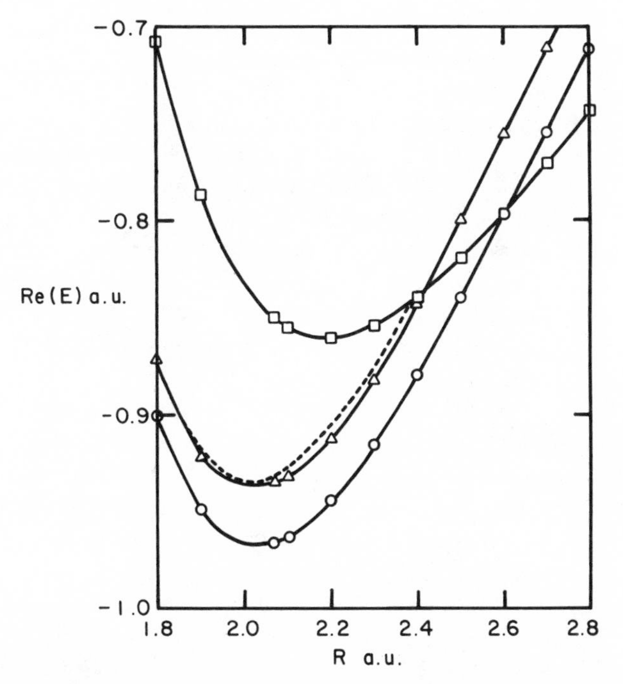

FIGURE 14. The SCF energies for N_2 and N_2^-:(□) CSCF for Re(E_{res}) for N_2^-, (○) SCF N_2, (Δ) Re($E_{res} - \varepsilon_{\pi_g}$) from CSCF on N_2^-, (---) position of the N_2 ground state relative to N_2^- from Ref. (49). Scale is relative to -180 a.u.

results are summarized in Table 1 and compared with other *ab initio* calculations.

McNutt and McCurdy(54) performed CSCF calculations on the ^2P state in a (7*s*/8*p*) basis of Slater functions with the four most diffuse *p* functions having complex exponents. The CSCF calculations showed remarkable

TABLE 1

Ab Initio Calculations on the Be–^2P Resonance

Calculation	Method	E_r(eV)	Γ(eV)
Rescigno *et al.*(18)	Static exchange	0.76	1.11
McCurdy *et al.*(33)	Complex self-consistent-field	0.70	0.51
McNutt and McCurdy(54)	Complex configuration interaction (CSCF start) (singles and doubles)	0.58	0.38
	(Singles, doubles, and triples)	0.323	0.296
Donnelly and Simons(53)	2nd-order electron propagator + coord. rotation	0.57	0.99
Junker(64)	Complex stabilization (CI)		0.40

stability with respect to the scaling parameter $e^{i\theta}$ of the complex Slater exponents. Using the occupied CSCF orbitals as part of the orbital basis, complex CI calculations were performed in a $(7s/5p/4d)$ basis of orthogonal orbitals. The CI calculations were performed first with all single and double excitations from the $2s$ and $2p$ orbitals and then with all single, double, and triple excitations from these orbitals included. Analogous calculations were performed on the Be neutral ground state.

The results show that the effects of correlation on ^2P resonance states are dramatic. The best CI estimates for the position and width are 0.323 and 0.296 eV, respectively, in comparison with 0.70 and 0.51 eV for the position and width at the CSCF level. This remarkable sensitivity of the width to correlation is due to the near degeneracy of the $2s$ and $2p$ orbitals and is unexpected for internuclear distances away from the crossing point in a molecular shape resonance.

7. QUESTIONS STILL UNANSWERED

7.1. *Extensions and Limitations of the Complex Basis Function Method*

In the first book he edited on the subject of autoionization,[55] Temkin wrote an article entitled "The Art and Science of Calculating Autoionization." There is still a substantial amount of "art" in most procedures for computing the positions and lifetimes and metastable states, and the principal reason is that the width of the resonance and energy shift in its position can be sensitive functions of electron correlation. This state of affairs is the central theme of the present chapter, which is that a complex variational principle allows us to convert the resonance problem to a problem such as the computation of a bound-state wave function where the wave function in question is square integrable. Particularly for molecular resonance, it seems that this path may eventually make calculating complex resonance energies as routine as calculating the energies of bound electronic states.

Even if that goal is reached, it leaves a major question unanswered. Temkin[55] raised this question in 1967 by asking what it means in terms of the phase shift behavior when we know where there is a resonance whose width is not small. Direct methods of computing the position of a resonance pole do not address the question of its interference with background scattering. Nesbet[42] has raised the question of whether or not this can be a serious issue in calculating vibrational excitation cross sections. In fact, the notion that vibrational excitation or dissociative-attachment cross sections can be calculated from the knowledge of only the *local* complex potential in Eq. (1) is certainly a simplification. Cederbaum and Domcke[56] have critically evaluated the local complex potential model in the context of an exactly soluble model and pointed out circumstances when the boomerang model,[5,6] which is based on the local potential, can fail. On the other hand, the boomerang model has the considerable advantage of simplicity, and McCurdy and Turner[57] have given a wave packet version that is more conveniently applicable to the polyatomic case. Ignoring these problems momentarily, because we know at least one system ($e + N_2$) in which the local complex potential model is able to provide the vibrational excitation cross section, there are numerous other questions about complex basis function calculations on resonances.

For example, how well would a complex multiconfiguration SCF (MCSCF) calculation work for a Feshbach resonance? An example of the trial function for such a calculation is given in Eq. (37). The idea has appeal because in a sense it corresponds to variationally determining the Feshbach Q-space and P-space parts of the resonance wave function at the same time. To our knowledge, there has been no complex MCSCF calculation ever attempted.

Also, what can we do to recover a meaningful version of Koopmans's theorem in a CSCF calculation? One method is to try to employ a variational function of the form

$$\int A(\psi^*_{\text{target}}\phi_{\text{res}})\hat{H}A(\psi_{\text{target}}\phi_{\text{res}})\,d\tau \tag{42}$$

where only the target orbitals are complex conjugated, while the resonance orbital is not. The original complex basis function approach seemed no less bizarre than this when it was first proposed.

How can we use the wave functions from a complex basis function calculation on resonances, say, for an autoionizing state of an atom, to calculate an optical oscillator strength for the transition to that state in an optical spectrum? In other words, how do we calculate the matrix element

$$d = \langle \psi_0 | \mu | \psi_{\text{complex}} \rangle \tag{43}$$

where μ is the dipole operator, ψ_0 is a bound state, and ψ_{complex} is a complex basis function approximation to an autoionizing state? If we did not use explicit complex coordinates to obtain ψ_{complex}, what contour is implicit in this matrix element? See Ref. (29) and references therein for more on this question.

Isaacson and Miller[58] have employed the Siegert basis function method, a different complex basis function procedure than those described in this chapter, to compute the complex potential function for the Penning ionization of $He(2^1s) + H$ and $He(2^3S) + H$. They were the first of the complex basis function calculations on a molecular system, but the method was abandoned and never used for a molecule again to our knowledge. Can this method or some variation of it be made into an efficient procedure for calculations on other molecular Feshbach resonances?

7.2. *Using the Complex Variational Principle without Complex Basis Function Calculations*

A final point concerns the possibility that we can use the complex variational principle in Eq. (17) without performing any complex basis function calculations at all. In the CI calculations on Be^- discussed earlier, some subset of the basis function exponents were scaled by $\eta = e^{i\theta}$, and after performing calculations for several discrete values of η, we used a rational fraction to find the stationary point

$$\left[\frac{dE(\eta)}{d\eta}\right]_{\eta_{\text{stationary}}} = 0 \tag{44}$$

and thus the resonance energy $E(\eta_{\text{stationary}})$. The question is, can we compute $E(\eta)$ for *real* η (a stabilization plot) and analytically continue to complex values of η to find the complex stationary point?

McCurdy and McNutt[59] have shown that it is indeed possible to do so. In Fig. 15, we show a stabilization graph for a model potential-scattering

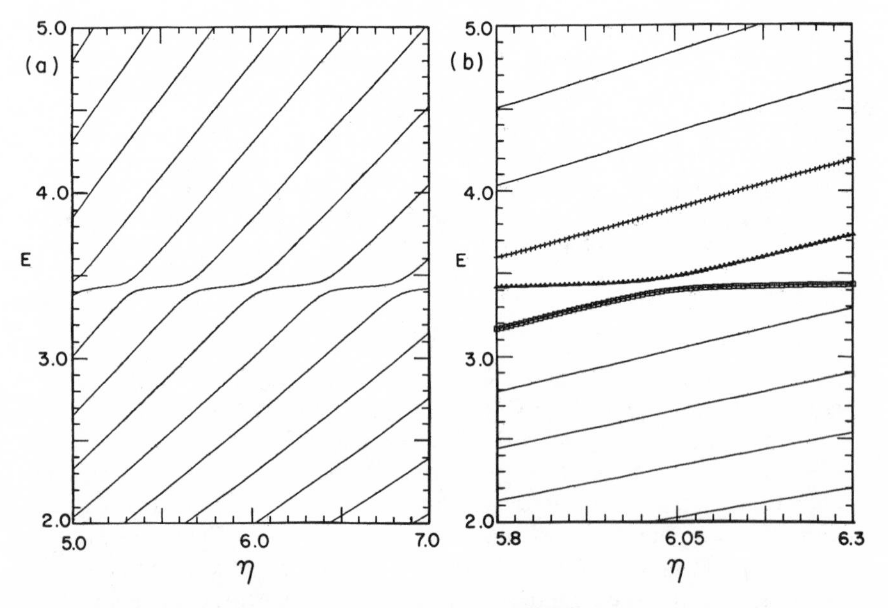

FIGURE 15. Stabilization graphs for the potential in Eq. (45) in a basis of Laguerre functions. The η is a real-valued nonlinear parameter in the basis functions as in Ref. (59). (a) shows a series of avoided crossings that have associated branch points and (b) shows enlargement with avoided crossing eigenvalues used in the calculations in Table 2 emphasized.

problem with the potential[59]

$$V(r) = \tfrac{15}{2} r^2 e^{-r} \tag{45}$$

(which supports a barrier-tunneling resonance) and the mass of the scattered particle equal to that of an electron. The basis consists of Laguerre functions, and η in these plots is a *real-valued* nonlinear parameter in the basis functions. The avoided crossings and "stabilization" near the resonance energy are typical of such stabilization graph.[60] Thompson and Truhlar[61] analytically continued one of the $E(\eta)$ curves in stabilization plots like those shown in Fig. 15 but found that by doing so they could obtain results reliable to only one or two significant figures in the width.

The reason for that inaccuracy and a way around the problem were given by McCurdy and McNutt.[59] The avoided crossings in Fig. 15 indicate that there are branch points in the eigenvalue function $E(\eta)$ at complex η where pairs of $E(\eta)$ curves are actually equal. The solution McCurdy and McNutt proposed is effectively to continue *two or more* of the $E(\eta)$ curves simultaneously. This can be done by using the stabilization plots to manufacture a part of the characteristic polynomial of the real matrix whose eigenvalues form the stabilization plot in the first place. McCurdy and McNutt then analytically continued the coefficients of that characteristic polynomial to complex η and found its complex roots.

A simpler and perhaps even more stable procedure for two-root continuation that clearly indicates the spirit of this approach is to use the expression

$$E_{\pm}(\eta) = \tfrac{1}{2}\{f_1(\eta) \pm [f_2(\eta)]^{1/2}\} \tag{46}$$

where

$$f_1(\eta) = E_1(\eta) + E_2(\eta)$$
$$f_2(\eta) = [E_1(\eta) - E_2(\eta)]^2 \tag{47}$$

and $E_1(\eta)$ and $E_2(\eta)$ are two of the $E(\eta)$ eigenvalue curves involved in an avoided crossing in Fig. 15. The $E_1(\eta)$ and $E_2(\eta)$ have branch points, but $f_1(\eta)$ and $f_2(\eta)$ are *analytic functions* in the neighborhood of those branch points. Equation (46) simply reproduces $E_1(\eta)$ and $E_2(\eta)$, of course, but it forms the basis of a method to analytically continue them

1. Construct $f_1(\eta)$ and $f_2(\eta)$ at *real* η, and fit these functions with continued fractions or Padé approximates.
2. Use these continued fraction representations to evaluate Eq. (46) for complex η.
3. Find the stationary points of $E_+(\eta)$ or $E_-(\eta)$ according to Eq. (44).

Results of this method[62] are shown in Table 2 and compared with the

TABLE 2
Results of Analytical Continuation of the Stabilization Graphs in Fig. 15[a]

J	E_R(a.u.)	Γ(a.u.)
10	3.42631	0.02594
20	3.42641	0.02556
30	3.42641	0.02555
40	3.42640	0.02556
Exact value	3.42639	0.02555

[a] The basis set consists of 50 Laguerre functions like those used in Ref. (59); J is the number of real-valued points in the analytic continuation using Eqs. (46) and (47).

exact result for this potential. This approach, as expressed in Eqs. (46) and (47), has also been used to compute resonance positions and widths in a model molecular vibration problem by Bai *et al.*[(63)] with excellent results. The validity of the entire procedure rests on the complex variational principle, but it is the ultimate application of the idea. *No complex basis function calculations are necessary.* Applications to molecular Feshbach resonances are underway in our laboratory.

Acknowledgements

I would like to thank the experts in this field with whom I have had many stimulating and enlightening conversations: Bill Reinhardt, Barry Simon, Tom Rescigno, Bob Junker, Howard Taylor, Andy Hazi, Aaron Temkin, and others. My work in this area was supported by National Science Foundation Grants CHE-7907787 and CHE-8217439 and by the Sloan and Dreyfus Foundations.

REFERENCES

1. See, for example, G. J. Schulz, *Rev. Mod. Phys.* **45** 378, 423 (1973); G. J. Schulz, in *Principles of Laser Plasmas*, (Bekefi, ed.), p. 33, Wiley, New York (1976).
2. K. D. Jordan and P. D. Burrow, *Acc. Chem. Res.* **11**, 341 (1978).
3. K. Fukui, *Acc. Chem. Res.* **4**, 57 (1971).
4. P. D. Burrow, J. A. Michejda, and K. D. Jordan, unpublished; for another example, see K. D. Jordan, J. A. Michejda, and P. D. Burrow, *J. Am. Chem. Soc.* **98**, 1295 (1976).
5. D. T. Birtwistle and A. Herzenberg, *J. Phys. B* **4**, 53 (1971).

6. L. Dubé and A. Herzenberg, *Phys. Rev. A* **20**, 194 (1979).
7. R. K. Nesbet, *Phys. Rev. A* **19**, 551 (1979).
8. J. Aguilar and J. Combes, *Commun. Math. Phys.* **22**, 269 (1971); E. Balslev and J. Combes, *Combes, Commun. Math. Phys.* **22**, 280 (1971).
9. See, for example, J. R. Taylor, *Scattering Theory*, (p. 221, Wiley, New York 1972).
10. See, for example, C. W. McCurdy, in *Electron–Molecule and Photon–Molecule Collisions* T. N. Rescigno, V. McKoy, and B. Schneider eds.), p. 229, Plenum, New York (1979).
11. G. D. Doolen, M. Hidalgo, J. Nuttall, and R. W. Stagat, in *Atomic Physics* (S. J. Smith and G. K. Walters, eds.), p. 257, Plenum, New York (1973); G. D. Doolen, J. Nuttall, and R. W. Stagat, *Phys. Rev. A* **10**, 1612 (1974).
12. G. D. Doolen, *J. Phys. B* **8**, 525 (1975).
13. E. Brändas and P. Froelich, *Phys. Rev. A* **16**, 2207 (1977); and E. Brändas, P. Froelich, and M. Hehenberger, *Int. J. Quantum Chem.* **14**, 419 (1978).
14. R. Yaris and P. Winkler, *J. Phys. B* **11**, 1475 (1978).
15. P. R. Certain, *Chem. Phys. Lett.* **65**, 71 (1979).
16. N. Moiseyev, P. R. Certain, and F. Weinhold, *Int. J. Quantum Chem.* **65**, 727 (1978); N. Moiseyev, P. R. Certain, and F. Weinhold, *Mol. Phys.* **36**, 1613 (1978); and N. Moiseyev, S. Friedland, and P. R. Certain, *J. Chem. Phys.* **74**, 4739 (1981).
17. Y. K. Ho, *Phys. Rev. A* **23**, 2137 (1981); and Y. K. Ho, *Phys. Rev. A* **19**, 2349 (1979).
18. T. N. Rescigno, C. W. McCurdy, and A. E. Orel, *Phys. Rev. A* **17**, 1931 (1978).
19. B. R. Junker and C. L. Huang, *Phys. Rev. A* **18**, 313 (1978).
20. B. Simon, *Phys. Lett.* **71A**, 211 (1979).
21. B. R. Junker, *Phys. Rev. Lett.* **44**, 1487 (1980); and B. R. Junker, *Int. J. Quantum Chem.* **14S**, 55 (1981).
22. D. R. Herrick, F. H. Stillinger, *J. Chem. Phys.* **62**, 4360 (1975); D. R. Herrick, *J. Chem. Phys.* **65**, 3529 (1981). I am indebted to W. P. Reinhardt for having pointed out these references to me.
23. E. Brändas, P. Froelich, C. H. Obcemea, N. Elander, and M. Rittby, *Phys. Rev. A* **26**, 3656 (1982).
24. J. Turner and C. W. McCurdy, *Chem. Phys.* **71**, 127 (1982).
25. C. W. McCurdy and T. N. Rescigno, *Phys. Rev. Lett.* **41**, 1364 (1978).
26. N. Moiseyev and C. T. Corcoran, *Phys. Rev. A* **20**, 814 (1979).
27. C. W. McCurdy, *Phys. Rev. A* **21**, 464 (1980).
28. J. D. Morgan and B. Simon, unpublished.
29. C. W. McCurdy and T. N. Rescigno, *Phys. Rev. A* **21**, 1499 (1980).
30. C. C. J. Roothaan, *Rev. Mod. Phys.* **32**, 179 (1960).
31. E. R. Davidson and L. Z. Stenkamp, *Int. J. Quantum Chem.* **105**, 21 (1976).
32. C. C. J. Roothaan and P. S. Bagus, *Meth. Comput. Phys.* **2**, 47 (1976); see also Ref. 30.
33. C. W. McCurdy, T. N. Rescigno, E. R. Davidson, and J. G. Lauderdale, *J. Chem. Phys.* **73**, 3268 (1980).
34. R. Lefebvre, in *Modern Quantum Chemistry*, part I (O. Sinanoglu, ed.), p. 125, Academic, New York (1965).
35. For a review, see A. C. Wahl and G. Das, in *Methods of Electronic Structure Theory, Modern Theoretical Chemistry* (H. F. Schaefer, ed.), p. 125, Academic, New York (1976).
36. L. Schlessinger, *Phys. Rev.* **167**, 1411 (1968). Schlessinger's algorithm is for constructing a Thiel continued fraction. For more discussion and references, see B. Jones and W. J. Thron, *Continued Fractions*, p. 393, Addison-Wesley, Reading, MA (1980).
37. M. Allan and S. F. Wong, *Phys. Rev. Lett.* **41**, 1791 (198).
38. C. W McCurdy and R. C. Mowrey, *Phys. Rev. A* **25**, 2529 (1982).
39. J. M. Wadehra and J. N. Bardsley, *Phys. Rev. Lett.* **41**, 1795 (1978); J. N. Bardsley and J. M. Wadehra, *Phys. Rev. A* **20**, 1398 (1979).

40. J. C. Y. Chen and J. L. Peacher, *Phys. Rev.* **167**, 30 (1968).
41. R. J. Bieniek and A. Dalgarno, *Astrophys. J.* **228**, 635 (1979).
42. R. K. Nesbet, *Comments At. Mol. Phys.* **11**, 25 (1981).
43. H. S. Taylor and F. E. Harris, *J. Chem. Phys.* **39**, 1012 (1963); and I. Eliezer, H. S. Taylor and J. K. Williams, *J. Chem. Phys.* **47**, 2165 (1967).
44. J. G. Lauderdale, C. W. McCurdy, and A. U. Hazi, *J. Chem. Phys.*, **79**, 2200 (1983).
45. W. Kolos and L. Wolniewicz, *J. Chem. Phys.* **43**, 2429 (1965).
46. T. N. Rescigno, A. E. Orel, and C. W. McCurdy, *J. Chem. Phys.* **73**, 6347 (1980).
47. N. Chandra and A. Temkin, *Phys. Rev. A* **13**, 188 (1976).
48. See. A. Temkin, in *Electron–Molecule and Photon–Molecule Collisions* T. N. Rescigno, V. McKoy and, B. Schneider, eds.), p. 173, Plenum, New York (1979).
49. A. U. Hazi, T. N. Rescigno, and M. Kurilla, *Phys. Rev. A* **23**, 1089 (1981).
50. B. I. Schneider, M. LeDourneuf, and Vo Ky Lan, *Phys. Rev. Lett.* **43**, 1926 (1979).
51. D. A. Levin and V. McKoy, unpublished results reproduced in Ref. 49.
52. P. J. Hay, *J. Chem. Phys.* **76**, 502 (1982).
53. R. A. Donnelly and J. Simons, *J. Chem. Phys.* **73**, 2858 (1980).
54. J. F. McNutt and C. W. McCurdy, *Phys. Rev. A* **27**, 132 (1983).
55. A. Temkin, in *Autoionization* (A. Temkin ed.), p. 55, Mono, Baltimore (1966).
56. L. S. Cederbaum and W. Domcke, *J. Phys. B* **14**, 4665 (1981).
57. C. W. McCurdy and J. L. Turner, *J. Chem. Phys.* **78**, 6773 (1983).
58. A. D. Isaacson and W. H. Miller, *Chem. Phys. Letts.* **62**, 394 (1979).
59. C. W. McCurdy and J. F. McNutt, *Chem. Phys. Letts.* **94**, 306 (1983).
60. A. U. Hazi and H. S. Taylor, *Phys. Rev. A* **1**, 1109 (1970), give a detailed treatment of the stabilization phenomenon.
61. T. C. Thompson and D. G. Truhlar, *Chem. Phys. Letts.* **92**, 71 (1982).
62. R. C. Mowrey and C. W. McCurdy, unpublished results.
63. Y. Y. Bai, G. Hose, C. W. McCurdy, and H. S. Taylor, *Chem. Phys. Letts.*, **99**, 342 (1983).
64. B. R. Junker, private communication (1982).

CHAPTER SIX

DIAGNOSTICS OF SOLAR AND ASTROPHYSICAL PLASMAS DEPENDENT ON AUTOIONIZATION PHENOMENA

GEORGE A. DOSCHEK

1. INTRODUCTION

This chapter shows atomic physicists how calculations involving autoionization effects are useful to astrophysicists. The calculations are important when incorporated into spectroscopic diagnostic tools used to determine physical conditions in astrophysical plasmas. The discussion here is limited to the interpretation of spectral line intensities at ultraviolet and shorter wavelengths, and, therefore, the corresponding emphasis is on calculations involving highly ionized atoms.

Because of absorption by the earth's atmosphere, x-ray, XUV, and UV spectra of astrophysical objects have become available only since the advent of space research. During the last 20 years, x-ray and UV spectrometers and spectrographs of increasing sophistication have been flown on rockets and unmanned and manned spacecraft, and the bewildering and large amount of data obtained by these instruments has opened up exciting new areas in astrophysics. Interpretating the data has also led to new areas of research in the atomic physics of highly ionized atoms, which produce spectral transitions at x-ray, XUV, and UV wavelengths. Accurate collision strengths, radiative and autoionization transition probabilities, and wavelengths are needed for allowed and forbidden transitions for almost all the ions of the cosmically abundant elements. For some of the solar observations, which are of very high quality, accuracies of 20% or better in atomic data are desirable.

This chapter is concerned primarily with the most recent spectral and solar observations, for which the data are of such high quality that state-of-

GEORGE A. DOSCHEK ■ E. O. Hulburt Center for Space Research, Naval Research Laboratory, Washington, D.C. 20375.

the-art computational techniques in atomic physics are often required to interpret the spectra satisfactorily. We attempt to show how the atomic data are used to calculate spectral line intensities and how this information is related to properties of the emitting gas such as temperature, density, ionization balance, and atmospheric structure and dynamics. Recent outstanding spectral observations from nonsolar astrophysics are also briefly discussed.

A tabulation of some of the recent space experiments, upon which much of the discussion to follow is based, is given in Table 1. A listing or discussion of earlier experiments, i.e., usually lower spectral or spatial resolution, or uncalibrated, is given in Walker[57] and in Doschek.[58] The experiments listed in Table 1 have high spectral resolution as a common denominator, although in some instances, particularly in the wavelength region between about 300 and 1100 Å, only the best available resolution experiment is listed. Many of the spectral lines of diagnostic value are weak lines, because they arise from metastable levels, and therefore high spectral resolution is needed to resolve them from the much stronger and numerous lines due to electric dipole transitions.

Autoionization is important for many of the calculations that relate spectral line intensities—obtained from the instruments in Table 1—to macroscopic gas parameters such as temperature and density. The intensity of a spectral line depends in part on the abundance of the ion in which the transition occurs and the cross section for exciting the upper state of the transition. In a low density situation, in which local thermodynamic equilibrium is not valid, the abundance of the ion depends on whether or not the plasma is in ionization equilibrium. If the plasma is assumed to be in equilibrium, the ion abundance will depend on ionization and recombination rate coefficients, which in turn are influenced by autoionization processes. For example, electron impact ionization of an ion X^{+z} with charge z can occur directly by the process,

$$X^{+z} + e \rightarrow X^{+z+1} + e' + e''$$

However, in some cases, ionization via autoionization can be as important or more important than the direct process, i.e.,

$$X^{+z} + e \rightarrow (X^{+z})^{**} + e'$$
$$(X^{+z})^{**} \rightarrow X^{+z+1} + e''$$

The symbol ( )** denotes an excited state in the continuum of the ion X^{+z}. This state can either stabilize radiatively, in which case no ionization occurs; or autoionization can occur, in which case the ion X^{+z} is ionized to the ion X^{+z+1}, as indicated in the preceding equations.

In low-density plasmas, dielectronic recombination is, in many cases, more important than radiative recombination and is the competing process to electron impact ionization in determining the fractional abundance of an ion.[59,60] Dielectronic recombination first involves dielectronic capture into an autoionization continuum state of the potentially recombined ion. The rate of this process is proportional to the autoionization rate from this continuum state back into the initial state consisting of the recombining ion and an unbound electron. After dielectronic capture occurs, the system can either autoionize into one of a number of states consisting of the recombining ion plus an unbound electron, or the system can stabilize by radiative transitions into bound states of the recombined ion, in which case dielectronic recombination has occurred. The process is

$$X^{+z} + e \rightleftharpoons (X^{+z-1})^{**}$$
$$(X^{+z-1})^{**} \rightarrow X^{+z-1} + \sum_k h\nu_k$$

Either one or several photons might be emitted during the stabilization, depending on which upper bound states of X^{+z-1} are formed. For some lines in the x-ray spectrum of low-density plasmas, dielectronic recombination is not only important in determining ion abundances, it is also the main mechanism by which the upper levels of observed spectral lines are excited.[61] Such lines are particularly important in high-temperature plasmas ($> 10^7$ K), such as solar flares, but there are also some instances where dielectronic recombination can be important in relatively more benign plasmas, such as planetary nebulas.[62]

In many of the cases we shall deal with, spectral lines are produced by electron impact excitation. The direct process is

$$X^{+z} + e \rightarrow (X^{+z})^{*} + e'$$
$$(X^{+z})^{*} \rightarrow X^{+z} + \sum_k h\nu_k$$

where ( )* here denotes a bound excited state of the ion X^{+z}. However, dielectronic captures can, in many important instances, lead to substantial (factors of two to five) increases in electron impact excitation rate coefficients due to such processes as

$$X^{+z} + e \rightleftharpoons (X^{+z-1})^{**}$$
$$(X^{+z-1})^{**} \rightarrow (X^{+z})^{*} + e$$
$$(X^{+z})^{*} \rightarrow X^{+z} + \sum_k h\nu_k$$

These "resonances" can occur within only discrete kinetic energy intervals of the unbound electron. Nevertheless, at these energies, the cross section can be many times greater than the direct excitation cross section, and therefore the rate coefficients, i.e., the integrals of the cross section over an assumed electron velocity distribution, such as a Maxwellian, can be substantially increased.

The processes we have briefly outlined are discussed more extensively in Section 3, where specific calculations are considered and compared to available data. In Section 2, we summarize some of the recent findings concerning the solar atmosphere and briefly mention some recent observations of other astrophysical objects. Much of our knowledge about physical conditions in the solar atmosphere, described in Section 2, has been gained by applying the spectroscopic diagnostics discussed in Section 3. Similar progress concerning our knowledge of physical conditions in some stellar atmospheres, and in other astrophysical objects, can be expected in the near future when observations obtained with ever more powerful instrumentation enable solar diagnostics to be applied to these objects.

2. SOLAR AND ASTROPHYSICAL PLASMAS

2.1. The Solar Atmosphere

The sun has an extended outer atmosphere called the corona. Because the visible light from the corona is about five orders of magnitude weaker than the radiation from the solar disk, the disk must be occulted in some fashion, either artificially or naturally by an eclipse, in order to observe it. The coronal visible light is actually solar disk radiation, Thomson scattered by the free electrons in the corona. The free electrons are produced by the high coronal temperature, which under so-called quiet conditions, i.e., no flares or other activity, is about 1.5×10^6 K. At this temperature, the major constituent of the coronal gas, hydrogen, is almost completely ionized. Other elements, such as C, O, Si, and Fe, are highly ionized, and several different ions of these elements can exist with appreciable abundance at coronal temperatures.

The discovery that the temperature in the corona is above 10^6 K was based on spectroscopic studies of the energy levels of highly ionized iron. Grotrian,[(63)] using energy level splittings for Fe X and Fe XI determined by Edlén, identified forbidden lines in coronal spectra that were emitted by these ions at 6374 and 7892 Å. Subsequently, Edlén[(64)] identified 17 other coronal emission lines, all due to forbidden transitions in highly ionized atoms. The existence of these ions in the coronal gas implied a very high temperature, on the order of a million degrees.

Since the temperature at the visible surface of the sun is about 5700 K, some mechanism or mechanisms must continuously heat the coronal gas, to offset the energy lost by radiation, by conduction and mass flow back into the lower temperature atmosphere and by the solar wind. The coronal heating mechanism(s) is still not known, and identifying it constitutes one of the most intriguing and interesting problems in solar physics.

The density of the corona can be found from coronal visible light observations and is quite low; the electron density just above the surface, i.e., a few thousand kilometers, ranges between 10^8 and $10^9\ \mathrm{cm}^{-3}$ in quiet regions. The ionic composition of the plasma is determined by balancing electron impact ionization against radiative and dielectronic recombinations. The ionization and recombination rates are only weakly dependent on density, and the fractional ion abundances depend primarily on electron temperature. At a temperature of about 10^6 K, the most abundant ions are such ions as O VII, Fe IX, Fe X, and Fe XI. The lightest abundant elements, such as oxygen, are ionized up to the He-like, H-like, and totally stripped ionization stages, while such ions as iron have about half of their electrons removed.

Although the corona was first observed in visible radiation, the resonance lines of the abundant ions just mentioned fall in the EUV and x-ray wavelength regions. Since the solar disk is totally dark at x-ray wavelengths, there is no need for an occulter, and the corona can be studied down to, and against, the solar surface by using an x-ray telescope placed outside the earth's atmosphere. Photographs of the corona in visible and x-ray wavelengths are shown in Fig. 1. The x-ray photograph has been overlayed on the white light photograph of the outer corona. The x-ray photograph was obtained on 30 June 1973 at 1145 UT using the American Science and Engineering (AS&E) x-ray telescope on Skylab. The white-light photograph was obtained at about the same time by the High Altitude Observatory (HAO).

One outstanding aspect of these photographs is the inhomogeneous nature of the corona. It is difficult to find any region of the corona that could be called typical or average, and furthermore, the x-ray photograph shows one of the main results from Skylab, namely, that many of the coronal structures appear in the form of loops or arches. These loops and arches are the outlines of magnetic field lines. The footpoints of the loops are connected to lower temperature regions of opposite magnetic polarity. The loops appear bright compared to their surroundings, probably because the density within the loops is higher than the ambient density outside the loops. The plasma is confined to the loops because the magnetic pressure $B^2/8\pi$ equals or exceeds the gas pressure $2N_e kT_e$ (B is the field strength, and N_e and T_e are the electron density and temperature, respectively).

Apparently, the very existence of a substantial corona depends on the loops or, more generally, closed magnetic structures. Over other areas of the sun, called coronal holes, the field lines are not closed, but open. (They are not strictly open, but close at enormous distances from the sun, and therefore near the surface, they can be regarded as open.) The x-ray emission from these

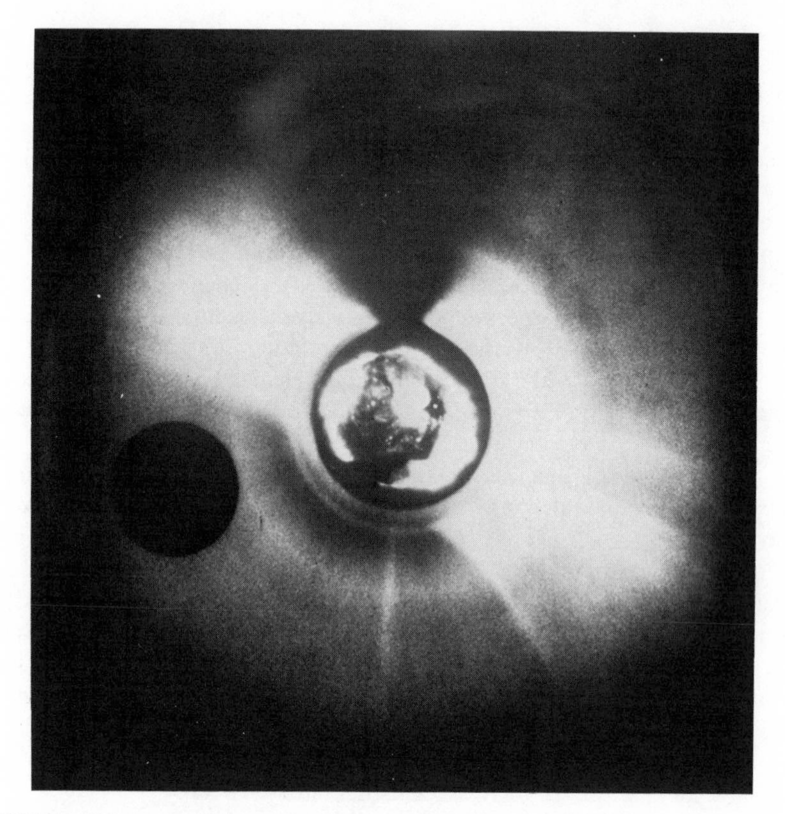

FIGURE 1. Soft x-ray and white-light photographs of the sun. The solar disk is dark in X rays but so bright at visible wavelengths that it must be occulted in order to observe the faint solar corona. The atmospheric structures in these photographs are determined by magnetic fields. X-ray photograph courtesy of American Science and Engineering (AS&E, from Skylab). White-light Photograph courtesy of the Naval Research Laboratory (NRL, from Skylab).

regions is negligible compared to closed field regions (see the large dark areas in the x-ray photograph in Fig. 1). We now know that the temperature in coronal holes is $\lesssim 10^6$ K close to the solar surface and the density is less than in quiet sun regions.[65,66] A further important result from Skylab is that high-speed solar wind streams originate from the coronal hole regions, at least during times of minimum activity in the solar cycle.[67]

Opposite in character to the coronal hole regions are the regions above sunspots, called active regions, where the magnetic field can be much stronger than in quiet sun or coronal hole regions. The active-region corona appears in the form of bright loops, as shown in the Naval Research Laboratory (NRL) photograph in Fig. 2, and both the temperature and density are greater in the active-region loops than in the quiet-sun loops.[68] The photograph is an

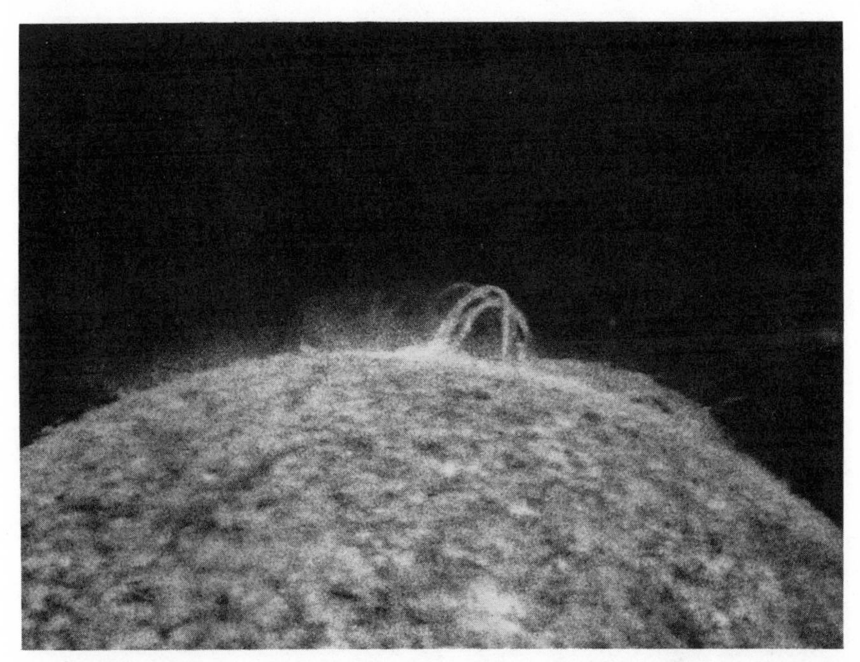

FIGURE 2. Magnetic flux tubes in the solar corona. Tubes appear brighter than the surrounding ambient corona because the confined plasma in the tubes is more dense than the ambient corona. Photograph courtesy of the Naval Research Laboratory (NRL, from Skylab).

image in He II 304 Å radiation obtained by NRL's slitless spectrograph on Skylab. Solar flares usually occur in or near active regions. Apparently, the probability of a flare is related to the complexity of the magnetic field geometry, but the relationship is not understood. In addition to solar flares, which are highly energetic and explosive phenomena, other violent phenomena, such as surges, sprays, and eruptive prominences, can occur in and near active regions. Surges and sprays are plasma ejected at high velocity ($\simeq 100$–$500\ \mathrm{km\ s^{-1}}$) into the corona from sites near or at the locations of flares and in many cases, appear to be one of the secondary effects of a flare.[69] A prominence is a complex of magnetic flux loops that contains cool plasma ($< 10^5$ K) that can extent fairly high into the corona.[70] For reasons not understood, these prominences can sometimes virtually explode (eruptive prominence), spewing cold hydrogen into the corona to heights of many solar radii, producing a so-called coronal transient. An NRL He II 304 Å image of an eruptive prominence is shown in Fig. 3. A coronal transient accompanied by a large eruptive prominence is shown in Fig. 4. The images in Fig. 4 were obtained by an NRL white-light coronagraph flown on the Air Force P78-1 spacecraft.[71]

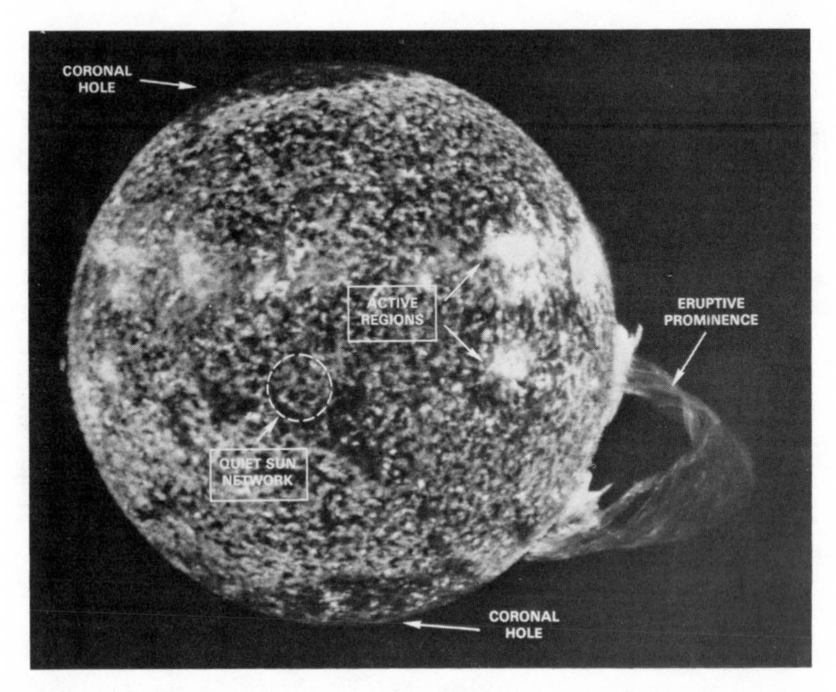

FIGURE 3. An erupting solar prominence as it appears in the resonance line of helium radiation (304 Å). The cause of such eruptions is not known. Other regions of the solar atmosphere are indicated. Photograph courtesy of NRL (from Skylab).

The hot corona in flux tubes must in some fashion couple to cooler atmospheric regions, i.e., the solar surface. Since the corona is much hotter than the surface, energy will be conducted downward from the corona into cooler regions, unless there is an inhibiting mechanism. The conductive flux along magnetic field lines is much larger than across field lines, so energy transport by conduction is primarily along the field lines. The energy conducted into the lower temperature regions is radiated away and perhaps is dissipated mechanically as well. In part because the classical coefficient of conductivity varies at $T_e^{5/2}$, a flux tube is quite hot ever much of its length, i.e., at coronal temperatures. In regions very close to the flux tube footpoints, the temperature drops almost discontinuously from coronal values to temperatures of about 2×10^4 K, which is the region of the atmosphere known as the chromosphere. The intermediate region, between 2×10^4 and about 10^6 K, is called the transition region. Its spatial extent along the axis of a coronal flux tube is extremely small. Images of the sun obtained in spectral lines formed at transition region temperatures reveal highly inhomogeneous structures, presently of indeterminate form, with the brightest regions concentrated in the

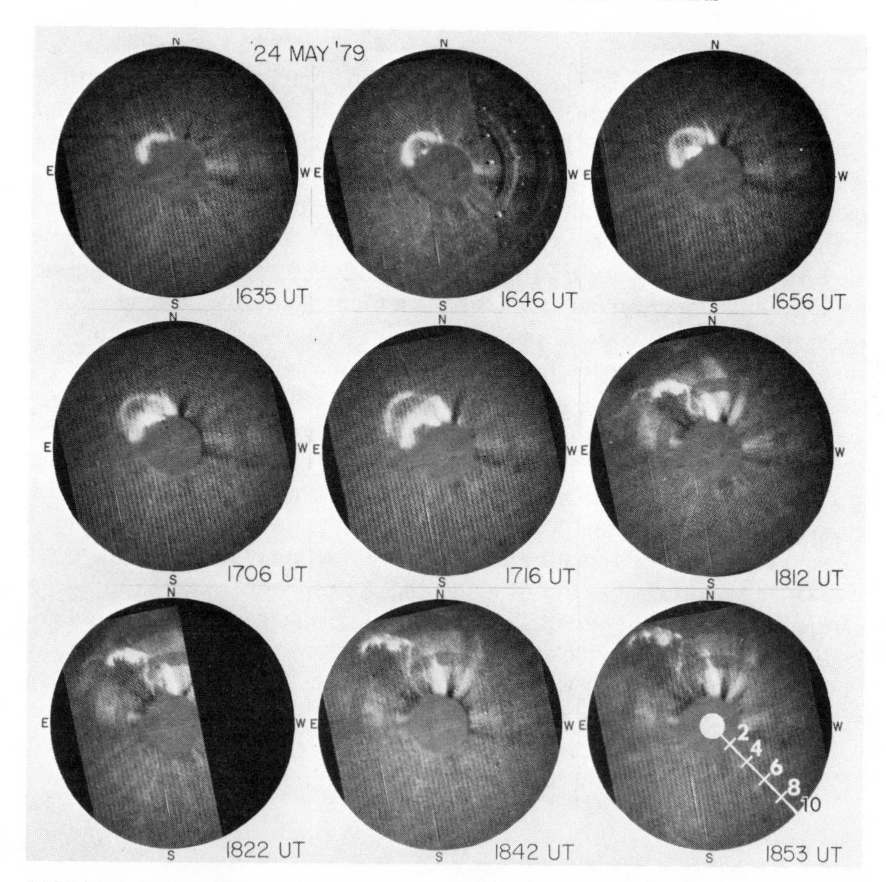

FIGURE 4. An erupting prominence and consequent coronal transient, observed by an NRL white-light coronagraph on the P78-1 spacecraft [see Ref. (71)]. The looplike structure seen in the first few frames is coronal plasma at temperatures near 10^6 K. The complicated extended structure seen later (e.g., 1842 UT) is cold prominence material at a temperature near 10^4 K. Both the cold and hot components are moving outward from the sun at a speed of $\simeq 700\,\mathrm{km\,s^{-1}}$. The scale shown in the 1853 UT photo is in units of solar radii. The size of the sun is indicated; the occulting disk is 2.5 solar radii in apparent radius.

so-called supergranule network.(72) This network is quite apparent in the He II 304 Å image shown in Fig. 3, and is the result of convection beneath the solar surface, which concentrates magnetic flux into the boundaries of large aggregations of convective cells. These boundaries form a lacework pattern and hence the name network.

However, it is in fact quite possible that most of the observed transition region emission does not arise from the extremely thin zones within coronal flux tubes, such as those described, but instead arises in magnetic structures that contain only cool transition region plasma.[73,74] The coronal flux tubes must still have transition regions, but these regions may not contribute most of the observed emission, which in this picture would arise in transition region loops magnetically isolated from the corona. In either case, the transition region is one of the most interesting parts of the solar atmosphere and may play a crucial role in coronal heating. Between the transition region and the solar surface, i.e., the photosphere, is the region called the chromosphere ($T_e \simeq 10^4$ K); we shall not deal significantly with this region in this chapter.

In summary, although much is known today about the structure of the solar atmosphere and the physical conditions within it, the basic problems, such as its origin and the mechanism or mechanisms by which it is maintained, still remain unsolved. The atmosphere may be heated by energy carried by acoustic or MHD (magnetohydrodynamic) waves generated from within the convection zone[75] (although acoustic heating is less likely[76]), by the ejection of mass (known as spicules) into the atmosphere,[77] or perhaps by the electrodynamic coupling between photospheric and coronal regions.[78] Certain problems also remain concerning the structure of the atmosphere. Of particular importance is the fact that the atmospheric regions cannot be characterized by plane-parallel layers. The atmosphere is highly inhomogeneous, and it is not possible to specify a unique temperature at a given height above the photosphere. A similar condition must hold for many stellar atmospheres as well. The physical association of the different temperature regions of the atmosphere, such as the observed chromosphere, transition region, and corona, is still uncertain.

2.2. *Solar Flares*

Solar flares are intrinsically fascinating phenomena, particularly from the diagnostic point of view. Flares are violent eruptions in the solar atmosphere that result in plasma being heated to temperatures of about 20×10^6 K, the acceleration of electrons to hundreds of kilovolts, the expulsion of plasma into the interplanetary medium, and substantial chromospheric and perhaps even photospheric heating.[69,79] A flare occupies only a small region of the solar surface, usually < 2 arc min square in area (1 arc sec $=$ 725 km), but over its lifetime, it can release between 10^{28} and 10^{33} ergs of energy. Virtually every region of the solar atmosphere is locally affected by the flare, and flares are observable throughout the spectrum from radio wavelengths down to gamma-ray emission. The causes of a flare are still unknown, although we know that the local magnetic field is intimately involved with the energy release and energy transport in the atmosphere after the commencement of a flare, and perhaps it is the flare's ultimate energy source.

Of particular interest to spectroscopy is the soft x-ray emission from flares, which arises in plasma at temperatures between about 2×10^6 and

20×10^6 K. The plasma responsible for this emission is also quite important from the point of view of flare energetics, since a fair fraction of the total flare energy released results in the soft x-ray plasma and its radiation. A flare is usually accompanied by a rapid rise in x-ray emission followed by a slower decline.

Observations obtained from Skylab have shown that the soft x-ray emission lines from such ions as Fe XXIV arises in small magnetic flux tubes, usually less than 1 arc min in length and frequently no longer than 10 arc sec, i.e., about 7.3×10^3 km.(80–82) An example from the NRL Skylab observations is shown in Fig. 5. Figure 5 is a monochromatic image of a flare loop in the radiation of Fe XXIV ($1s^2 2s\ ^2S_{1/2}$–$1s^2 2p\ ^2P_{1/2}$) at 255 Å and in the radiation of He II at 256 Å. The images are not spatially coincident because they have been dispersed by the spectrograph. The flux tube spans the gap between two regions of opposite magnetic polarity. These regions are separated by the indicated "neutral line" in Fig. 5. The gas confined within the flux tube (or perhaps unresolved tubes) has a temperature on the order of 18×10^6 K(83,84) and can remain this hot for time intervals ranging from a few

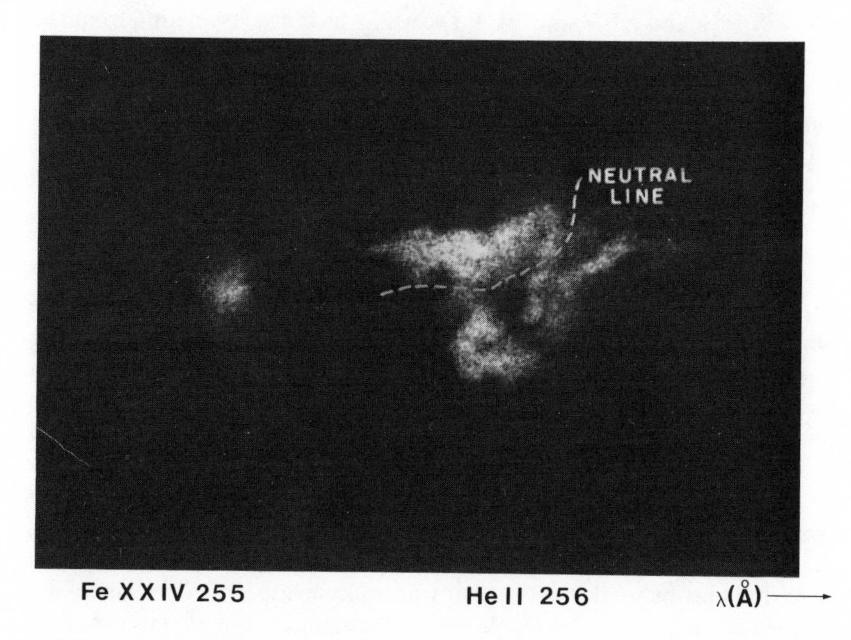

FIGURE 5. Photograph of a solar flare recorded by the NRL slitless spectrograph on Skylab [see Ref. (79)]. The spectrograph records images of the flare in monochromatic wavelengths. The right-most image is in He 256 Å radiation, while the left-most image shows the appearance of the flare in Fe XXIV radiation at 255 Å. The high-temperature image straddles the magnetic neutral line and has a looplike shape. The footpoints of the loop are located in two regions adjacent to the neutral line of opposite magnetic polarity.

minutes up to several hours. Recent observations from the P78-1 and SMM spectrometers have demonstrated rather conclusively that for many flares, continual energy input into flare flux tubes is necessary to explain the longevity of the emission.[85]

Solar flares also produce hard x-ray emission, which extends up to several hundred kilovolts.[86] This emission can be much more impulsive than the soft x-ray emission, with rapid intensity variations occurring on the order of seconds or even milliseconds.[87] The hard X rays are produced by electrons of similar energies. The number of these electrons is quite small in comparison to the thermal electrons that produce the 20×10^6 K plasma but nevertheless much larger than expected from a plasma that is no hotter than 20×10^6 K. Evidently, plasma instabilities produce the most energetic electrons. Recent evidence indicates, contrary to some theories, that these instabilities are not the actual flare trigger mechanisms but are instead simply another by-product of the overall energy release process.[88] From the spectroscopic standpoint, it is interesting to consider whether line ratios sensitive to non-Maxwellian velocity distributions can be identified.

In summary, flares are violent eruptions in the solar atmosphere that produce X rays and XUV and EUV spectra, which are fascinating from the standpoint of spectroscopic diagnostics, both from a fundamental atomic physics point of view and from their value in aiding our understanding of the nature of solar flares. Many of these diagnostics have similar uses in interpreting spectra from tokamak plasmas.

2.3. Stellar Atmospheres and Other Astrophysical Sources

Recent observations from the HEAO-2 x-ray observatory (Einstein) and the International Ultraviolet Explorer (IUE) have, in several cases, resulted in the application of spectroscopic diagnostics developed for the sun to more exotic astrophysical sources.[89,90] In particular, x-ray emission line-spectra between about 12 and 21 Å have been obtained by Einstein for certain supernova remnants, such as Puppis A.[56,90] In these remnants, the interstellar medium is shock heated to temperatures of a few million degrees. The supernova remnant spectra can be compared to solar spectra in the same wavelength range. Similarly, IUE observations between ~ 1100 and 3200 Å of cool stars (spectral-type F, G) have shown that many of these stars have transition regions that may be similar to the solar transition region.[89,91,92] Diagnostics developed to analyze the NRL solar spectra obtained from Skylab can, with appropriate assumptions and caveats, sometimes be applied to the stellar spectra. Also, high-quality IUE emission line spectra of planetary and other nebulas have been obtained.[93-95] The analysis of these spectra also benefits greatly from the most recent and accurate atomic data calculations. There are many other astrophysical areas in addition to these just mentioned that can potentially make use of the new atomic data base for highly ionized atoms.

3. PLASMA DIAGNOSTICS AND AUTOIONIZATION

3.1. *Theory of Spectral Line Formation*

For most cases discussed here, we consider a low-density plasma where the number density N of different ion species is determined by electron impact ionization and radiative and dielectronic recombinations. We neglect three-body recombination and photoionization, although in some cases photoionization can be important. The ion number densities for an element with nuclear charge Z are determined by solving the set of coupled-rate equations $(0 < z < Z)$,

$$\frac{\partial N_z}{\partial t} + \mathbf{U}\cdot\nabla N_z = N_e(N_{z-1}Q_{z-1} + N_{z+1}\alpha_{z+1}) - N_eN_z(Q_z + \alpha_z) \tag{3.1}$$

with the additional condition

$$\sum_0^Z N_z = N_T \tag{3.2}$$

where Q and α are ionization and total recombination rate coefficients (cm^3 sec^{-1}), z is the ionic charge, N_e is the electron density, N_T is the number density of the element, and U is the plasma fluid velocity. The solutions of eq. (3.1) are obviously model dependent for any realistic calculation. Frequently, ionization equilibrium is valid or at least assumed, in which case eq. (3.1) becomes $(0 < z < Z)$

$$N_zQ_z = N_{z+1}\alpha_{z+1} \tag{3.3}$$

The flux F_{ij}($erg\,cm^{-2}\,sec^{-1}$) at earth in an optically thin spectral line (usually the case encountered) is given by

$$F_{ij} = \frac{\Delta E_{ij}}{4\pi R^2} A_{ji} \int_{\Delta V} n_j dV \tag{3.4}$$

where ΔE_{ij} is the energy of the transition between upper-level j and lower-level i, A_{ji} is the radiative transition probability, ΔV is the volume of plasma where the line is formed, R is the earth-to-object distance, and n_j is the number density of the ions excited in level j. The excitation processes within an ion are usually much faster than ionization and recombination time scales, particularly near ionization equilibrium, so excitation can usually be decoupled from ionization and recombination. That is, in many cases, excitation processes are dominant over ionization and recombination processes in producing an excited state. In this case, the quantities n_j can be found by using detailed balance between excitation and radiative processes alone, neglecting ionization and recombination processes. The result is

$$n_j\left(\sum_{i<j} A_{ji} + N_e \sum_{i>j} C_{ji}^e + N_e \sum_{i<j} C_{ji}^d\right) = N_e \sum_{i<j} n_i C_{ij}^e + N_e \sum_{i>j} n_i C_{ij}^d + \sum_{i>j} n_i A_{ij} \tag{3.5}$$

with

$$\sum_j n_j = N_z \tag{3.6}$$

where $C^e(C^d)$ are excitation (deexcitation) rate coefficients for electrons (and sometimes protons). Assuming a Maxwellian velocity distribution, we also have

$$C_{ji}^d = C_{ij}^e \frac{\omega_i}{\omega_j} \exp\left(\frac{\Delta E_{ij}}{kT_e}\right) \tag{3.7}$$

where $\omega_{i(j)}$ is the statistical weight of the $i(j)$ level and k is Boltzmann's constant. There are many cases (e.g., resonance lines of Li-like ions) for which only two levels are important, i.e., the ground level and an excited level. Furthermore, in many cases, almost all the ionic population is in the ground level. Then

$$N_e n_1 C_{12}^e = n_2 A_{21} \tag{3.8}$$

and

$$n_1 \simeq N_z \tag{3.9}$$

Combining Eqs. (3.4), (3.8), and (3.9) gives

$$F_{12} = \frac{\Delta E_{12}}{4\pi R^2} \int_{\Delta V} N_e N_z C_{12}^e dV \tag{3.10}$$

The excitation rate coefficient is related to the excitation cross section σ_{ij} and an assumed velocity distribution $f_m(v)$ by

$$C_{ij}^e = \int \sigma_{ij}(v) v f_m(v) dv \tag{3.11}$$

where v is the speed of the electron. Frequently, collision strengths Ω_{ij} are tabulated as a function of incident particle kinetic energy E rather than the cross sections σ_{ij}. The relationship between them is

$$\sigma_{ij} = \left(\frac{I_H}{E}\right)\left(\frac{\Omega_{ij}}{\omega_i}\right)\pi a_0^2 \tag{3.12}$$

where I_H is the hydrogen ionization potential (2.18×10^{-11} erg) and a_0 is the Bohr radius. If $f_m(v)$ is a Maxwellian distribution, then

$$f_m(v) = 4\pi\left(\frac{m}{2\pi kT_e}\right)^{3/2} \exp\left(\frac{-mv^2}{2kT_e}\right) v^2 \tag{3.13}$$

where m is the electron mass. Combining Eqs. (3.11)–(3.13) gives

$$C_{ij}^e = \frac{8.63 \times 10^{-6}}{\omega_i k T_e^{3/2}} \int_{\Delta E_{ij}}^{\infty} \Omega_{ij}(E) \exp\left(\frac{-E}{kT_e}\right) dE \tag{3.14}$$

In some cases, only the value of the collision strength at a particular energy is available. If the collision strength is assumed constant, the excitation rate coefficient given by Eq. (3.14) becomes

$$C_{ij}^e = \frac{8.63 \times 10^{-6} \bar{\Omega}_{ij}}{\omega_i T_e^{1/2}} \exp\left(\frac{-\Delta E_{ij}}{kT_e}\right) \tag{3.15}$$

If resonances are small, and the value of $\bar{\Omega}_{ij}$ is taken to be the value of Ω_{ij} at about 1.5 ΔE_{ij}, then C_{ij} should be accurate to within a factor of two.

The excitation processes just discussed apply to electron or proton impact excitation of level j from level i. Autoionization is important in calculating $\Omega_{ij}(E)$, due to the existence of resonances, as outlined in Section 1. However, as we shall see in Section 3.2, some excited states of x-ray transitions are produced not by collisional excitation, but instead directly by dielectronic capture. Since dielectronic capture and autoionization are inverse processes, they are related using the principle of detailed balance and thermodynamic (Saha) equilibrium. Specifically,

$$\alpha_{nj}^{dc}(n \to j) = \frac{h^3}{2(2\pi mkT_e)^{3/2}} \frac{\omega_j}{\omega_n} A_{jn}^a \exp\left(\frac{-E_{nj}}{kT_e}\right) \tag{3.16}$$

where α_{nj}^{dc} is the rate coefficient for dielectronic capture of an unbound electron by an ion of charge $z + 1$ in state n into a doubly excited state j of an ion of charge z; E_{nj} is the energy difference between states j and n; and A_{jn}^a is the autoionization probability for the $j \to n$ transition. The dielectronic capture can be followed by a stabilizing radiative transition in the z-charge ion. If the lower state is defined as state i, the rate coefficient α_{ij}^d for the entire process of capture followed by radiative stabilization is given by

$$\alpha_{ij}^d = \alpha_{nj}^{dc}\left(\frac{A_{ji}}{\sum_{n'} A_{jn'}^a + \sum_{i'} A_{ji'}}\right) \tag{3.17}$$

where the sum over n' is over all possible states consisting of ion $z + 1$ and an unbound electron into which state j can autoionize, and the sum over i' is over all lower states in ion z. The quantity in parentheses in Eq. (3.17) is simply a branching ratio. A more detailed derivation of Eqs. (3.16) and (3.17) is given by Seaton and Storey.[60]

The flux in a spectral line produced by dielectronic recombination is given by

$$F_{ij} = \frac{\Delta E_{ij}}{4\pi R^2} \int N_{z+1} N_e \alpha_{ij}^d \, dV \tag{3.18}$$

In cm g sec units, the quantity α_{ij}^d becomes

$$\alpha_{ij}^d = 2.07 \times 10^{-16} \frac{\omega_j A_{jn}^a A_{ji}}{\sum_{n'} A_{jn'}^a + \sum_{i'} A_{ji'}} \frac{\exp(-E_{nj}/kT_e)}{\omega_n T_e^{3/2}} \tag{3.19}$$

In order to evaluate Eqs. (3.18) and (3.4), it is necessary to know either N_z or N_{z+1}. Frequently, ionization equilibrium is assumed, and these quantities are given by Eq. (3.3). In this case, N_z (or N_{z+1}) can be expressed as an identity

$$N_z = \frac{N_z}{N_T} \frac{N_T}{N_H} \frac{N_H}{N_e} N_e \tag{3.20}$$

where N_H is the hydrogen number density. We further define

$$\frac{N_z}{N_T} = f(T_e) \quad \text{and} \quad \frac{N_T}{N_H} = A_H \tag{3.21}$$

In fully ionized astrophysical plasmas, $N_H/N_e \simeq 1$; in the solar atmosphere, 0.8 is frequently adopted.

3.2. *Electron Temperature Measurements*

3.2.1. Methods. There are two methods for determining electron temperature. The more commonly known method involves using the Boltzmann factors in excitation rate coefficients. Consider an isothermal plasma of density N_e and volume V. The flux ratio of two lines from an ion, originating from excited levels 3 and 2 and terminating on lower (say, the lowest) level 1, is given by Eq. (3.10)

$$\frac{F_{13}}{F_{12}} = \frac{\Delta E_{13}}{\Delta E_{12}} \frac{C_{13}^e}{C_{12}^e} \tag{3.22}$$

assuming for simplicity that levels 3 and 2 can decay to only level 1. Using Eq. (3.15), Eq. (3.22) can be rewritten as

$$\frac{F_{13}}{F_{12}} = \frac{\Delta E_{13}}{\Delta E_{12}} \frac{\bar{\Omega}_{13}}{\bar{\Omega}_{12}} \exp\left(\frac{\Delta E_{12} - \Delta E_{13}}{kT_e}\right) \tag{3.23}$$

where $\bar{\Omega}$ is a mean collision strength. The flux ratio in Eq. (3.23) is temperature sensitive if $(\Delta E_{13} - \Delta E_{12})/kT_e \gg 1$. The method is independent of ionization equilibrium but does depend on the atmospheric structure. We assumed in deriving Eq. (3.23) an isothermal plasma of constant density. Most real plasmas are not isothermal, and the density may vary with temperature. In this case, the excitation rate coefficients cannot be removed from the integral in Eq. (3.10), and the flux ratio depends on atmospheric structure in addition to temperature.

Another method of deducing electron temperature is simply to assume ionization equilibrium [Eq. (3.3)].(34) There are many highly ionized atoms that are formed in ionization equilibrium over a very narrow temperature range.(96,97) This is particularly true for most ions formed in the transition region between 2×10^4 K and several hundred thousand degrees. It is *not* true for most Li-like, He-like, and H-like ions, which are formed over very broad temperature ranges.(98,99) However, if a line of an ion, such as C III or O III, is observed, for example, and ionization equilibrium is assumed, then from the equilibrium calculations, the temperature is known (i.e., $T_e \simeq 6 \times 10^4$ K for these ions).

Finally, a rather elegant method for determining temperature in high-temperature low-density plasmas has been developed by Gabriel and Jordan(61,100,101) and independently by Vainstein and his colleagues.(102)

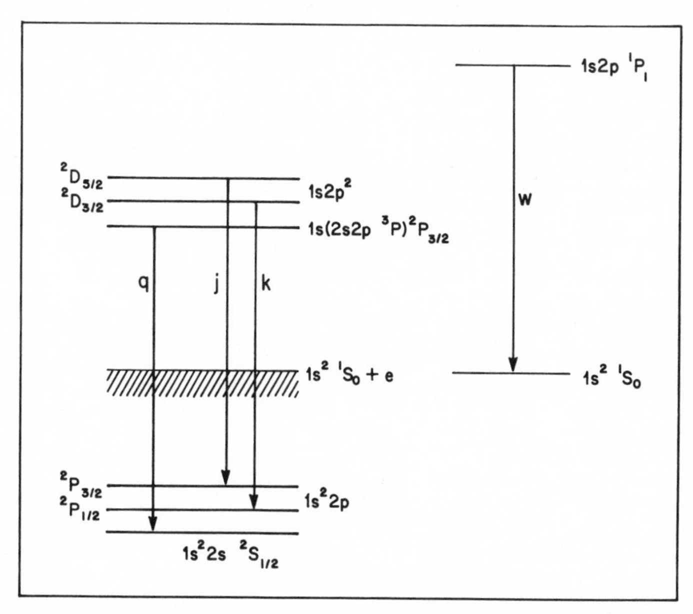

FIGURE 6. Energy-level diagrams showing x-ray lines formed by collisional excitation and dielectronic recombination. (Courtesy *Astrophys. J.*)

This method uses the different dependence on temperature of dielectronic recombination and electron impact excitation. We choose a line formed by only dielectronic recombination and a line formed primarily by electron impact excitation. As an example, consider the Li-like ion dielectronic x-ray line is $^2 2p^2P_{3/2}$–$1s2p^{2\,2}D_{5/2}$ and the He-like ion resonance line $1s^{2\,1}S_0$–$1s2p\,^1P_1$ excited by electron impact (see Fig. 6). The upper level $1s2p^{2\,2}D_{5/2}$ is formed by dielectronic capture in low-density plasmas because almost all of the Li-like ions are in the ground level $1s^22s\,^2S_{1/2}$, and the cross section for the electron impact excitation rate $1s^22s\,^2S_{1/2} \rightarrow 1s2p^{2\,2}D_{5/2}$ is very small and negligible compared to the dielectronic process. If for simplicity we again consider an isothermal plasma of constant density and use Eqs. (3.10) and (3.18), the flux ratio of the dielectronic line (F_D) to the resonance line (F_R) is given by

$$\frac{F_D}{F_R} = \frac{\Delta E_D}{\Delta E_R} \frac{\alpha_D^d}{C_R^e} \tag{3.24}$$

where ΔE_R and ΔE_D are the resonance line and dielectronic line energies, respectively. In the x-ray region, these energies are nearly equal. Since the resonance line is formed in an ion once more ionized than the ion where the dielectronic line is formed, $N_z = N_{z+1}$ when using Eqs. (3.10) and (3.18) to form the ratio given by Eq. (3.24). Using Eqs. (3.19) and (3.15) (with $\Delta E_{ij} \equiv \Delta E_R$) in Eq (3.24) gives

$$\frac{F_D}{F_R} = \beta \frac{\exp[-(E_{nj} - \Delta E_R)/kT_e}{T_e} \tag{3.25}$$

where β is a constant containing atomic factors, and β is given by

$$\beta = 2.4 \times 10^{-11} \frac{\Delta E_D \omega_j A_{jn}^a A_{ji}}{\Delta E_R \bar{\Omega}_R (\sum_{n'} A_{jn'}^a + \sum_{i'} A_{ji'})} \tag{3.26}$$

The ratio F_D/F_R in Eq. (3.25) can be quite temperature sensitive, and as we shall see, this diagnostic is very useful for flare plasmas.

Finally, we note that it is not usually possible to determine temperature from the Doppler-broadened profiles of optically thin lines, as might be thought. The reason is that most of the lines in solar, and many other nonsolar astrophysical spectra, are far wider than expected assuming ionization equilibrium.[38,103] These excess widths are usually attributed to random mass motions (or in some cases, they are due to stellar rotation), but in any event, they are not simply dependent on temperature.

3.2.2. The Solar Transition Region. As discussed in Section 2, the transition region is defined as that region of the solar atmosphere at a temperature between about 2×10^4 and 10^6 K. Most of the important spectral

TABLE 1
Recent High-Resolution Astrophysical Spectrometers and Spectrographs

Source	Flight mode	Instruments	References
Leicester Solar Physics Group	A series of rockets	Solar x-ray spectrometers ($\lambda < 25$ Å)	1, 2
P. N. Lebedev Physical Institute	Vertical-series rockets and unmanned satellites (Intercosmos series)	Solar x-ray spectrometers ($\lambda < 25$ Å)	3–6
Naval Research Laboratory	Unmanned satellite (P78–1)	Solar x-ray spectrometers ($\lambda < 8.5$ Å)	7,8
The Aerospace Corporation	Unmanned satellite (P78–1)	Solar x-ray spectrometers ($\lambda < 25$ Å)	9–11
Lockheed Corporation; Culham Laboratory; Mullard Space Science Laboratory	Unmanned satellite [Solar Maximum Mission (SMM)]	Solar x-ray spectrometers ($\lambda < 25$ Å)	12–14
Institute of Space and Aeronautical Science of the University of Tokyo	Unmanned satellite (Hinotori)	Solar x-ray spectrometers ($\lambda < 2$ Å)	15
Goddard Space Flight Center	Rockets	Solar grazing incidence spectrographs (60–770 Å)	16, 17
Air force, Cambridge Research Laboratory	Rocket	Solar grazing incidence spectrometer (50–300 Å)	18
Goddard Space Flight Center	Unmanned satellite [Fifth Orbiting Solar Observatory (0S0–5)]	Solar grazing incidence spectrometer (25–400 Å)	19
Astrophysics Research Unit; Culham Laboratory	A series of rockets	Solar high-resolution spectrographs (12–3000 Å)	20–23
University of Colorado	A series of rockets	Solar XUV and EUV spectrometers (200–700, 609–1272 Å)	24–26
Naval Research Laboratory	Skylab-manned space station	Solar XUV spectro-heliograph (170–600 Å)	27, 28
Harvard College Observatory	Skylab-manned space station	Solar EUV spectometers and spectroheliometer (280–1350 Å)	29
Harvard College Observatory	Rockets	Solar UV telescope spectrometer (1400–1875 Å)	30, 31
Culham Laboratory Harvard College Observatory; Imperial College, London; York University (Toronto)	March 1970; eclipse rocket	Solar EUV spectro-heliograph (977–2200 Å)	32–34

(*continued overleaf*)

TABLE 1 (continued)

Source	Flight mode	Instruments	References
Naval Research Laboratory	March 1970; eclipse rocket	Solar XUV spectroheliograph (1390–1945 Å)	35, 36
Naval Research Laboratory	Skylab-manned space station	Solar UV spectrograph (1150–4000 Å)	37, 38
Naval Research Laboratory	Rockets	Solar high-resolution telescope spectrograph (HRTS) 1150–1700 Å) and other instrumentation	39–42
University of Hawaii	Rockets	Solar UV echelle spectrographs (1800–2000 Å, 2700–2900 Å)	43, 44
University of Colorado	Unmanned satellite (0S0–8)	Solar UV telescope spectrometer (1150–2200 Å)	45, 46
Laboratoire de Physique Stellaire et Planetaire, CNRS	Unmanned satellite (0S0–8)	Solar UV telescope six-channel spectrometer (1025–3969 Å)	47, 48
NASA Marshall Space Flight Center; High-Altitude Observatory; University of Colorado; Goddard Space Flight Center	Unmanned satellite [Solar Maximum Mission (SMM)]	Solar ultraviolet telescope spectrometer and polarimeter (1170–3600 Å)	49, 50
Princeton University	Unmanned satellite (Third Orbiting Astronomical Observatory, Copernicus)	Nonsolar high-resolution telescope spectrometer (950–3000 Å)	51–53
National Aeronautics and Space Administration; European Space Agency; Space Research Council	Unmanned spacecraft [International Ultraviolet Explorer (IUE)]	Nonsolar echelle spectrometers (1150–3200 Å)	54
Masachusetts Institute of Technology	Unmanned spacecraft (High-Energy Astrophysical Observatory, HEAO–2, Einstein)	Nonsolar x-ray crystal spectrometer (4.1–62 Å)	55, 56

lines from ions formed in this region fall roughly between 300 and 4000 Å. It is well to keep in mind certain spectrographic instrumentation considerations. Because of the nature of the materials used in the construction of the instrumentation, and other similar considerations, the major instruments flown on space vehicles have recorded spectra in the wavelength windows of $\simeq 170-600$, $\simeq 300-1400$, $\simeq 1100-2000$, and $\simeq 2000-4000$ Å (see Table 1). Therefore, it is desirable to find lines that all fall in one or another of these windows. It is more difficult to apply diagnostics to spectra obtained from different instruments.

Only a few transition region lines fall in the 2000–4000 Å region. However, the Mg II h and k lines near 2800 Å due to the transitions $3s\,^2S_{1/2}$–$3p\,^2P_{1/2,3/2}$ are important diagnostics for both solar and stellar atmospheres. The Mg II lines are actually formed at the base of the transition region (in temperature space), a region that may be regarded as the upper chromosphere.

It is possible in certain instances to determine the temperature of the Mg II emitting region by considering two other Mg II lines, i.e., $3p\,^2P_{1/2}$–$3d\,^2D_{3/2}$ and $3p\,^2P_{3/2}$–$3d\,^2D_{5/2}$.[104] The upper levels $3d\,^2D_{3/2}$ and $3d\,^2D_{5/2}$ are not populated by excitation from the $3p$ levels, because at low densities, the $3p$ levels have a negligible population. Rather, the $3d$ levels are excited directly from the $3s$ levels. The $3s \rightarrow 3d$ excitation rates in Na-like ions have been calculated by Blaha,[105] and Flower and Nussbaumer.[106] The ratio of the $3p$–$3d$ lines to the $3s$–$3p$ lines is temperature sensitive because of the Boltzmann factors [see Eq. (3.23)]. The $3s$–$3d$ excitation energy is about twice the excitation energy of the $3s$–$3p$ lines, and the energy difference of about 35,800 cm^{-1} divided by kT_e is around 3.2, so there is considerable sensitivity in the line ratios to temperature. Furthermore, because the $3s$–$3p$ and $3p$–$3d$ transition energies are similar, the lines are close in wavelength, thereby minimizing instrument calibration difficulties (e.g., $^2S_{1/2}$–$^2P_{3/2}$ falls at 2795.523 Å, and $^2P_{3/2}$–$^2D_{5/2}$ falls at 2797.989 Å).

Figure 7 shows the temperature dependence of the $3p$–$3d$ to $3s$–$3p$ ratio R_C. Unfortunately, there are two difficulties associated with the Mg II ratio: the $3p$–$3d$ lines are quite weak compared to the $3s$–$3p$ lines, and radiative transfer in the $3s$–$3p$ lines is also important. This makes the ratio dependent on other factors, such as the electron pressure p_0 as well as on T_e. [The other two curves in Fig. 7 are the ratio R for two different values of p_0. The derivation of these curves is described in Feldman and Doschek[104].] So far, the ratio has been applied to observations only above the solar limb.[104] Results at heights of 8 and 12 Arc sec above quiet-sun (QS) and active regions (AR), respectively, are shown in Fig. 7. The deduced temperatures depend on the assumed electron pressure p_0 because of resonance scattering; but for the two quite plausible values of p_0 shown in Fig. 7, the deduced temperature is about

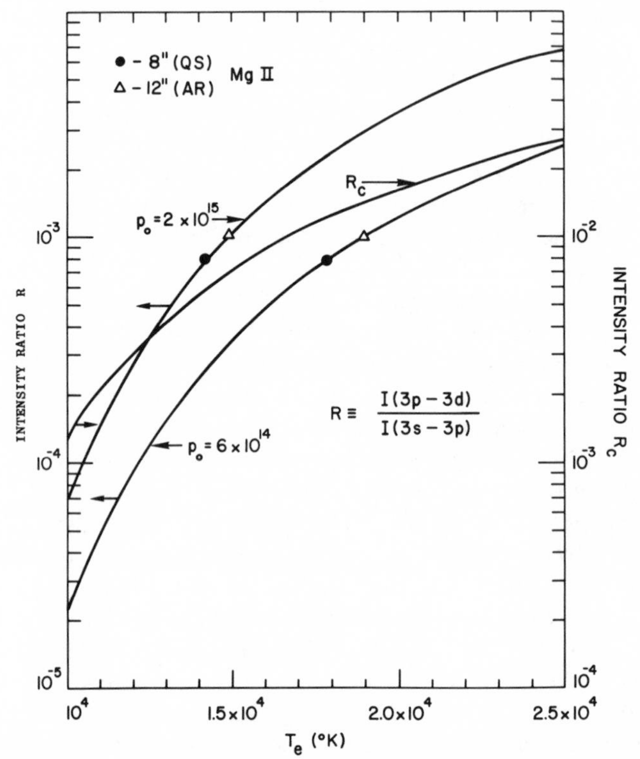

FIGURE 7. Temperature-sensitive line ratios of Mg II. The R_c is the ratio neglecting radiative transfer effects in the (3*s*–3*p*) lines. The R is the ratio at heights of 8–12 arc sec outside the limb and includes the effect of resonance scattering in the (3*s*–3*p*) lines (see text for other details). (Courtesy *Astrophys. J.*)

1.5×10^4 K, which is in good agreement with ionization equilibrium expectations.

The particular line ratios discussed for Mg II can be applied to higher temperature regions in the transition region by using more highly ionized ions of the Na I isoelectronic sequence. Atomic data for Si IV, S VI, Ca X, and FeXVI have been calculated by Flower and Nussbaumer[106] and compared to available observations. The theoretical temperature behavior for the Si IV and S VI lines is shown in Fig. 8. Also marked is T_M, the temperature of maximum ion abundance in ionization equilibrium. If the atmosphere is in equilibrium, the temperatures deduced for the line ratios should equal T_M. As can be seen, the line ratios are good temperature diagnostics. Although the wavelength separations are larger than for Mg II, resonance scattering in the $3s$–$3p$ lines is not important (in the solar case), and so the ratios depend on only temperature. Solar observations are available from Harvard College Observatory

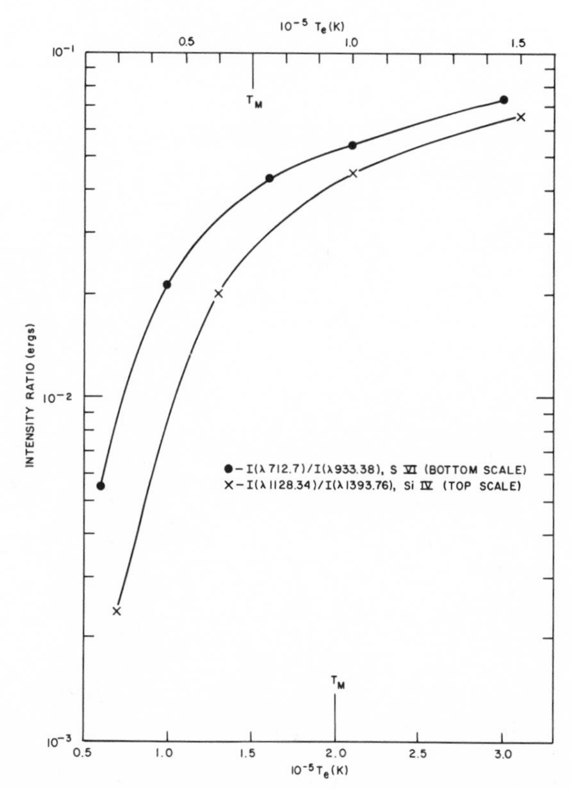

FIGURE 8. Theoretical temperature sensitive line ratios of Si IV and S VI. The sensitivity occurs primarily at temperatures lower than T_M, the temperature of maximum ion abundance in ionization equilibrium. Atomic data are from (Ref. 106.)

(HCO) spectrometers flown on various spacecraft and Skylab. Generally speaking, the observed ratios are too large to be explained by any reasonable temperature. However, the HCO spectrometers have poor spectral resolution, and the lines are weak. Blends may account for the discrepancies, e.g., the 1122 Å Si IV line is severely blended.[107] Nevertheless, it may be worthwhile to calculate the $3s \rightarrow 3d$ excitation rate more accurately. For these weak excitations, resonances may significantly enhance the rate coefficients.

There are many other temperature-sensitive line ratios applicable to the transition zone, particularly for temperatures between $\simeq 2 \times 10^4$ and 2×10^5 K. For example, most of the lines of ions such as O III, O IV, and O V, due to $2s^2 2p^k$–$2s^2 2p^{k+1}$ electric dipole transitions, fall between 500 and 650 Å, while intercombination transitions (i.e., transitions involving a change in spin) between these same configurations occur between 1100 and 2000 Å. Although many of these lines cannot be observed with a single instrument, it is possible

TABLE 2
Temperature-Sensitive Line Ratios[a]

Ratio	Observed	Theoretical
1218.35/629.73	0.16	0.094
1218.35/760[b]	2.1	1.4
1401.16/554	0.16	0.090
1401.16/790	0.20	0.095
1666.15/835	0.26 (blend)	0.15
1666.15/703	0.67	0.22
1666.15/525.80	1.5	2.3
1406.00/750	0.38	0.16
1406.00/661.42	0.64	0.33
1062.67/661.42	0.83	0.25

[a] Intensities measured in ergs (not photons).
[b] Wavelengths without decimal fractions refer to the intensity of the entire multiplet.

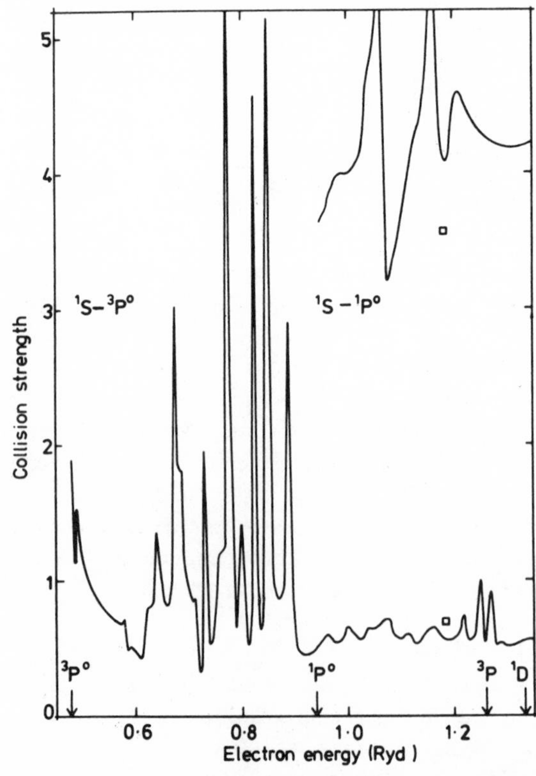

FIGURE 9. Resonance fine structure in the collision strength for the indicated C III transition. The resonances substantially increase the excitation rate coefficient. Data are from Ref. (109.)

to intercompare some of the data obtained with the HCO Skylab spectrometer (300–1400 Å) with the NRL Skylab spectra (1150–2000 Å). Table 2 shows some results.[108] The atomic data used in these calculations come from a variety of sources, and the results, particularly for the intercombination lines, depend strongly on resonances and hence on autoionization phenomena. For example, resonances in the cross sections [Eq. (3.12)] for the C III 1909 Å, O V 1218 Å, O III 1666 Å, and S IV 1406 Å intercombination lines have been calculated[109–112] (e.g., see Figs. 9–11). These resonances increase the excitation rate coefficients by factors between two and five. The results in Table 2 were obtained under the assumptions of ionization equilibrium and an atmospheric model. Note that in general, the theoretical ratios are smaller than the observed ones. It is possible that this discrepancy is at least partly due to Lyman continuum absorption at wavelengths less than 912 Å. In order to resolve this issue, which is quite important from the point of view of solar physics, it is essential that atomic data of the highest quality be available.

So far, there are very few solar observations of high quality available for transition region lines of the type $2s^2 2p^k - 2s^2 2p^{k-1} 3\ell$. The ratios of these lines to longer wavelength lines are extremely sensitive to temperature, and observations are desirable in spite of possible instrumental calibration difficulties.

Since many transition region ions are formed over a narrow temperature

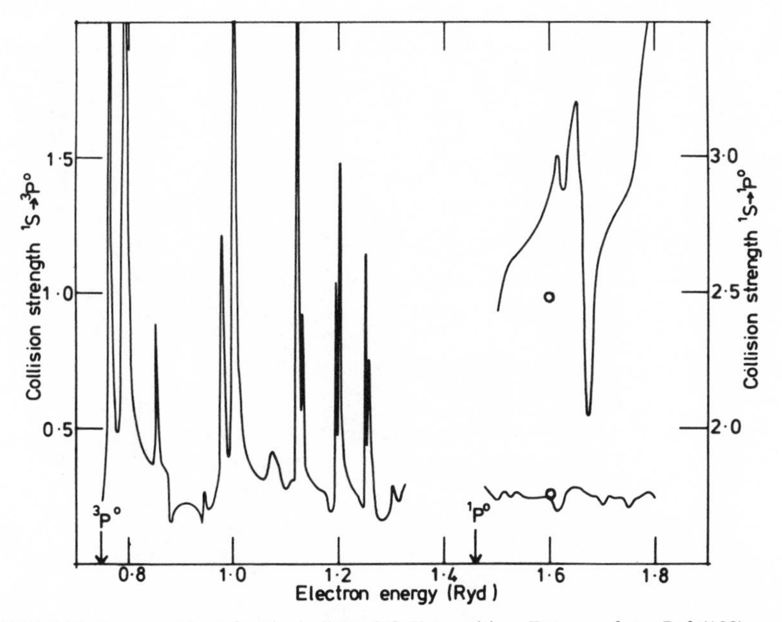

FIGURE 10. Same as Fig. 8 for the indicated O V transition. Data are from Ref. (109).

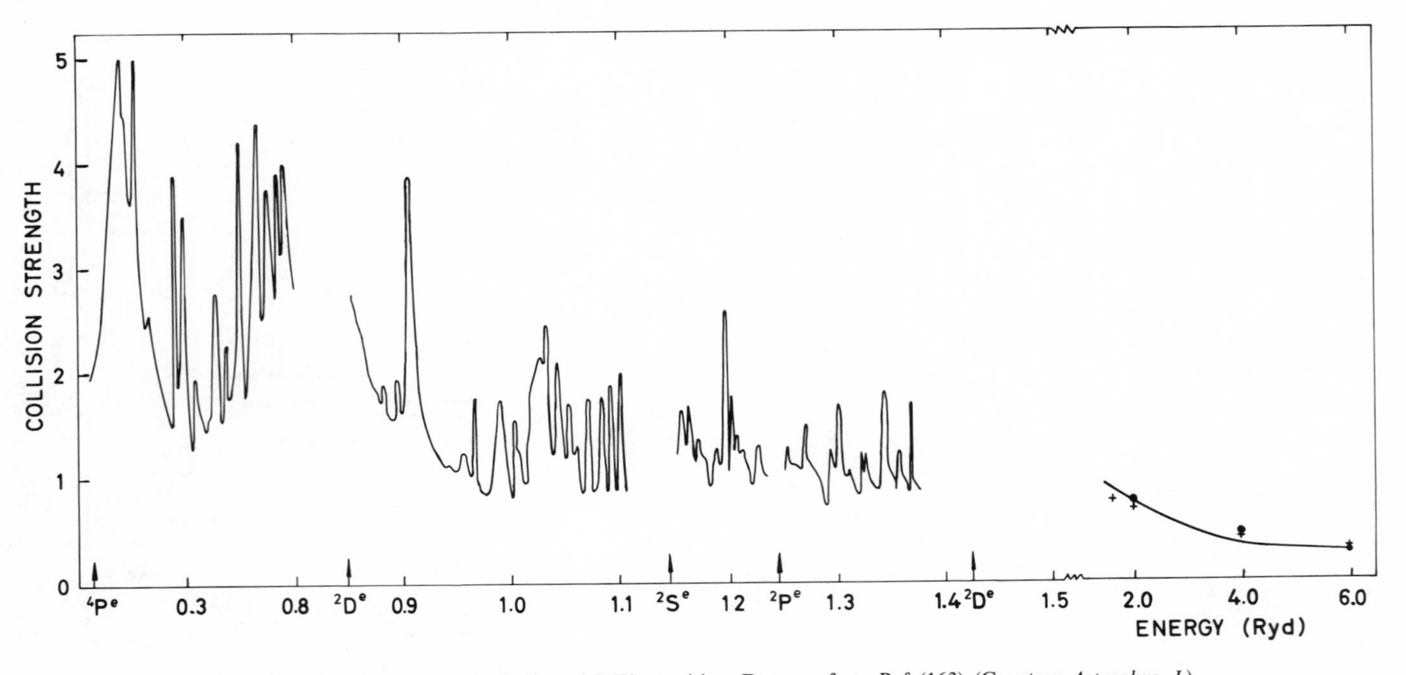

FIGURE 11. Same as Fig. 9 for the indicated S IV transition. Data are from Ref. (163). (Courtesy *Astrophys. J.*)

range, it may be questioned why temperature diagnostics are desirable. The reason is that in some atmospheric models, the temperature gradient in the transition region is quite steep. Under this condition, a small flow velocity of ions through the transition region, or *in situ* heating and cooling, or ion diffusion, or high-temperature particles propagating into low-temperature regions, might invalidate the assumption of ionization equilibrium and alter the line ratios from their expected equilibrium values. Although most solar observations appear to be consistent with the assumption of ionization equilibrium, at least in the lower transition region, this assumption may break down in the atmospheres of some stars, particularly those stars with large stellar winds. These points are discussed further in Section 4.

3.2.3. Solar Flares. As mentioned in Section 3.2.1, lines produced by dielectronic recombination can be used to determine electron temperature in solar flare plasmas [see Eq. (3.25)]. Recently, very high resolution solar flare spectra have become available,(7,10,14,15) as well as high-resolution tokamak spectra.(113–116) Two spectra of highly ionized iron lines are shown in Fig. 12,

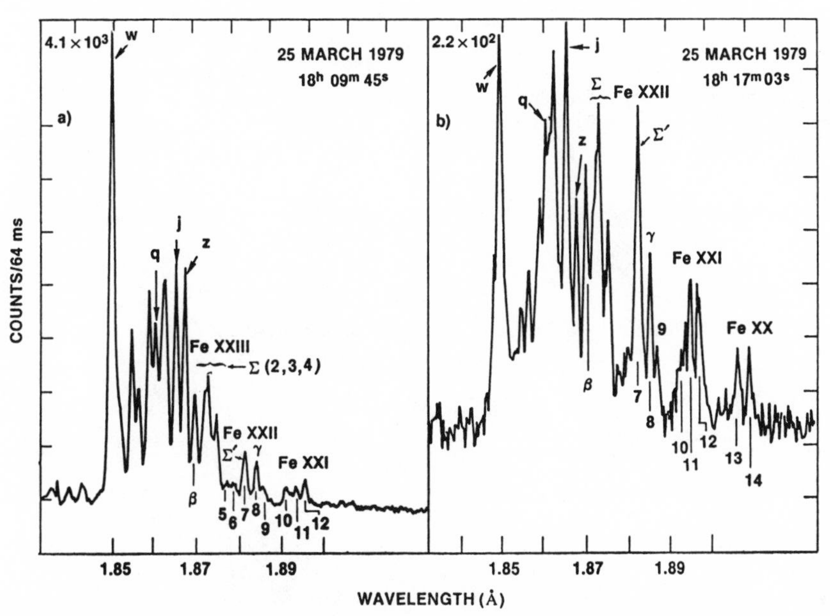

FIGURE 12. Iron line spectra recorded by an NRL spectrometer on the P78-1 spacecraft. The left-most spectrum was recorded near the flare maximum; the right-most spectrum was recorded during the decay phase. The numbers in the upper-left corners of each spectrum represent fullscale count rates for each spectrum. Most of the lines are formed by dielectronic recombination. The temperature determined from the j/w ratio is about 19×10^6 K for the left-most spectrum and about 13×10^6 K for the right-most spectrum (see Fig. 18).

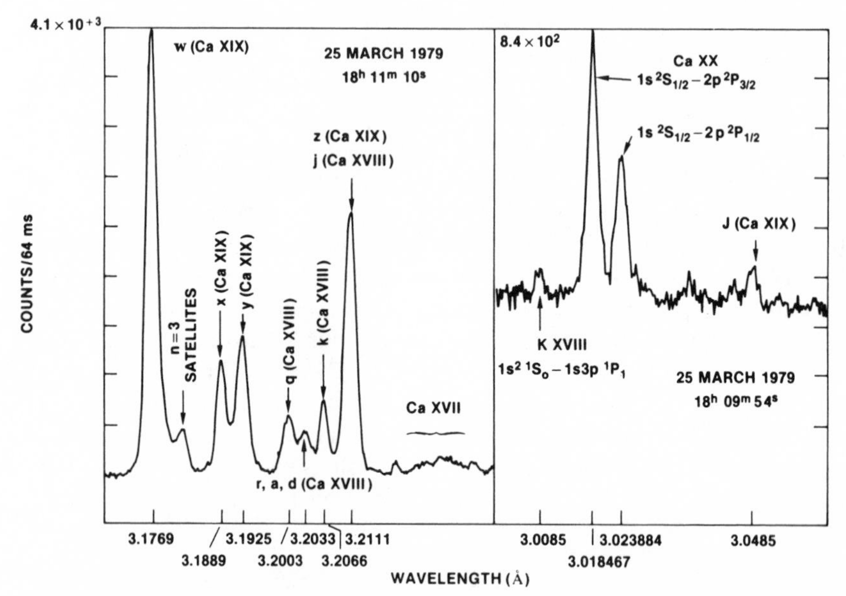

FIGURE 13. The NRL P78-1 x-ray spectra of calcium lines. The ratio k/w gives the electron temperature. Also shown are the two-fine structure transitions of Ca XX Lyman α and the dielectronic satellite line J. The shapes of the spectral lines are intrinsic, i.e., not due to instrumental resolving power or flare source size.

TABLE 3
Observed He-Like and Li-Like Calcium and Iron Lines

Key	Identification	Calcium λ(Å)[a]	Iron λ(Å)
w	$1s^2\,{}^1S_0$–$1s2p\,{}^1P_1$	3.1769	1.84992
x	$1s^2\,{}^1S_0$–$1s2p\,{}^3P_2$	3.1889	1.85519
y	$1s^2\,{}^1S_0$–$1s2p\,{}^3P_1$	3.1925	1.85947
z	$1s^2\,{}^1S_0$–$1s2s\,{}^3S_1$	3.2111	1.86801
$n=3$	$1s^23l$–$1s2p3l$	3.1822	
t	$1s^22s\,{}^2S_{1/2}$–$1s2p({}^3P)2s\,{}^2P_{1/2}$		1.8568
q	$1s^22s\,{}^2S_{1/2}$–$1s2p({}^1P)2s\,{}^2P_{3/2}$	3.2003	1.8610
r	$1s^22s\,{}^2S_{1/2}$–$1s2p({}^1P)2s\,{}^2P_{1/2}$	3.2033	1.8631
k	$1s^22p\,{}^2P_{1/2}$–$1s2p^2\,{}^2D_{3/2}$	3.2066	1.8631
j	$1s^22p\,{}^2P_{3/2}$–$1s2p^2\,{}^2D_{5/2}$	3.2111	1.8660

[a]See Ref. (124) for a discussion of the wavelengths.

and spectra of calcium are shown in Fig. 13. At temperatures in excess of 10^7 K, such elements as calcium and ion are very highly stripped, and the He-like iron is quite abundant. The lines of He-like iron and calcium in the spectra are listed in Table 3; the notation *w*, *x*, *y*, and *z* was introduced by Gabriel.(100)

All of the He-like lines are produced by electron impact excitation from the ground state $1s^2\,{}^1S_0$. However, most of the other lines shown in Figs. 12 and 13 are produced mainly by dielectronic recombination of He-like and lower-ionization-stage ions.(100,117,118) These so called satellite lines of lower ion stages are much stronger relative to the He-like lines for iron than for calcium. This is not entirely due to the fact that the abundances of ion stages lower than He-like are greater for iron than for calcium. To see this, consider

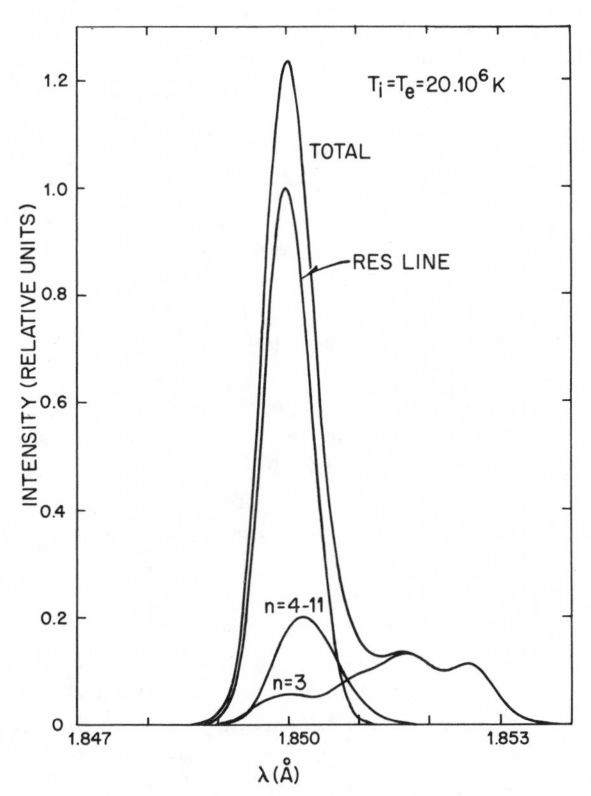

FIGURE 14. Theoretical calculations showing the effect of Fe XXIV dielectronic satellite lines due to transitions of the type $1s^2\,nl$–$1s2p\,nl$ on the total spectral intensity near 1.8500 Å, the wavelength of the Fe XXV resonance line (n, l = principal quantum number, orbital angular momentum). The calculation is from Ref. (120).

the ratio of the transitions discussed in Section 3.2.1, i.e.,

$$\frac{1s^2 2p\,^2P_{3/2} - 1s2p^2\,^2D_{5/2}}{1s^2\,^1S_0 - 1s2p\,^1P_1}$$

In the notation in Fig. 12, this is the line ratio j/w. An equivalent ratio for calcium (Fig. 13) is k/w, i.e, line k is $1s^2 2p\,^2P_{1/2} - 1s2p^2\,^2D_{3/2}$. (Line j of calcium is blended with line z.) From Eq. (3.26), these ratios are proportional to

$$B \equiv \frac{A^a A}{A^a + A} \tag{3.27}$$

where the subscripts are dropped for simplicity and where, for these particular transitions, the sums in the denominator are equal or nearly equal to the corresponding values in the numerator. For low Z ions, such as O VII, $A^a \gg A$ and $B \propto A$. Since A scales roughly as Z^4, and A^a is approximately constant with Z, the intensities of the Li-like satellite lines increase as Z^4 relative to the He-like lines. The satellite lines are so weak for O VII that they are not useful temperature diagnostics. For calcium, A^a is only slightly larger than A, and the Z^4 scaling is no longer valid. The satellite lines are sufficiently strong to be a useful temperature diagnostic for Mg XI and all heavier He-like ions. The existence of the satellite lines is very important for analyzing flare and active-region spectra, since the He-like ions are formed over extremely broad temperature ranges in ionization equilibrium, and it is not possible to assume a temperature simply because a particular He-like line is present.

The theory of the Li-like satellite lines, due to $1s$–$2p$-type transitions, their wavelengths, and intensities, is now well understood.[119–122] In addition to the $1s$–$2p$-type lines, transitions of the type $1s^2 3l - 1s2p3l$ also appear.[119,123–125] Figure 14 shows some calculations for these lines for iron, and Fig. 15 shows some of them resolved in tokamak iron spectra.[126] (Solar flare spectra appear to have lower spectral resolution than tokamak spectra, because the flare lines are considerably broadened by turbulence, while the tokamak lines are not.) Also, transitions of the type $1s^2 2l - 1s2l3p$ have been observed in solar and laser-produced plasmas (Fig. 16). These lines fall near the $1s^2\,^1S_0 - 1s3p\,^1P_1$ lines of He-like ions. In addition, there are transitions in Li-like ions from higher-principal quantum numbers, e.g., $1s^2 nl - 1s2pnl$. These lines eventually merge in wavelength with the resonance line $1s^2\,^1S_0 - 1s2p\,^1P_1$. The contribution of these lines to the w feature has been calculated by Bely-Dubau *et al.*,[120] and observed spectra can be corrected for this effect. The strongest lines of $1s$–$2p$ transitions in Li-like Ca XVIII and Fe XXIV are given in Table 3.

So far, we have considered only He-like and Li-like lines in Figs. 12 and 13; however, attempts to understand the remaining lines due to lower ionization stages have begun,[117,118,127] particularly for iron, where the lines are

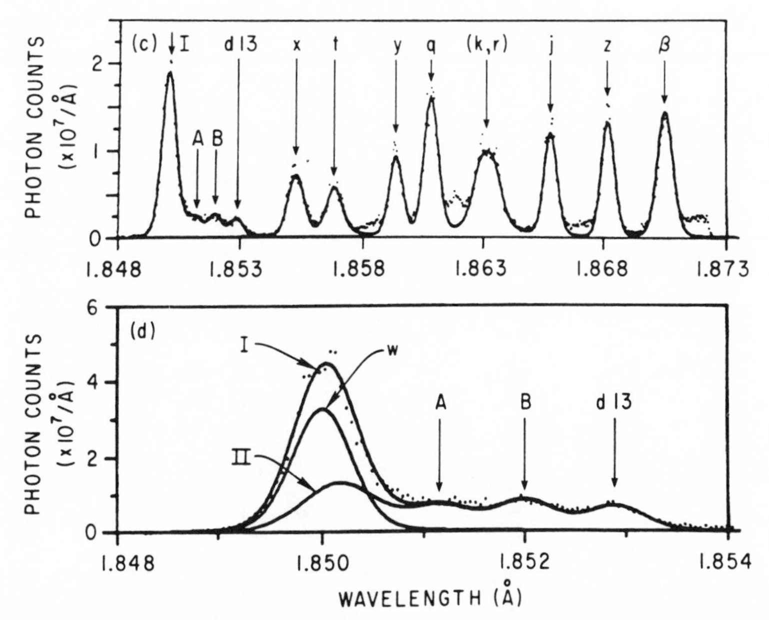

FIGURE 15. Tokamak spectra of Fe XXIV and Fe XXV lines. The lower panel shows an expanded version of the region around the Fe XXV resonance line. The effect of the $1s^2 3l$–$1s2p3l$ lines on the total intensity at the wavelength (1.8500 Å) of the resonance line can be clearly seen. Lines in the spectra are better resolved than in flare spectra because turbulence in the tokamak plasma is much less than in the flare plasma. Data are from Ref. (126.)

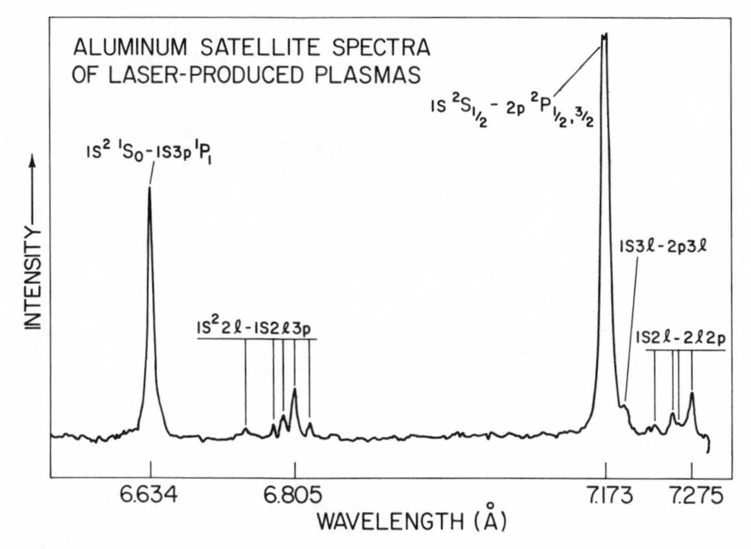

FIGURE 16. X-ray spectra of an aluminum laser-produced plasma obtained using the glass laser facility at NRL. (Courtesy *Astrophys. J.*)

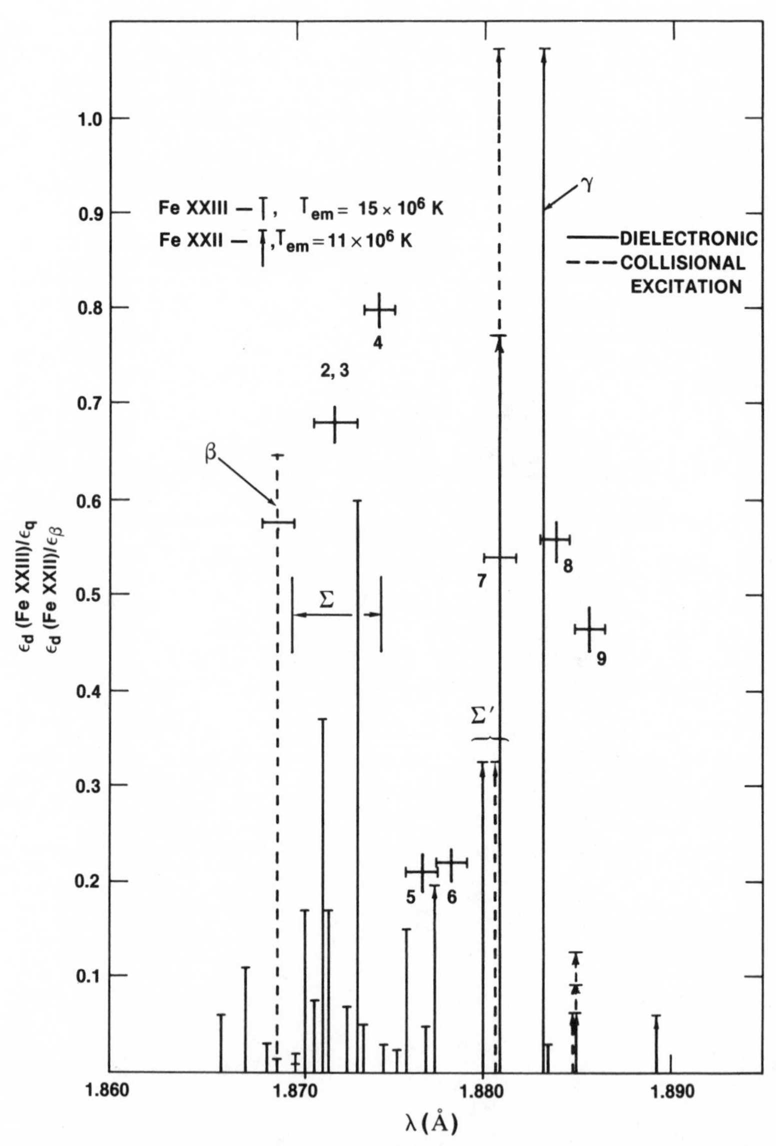

FIGURE 17. Calculations of wavelengths and intensities of the strongest x-ray lines of Fe XXII and Fe XXIII. Calculations are from Refs.(128–130). Features marked with Greek letters are shown in Fig. 12. A complete discussion of the figure is given in Ref. (117). (Courtesy *Astrophys. J.*)

quite strong and can provide useful temperature diagnostics. Figure 17 shows the region between 1.860 and 1.890 Å. In this region, *ab initio* wavelength and line intensity calculations by Cowan[117,128,129] show that most of the emission is due to dielectronic recombination lines of Fe XXIII and Fe XXII, formed by dielectronic capture onto Fe XXIV and Fe XXIII, respectively. Note that several lines blend together to form the feature marked Σ in Figs. 12 and 17, but the feature γ appears to be due to a single dielectronic line. The lines shown schematically in Fig. 17 are all due to $1s$–$2p$-type transitions. Transitions from higher quantum numbers are weaker, and calculations indicate that they do not severely influence interpretation of the spectra.[130] In general, the theoretically predicted spectrum for all of the most prominent features in Fig. 12 agrees fairly well with the observed spectrum.[118] Finally, we note that a rigorous theory of autoionization is discussed by Temkin and Bhatia[131] in this volume.

Although most of the emission in Fig. 12 from Fe XXIV and lower ionization stages is produced by dielectronic recombination, there are a few lines produced primarily by inner-shell excitation. A few of the prominent lines are

$$q[1s^2 2s^2 S_{1/2} - 1s(2s2p\,^3P)^2 P_{3/2}], \qquad \beta(1s^2 2s^2\,^1S_0 - 1s2s^2 2p^1 P_1)$$

and part of the feature marked Σ' is produced by inner-shell excitation of Fe XXII. These lines can be combined with dielectronic lines and used as

TABLE 4
Major Features in Iron Line Spectra Due to Fe XX–Fe XXIII[a]

Key	λ(Å)	Main contributor	Ion
β	1.8704	$1s^2 2s^2\,^1S_0$–$1s2s^2 2p\,^1P_1$	Fe XXIII
Σ(2)	1.8729	$1s^2 2s2p$–$1s2s2p^2$	Fe XXIII
Σ(3)	1.8735		
Σ(4)	1.8754		
5	1.8779		
6	1.8794	$1s^2 2s^2 2p\,^2P_{3/2}$–$1s2s^2 2p^2\,^2S_{1/2}$	Fe XXII
Σ'(7)	1.8824	$1s^2 2s^2 2p$–$1s2s^2 2p^2$	Fe XXII
γ(8)	1.8851	$1s^2 2s^2 2p^2\,^2P_{3/2}$–$1s2s^2 2p^2\,^2D_{5/2}$	Fe XXII
9	1.8867	$1s^2 2s^2 2p\,^2P_{3/2}$–$1s2s^2 2p^2\,^2D_{3/2}$	Fe XXII
10	1.8916	$1s^2 2s^2 2p\,^2P_{3/2}$–$1s2s^2 2p^2\,^4P_{5/2}$	Fe XXII
		$1s^2 2p\,^2P_{1/2}$–$1s2s^2\,^2S_{1/2}$	Fe XXIV
		$1s^2 2s^2 2p^2$–$1s2s^2 2p^3$	Fe XXI
11	1.8942	$1s^2 2s^2 2p^2$–$1s2s^2 2p^3$	Fe XXI
12	1.8966	$1s^2 2s^2 2p^2$–$1s2s^2 2p^3$	Fe XXI
		$1s^2 2p\,^2P_{3/2}$–$1s2s^2\,^2S_{1/2}$	Fe XXIV
13	1.9051	$1s^2 2s^2 2p^3$–$1s2s^2 2p^4$	Fe XX
14	1.9075	$1s^2 2s^2 2p^3$–$1s2s^2 2p^4$	Fe XX

[a] The features from β through 9 are shown in Fig. 17.

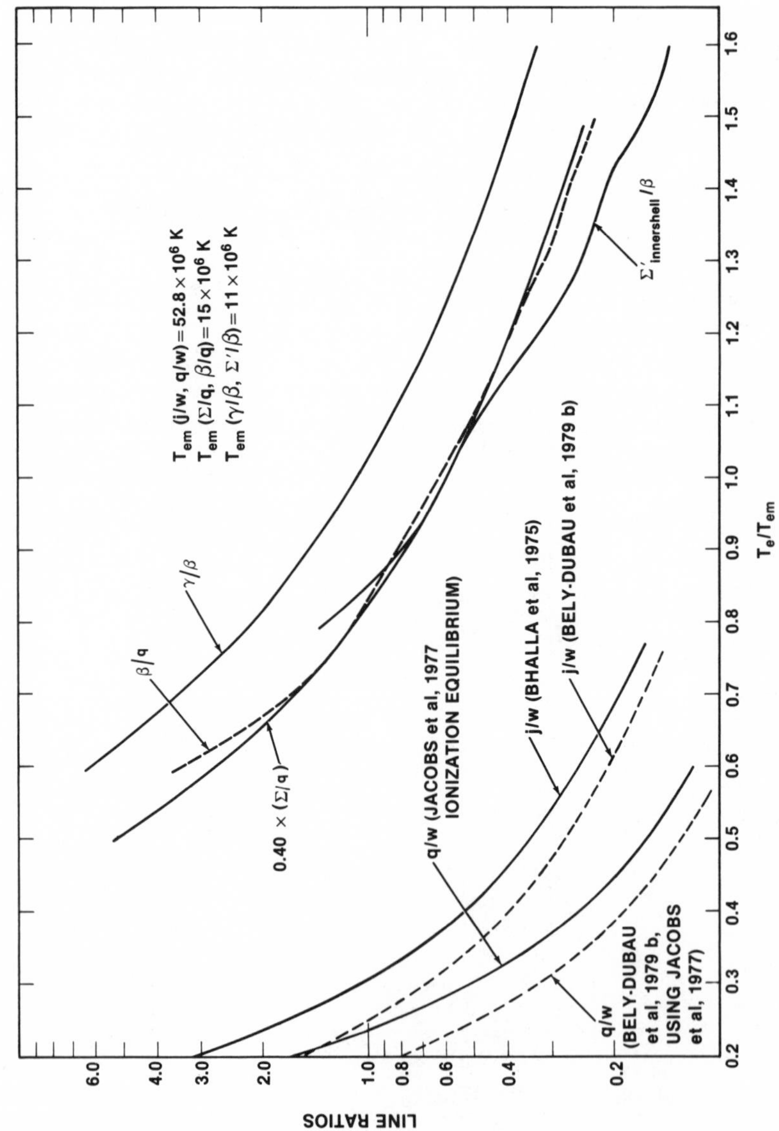

FIGURE 18. Theoretical calculations of intensity ratios as a function of temperature for the transitions indicated. See Ref. (117) for details. (Courtesy *Astrophys. J.*)

temperature diagnostics, completely analogous to the j/w and k/w ratios.[117] Additional possible temperature-sensitive ratios are Σ/q and γ/β (see Figs. 12 and 17). The behavior with temperature of the ratios we are discussing is shown in Fig. 18. An abbreviated list of the observed features of Fe XX–Fe XXIII is given in Table 4.

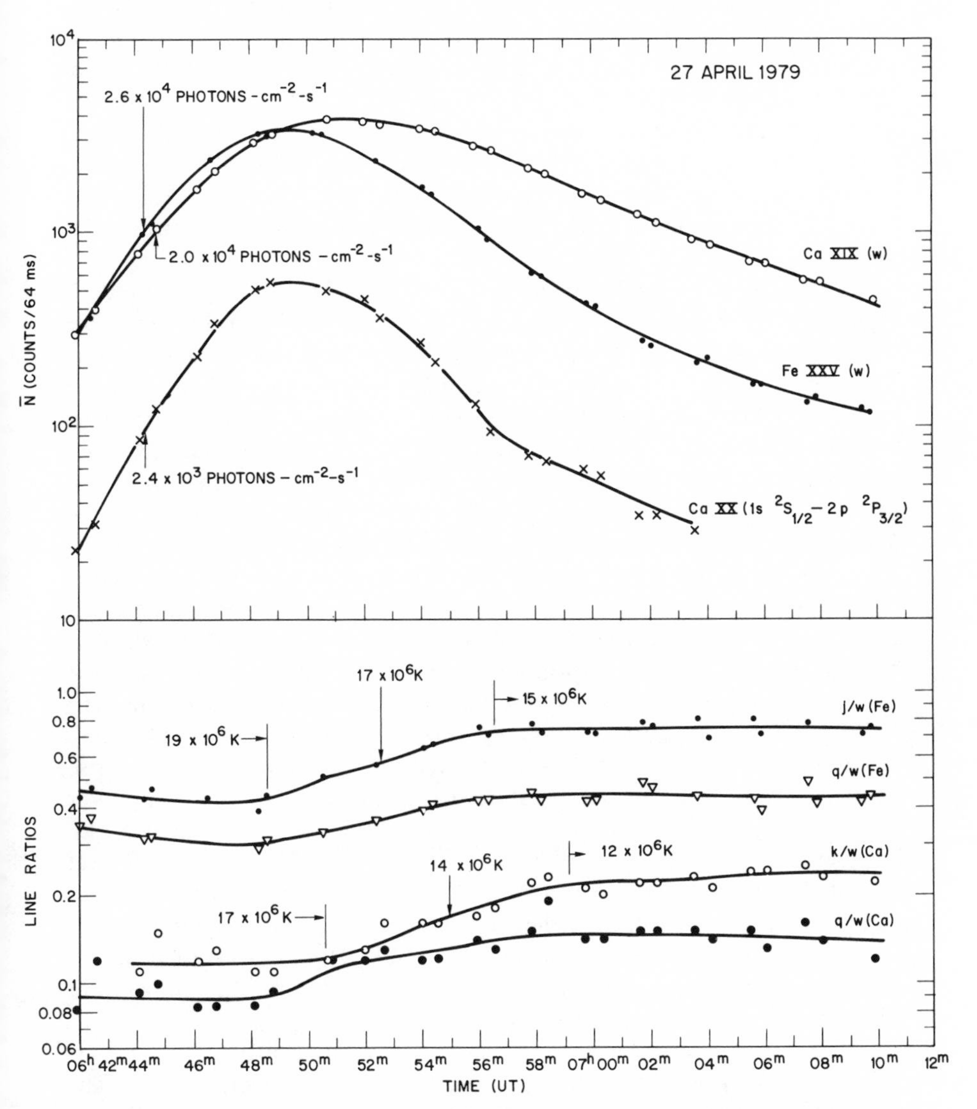

FIGURE 19. Light curves and intensity ratios for the x-ray lines indicated, for a large flare that occurred on 27 April 1979. The behavior of the quantities shown is typical of large flares. (Courtesy *Astrophys. J.*)

In addition to the spectra of Ca XVIII and Ca XIX, Fig. 13 shows spectra of the Ca XX Lyman-α lines near 3.0 Å, i.e., $1s\,^2S_{1/2}$–$2p\,^2P_{1/2,3/2}$. There are also satellite lines associated with these lines due to transitions of the type $1s2l$–$2p2l$. The strongest of these lines, called J, is due to the transition $1s2p\,^1P_1$–$2p^2\,^1D_2$ and appears in P78-1 flare spectra at 3.0485 Å.[124,132] The ratio J/Lyman-α is a temperature indicator analogous to the j/w ratio, and the theory for the formation of the Ca XIX satellites has been developed and applied to observations.[102,132] Extensive work has also been done on the satellite lines of Mg XII and Fe XXVI.[102,133–135] The Fe XXVI lines were observed by SMM (Solar Maximum Mission) spectrometers.[135] We stress that the dielectronic temperature indicator is independent of ionization equilibrium, which in itself is important to flare researchers, since there is no completely compelling *a priori* reason for assuming that ionization equilibrium must be valid in flares.

Figure 19 shows results for a large flare that occurred on 27 April 1979.[85] Shown are the fluxes in line w for Ca XIX and Fe XXV and the flux in the strong component of the Lyman-α line, as functions of time during the flare. Also shown are the ratios j/w for iron and k/w for calcium and temperatures deduced over certain time intervals. Several conclusions have been obtained from these and similar results for other flares.[8,85] Three of them are: (1) Flare heating is extremely rapid; by the time line intensities are strong enough to be observed, the temperature is already on the order of 16×10^6 K; (2) the temperature seldom exceeds $\simeq 20 \times 10^6$ K, and frequently this is the highest temperature reached *by the bulk* of the thermal plasma during large flares; (3) the temperature behavior is quite variable after peak flux in the lines is attained. The temperature in some flares decreases monotonically, while in other flares, such as those shown in Fig. 19, the temperature drops by about 5×10^6 K and then remains constant for time periods on the order of an hour or longer, indicating continuous energy input well into the decay phase.

Figure 20 shows results for another flare.[8] This flare had a rapid rise phase ($\simeq 2$ min), and the temperature declined rather monotomically after peak intensity was reached. Also shown in Fig. 20 is the flux in the Fe XXIV line due to the transition $1s^22p\,^2P_{3/2}$–$1s^24d\,^2D_{5/2}$ at 8.317 Å. The upper level is excited from $1s^22s\,^2S_{1/2}$, and for Li-like ions, this excitation rate is comparable to the $2s \rightarrow 4p$ rate.[136] The ratio of the 8 Å line to line q is temperature sensitive because of the Boltzmann factor (Fig. 21). The temperature behavior predicted from observed $8.317/q$ ratios agrees well with the temperature behavior deduced from the dielectronic lines.[8]

The fact that lines such as q and β are produced by electron impact excitation provides an important diagnostic in addition to temperature.[100] The ratio of line q to line w, for example, depends on the ratio R of ion populations N(Fe XXIV)/N(Fe XXV). In ionization equilibrium, this ratio depends primarily

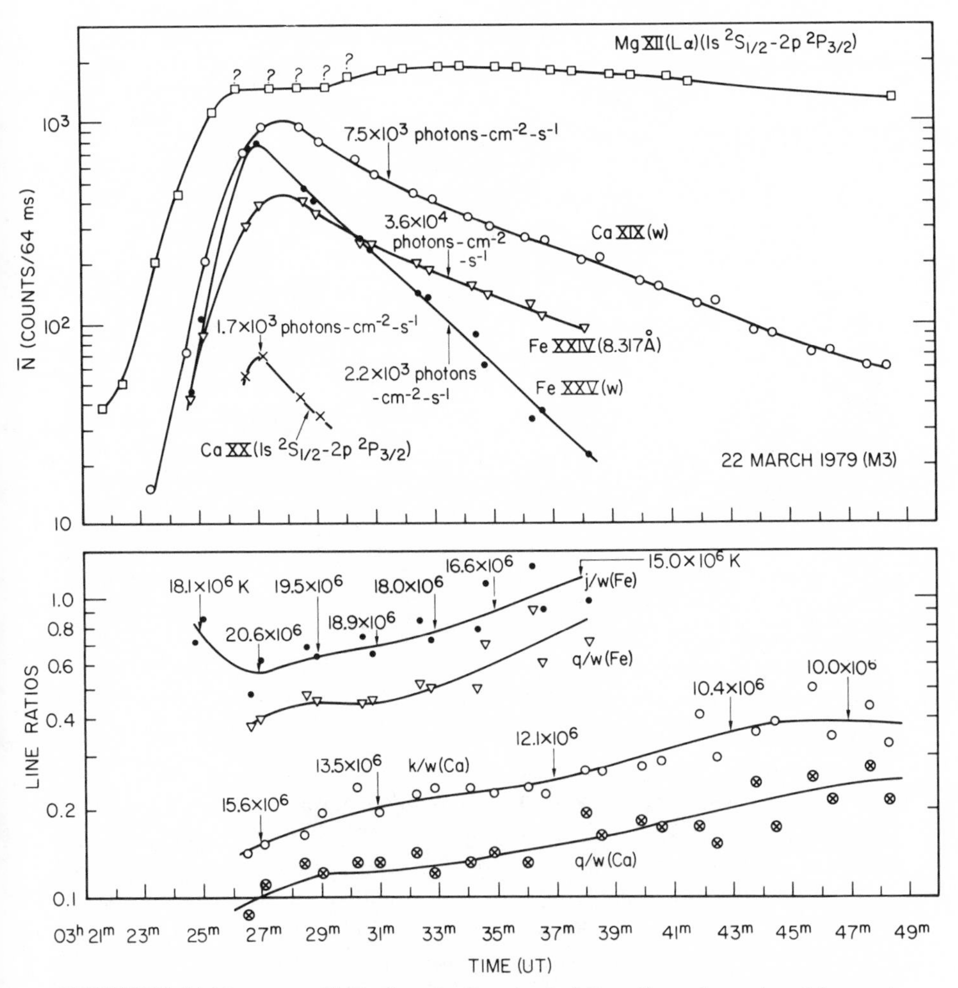

FIGURE 20. Light curves and intensity ratios for a somewhat smaller and more impulsive event than depicted in Fig. 19. These data are very important for understanding energy input and transport in the soft x-ray flare plasma. (Courtesy *Astrophys. J.*)

on electron temperature. Since the temperature is measured using the j/w or k/w ratios, the ratio R expected in ionization equilibrium at the measured temperature can be calculated, and from this ratio, the q/w intensity ratio can be calculated and then compared to observation. If the measured q/w ratio is different from the calculated q/w ratio, then nonequilibrium may exist in the plasma. If the measured q/w ratio is larger than predicted in equilibrium, then transient ionization is implied. Conversely, if the measured ratio is smaller

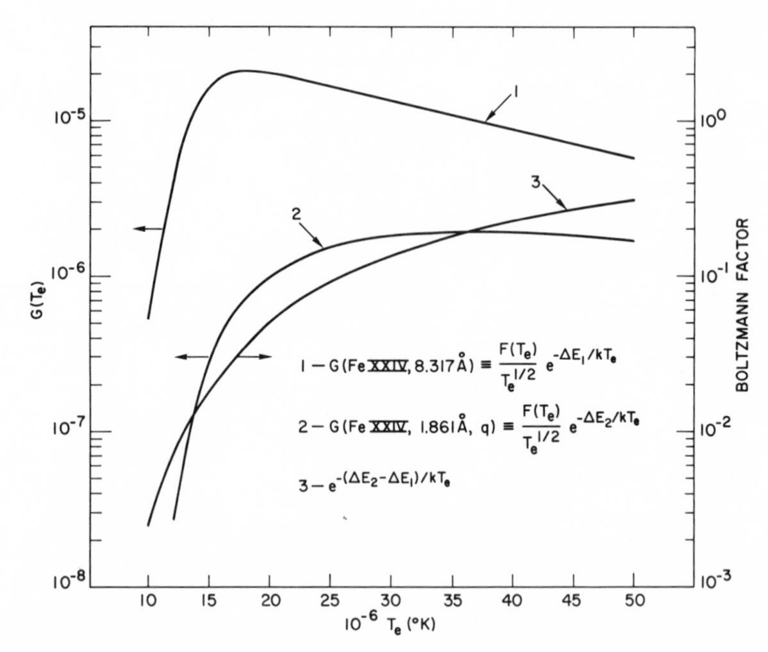

FIGURE 21. Contribution functions and Boltzmann factors for the indicated x-ray lines. There is considerable temperature sensitivity in these quantities at the low temperatures quite common in flares. (Courtesy *Astrophys. J.*)

than the calculated ratio, then transient recombination is implied.

The measured q/w ratios in flare spectra for both iron and calcium are always larger than expected in equilibrium.(85,124,137) These ratios are also shown in Figs. 19 and 20. Since the enhancement is almost always the same, and since it persists even into the decay phase of flares when cooling is expected, it appears that a transient ionization explanation is in fact unlikely. Rather, the result may reflect uncertainties in the ionization equilibrium calculations. Although flare plasmas are not isothermal (i.e. the k/w and j/w temperatures are slightly different in Figs. 19 and 20 because the calcium and iron lines arise in different regions at different temperatures), it does not appear that the q/w enhancement can be a result of the multithermal character of the source. Finally, we note that the q/w ratio is much larger in tokamak spectra than in flare spectra (Fig. 22), and in this case, ion circulation is a likely explanation for the enhancement.

Analysis of high-resolution flare spectra is continuing, and it is helping solar physicists understand better the energy transport processes in flaring flux tubes. Spectroscopic diagnostics that depend heavily on autoionization through dielectronic recombination are important aids in understanding high-energy phenomena in the solar atmosphere. In addition to the review by

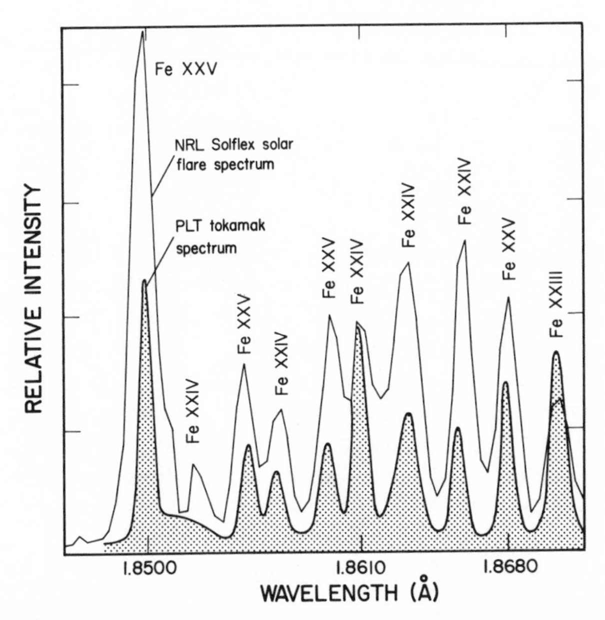

FIGURE 22. Solar flare and tokamak x-ray spectra compared. The PLT spectrum is from Ref. (113). Note the difference in the q/w ratio in the flare and tokamak spectrum. (Line q is at 1.8610 Å and line w is at 1.8500 Å.) The j/w ratios in the two spectra are about the same, implying nearly equal electron temperatures in the two plasmas. (Line j is at 1.8660 Å.)

Seaton and Storey[60] mentioned earlier, a more recent review of dielectronic recombination has been published by Dubau and Volonté.[138]

Before leaving this section, it is worth remarking that dielectronic recombination has also been an important line-forming mechanism in low-density and low-temperature astrophysical plasmas, such as planetary nebulas.[62] In these nebulas, typically $T_e \simeq 10^4$ K and $N_e \simeq 10^{3-4}\,\mathrm{cm}^{-3}$. Ions, such as C IV, are formed by photoionization from the ultraviolet continuum of a star embedded in the nebula. In the case of planetary nebulas, the star is centrally located in the nebula and the source of the gas. Storey[62] found that dielectronic recombination exceeds radiative recombination by factors of two to four, which significantly affects ionization equilibrium calculations and in addition can produce certain spectral lines, such as the C III line at 2297 Å due to the transition $2s2p\,{}^1P_1$–$2p^2\,{}^1D_2$. At low nebular temperatures and densities, the bound $2p^2\,{}^1D_2$ state is excited much more effectively by dielectronic capture followed by radiative cascade than by direct collisional excitation, i.e., the process could proceed as follows:

$$C^{+3}(2s\,{}^2S_{1/2}) + e \rightleftharpoons C^{+2}(2p4d\,{}^1F^0)$$
$$C^{+2}(2p4d\,{}^1F^0) \rightarrow C^{+2}(2p^2\,{}^1D_2^e) + h\nu$$

These processes severely affect derived element abundances calculated from IUE spectra of planetary nebula[139] and underline the importance of having a thorough understanding of atomic processes when analyzing astrophysical spectra.

3.3. Electron Density Measurements

3.3.1. Methods. Line ratios can be found that are sensitive primarily to electron density. If the temperature of formation of the lines is also known, then the local electron pressure ($N_e T_e$) of the plasma can be calculated. If density-sensitive line pairs can be found over a range of temperatures, then the pressure throughout the plasma can, in principle, be determined, and this result is important from the hydrodynamical standpoint.

Density sensitivity arises because of the existence of metastable levels. The transition probabilities for magnetic dipole transitions, and transitions that involve a change in spin (intercombination lines), are much less than similar electric dipole transition probabilities that do not involve a spin change. We shall define electric dipole transitions without spin change as "allowed" transitions.

The principle behind using metastable levels to measure density is illustrated in Fig. 23; two hypothetical atoms are shown in the figure. On the left, the three-level case, level 3, is defined as the metastable level, and level 1 is the ground level. The transition $3 \rightarrow 1$ might typically be an intercombination line, while the $2 \rightarrow 1$ transition might be an allowed line. At all densities of interest, collisional excitation and deexcitation out of level 2 is negligible compared to the radiative decay A_{21}. At low densities, most of the ion is at the ground level, and levels 2 and 3 are excited by electron impact from level 1. Every upward excitation is followed immediately by a radiative decay. In this situation, the line ratio $(3 \rightarrow 1)/(2 \rightarrow 1)$ is roughly proportional to the ratio of excitation rates C^e_{13}/C^e_{12}. In many cases, this ratio is proportional to the statistical weights of the upper levels and is of order unity. When the density is increased, eventually collisional deexcitation from the metastable level 3 into either level 2 or level 1 becomes comparable to A_{31} and A_{32}, i.e., $N_e(C^d_{32} + C^d_{31}) \approx A_{31}$. This results in decreasing the $(3 \rightarrow 1)/(2 \rightarrow 1)$ ratio, because the excitation $1 \rightarrow 3$ can now possibly result in a $2 \rightarrow 1$ transition. At very high densities, where collisional excitation and deexcitation out of level 2 are larger than A_{21}, the relative populations of levels 3 and 2 are in the ratio of their statistical weights, and in this case, the $(3 \rightarrow 1)/(2 \rightarrow 1)$ ratio is roughly proportional to A_{31}/A_{21}, which is $\ll 1$ [see Eq. (3.4)]. A schematic representation of the line behavior with density is also shown in Fig. 23. An example similar to the three-level case can be found in C III, one of the ions of the beryllium isoelectronic sequence, for nebular densities. A term diagram is

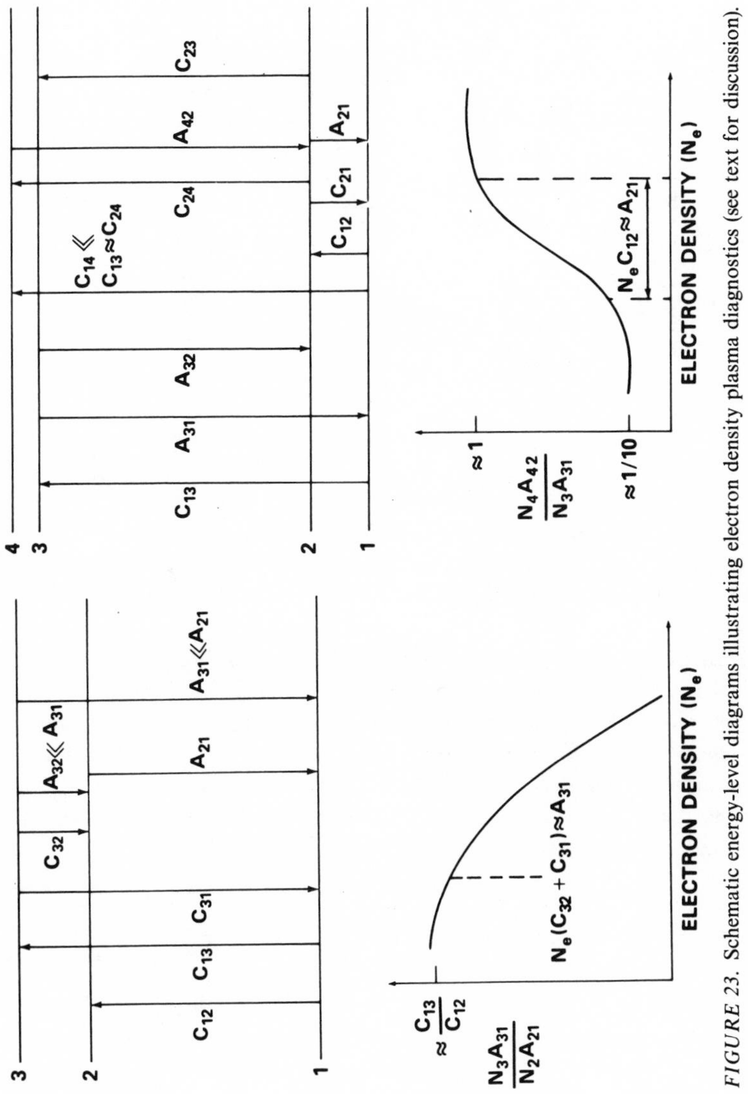

FIGURE 23. Schematic energy-level diagrams illustrating electron density plasma diagnostics (see text for discussion).

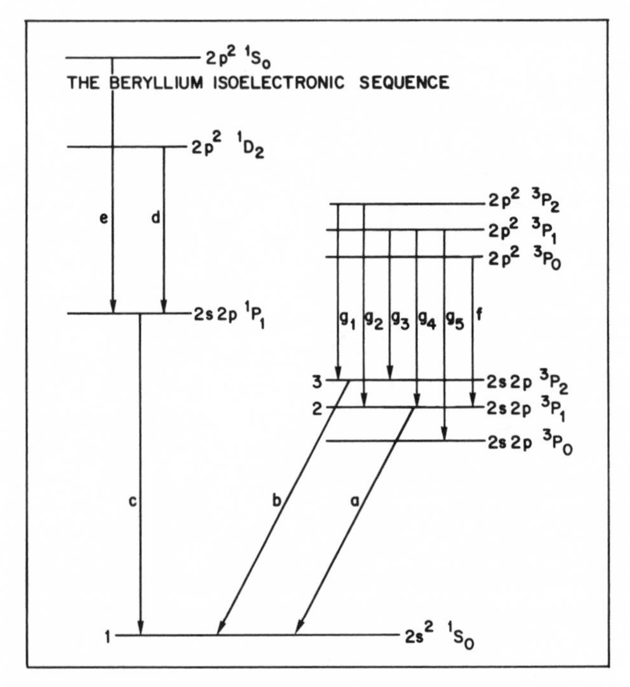

FIGURE 24. Energy-level diagram for the Be I isoelectronic sequence.

shown in Fig. 24 with the levels of Fig. 23 indicated. We shall return to the discussion of the C III lines in Section 3.3.2.

The discussion of the three-level case can be made more quantitative for densities of interest by considering Eqs. (3.4), (3.5), (3.7), and (3.15), and the approximations we have been assuming, i.e., $A_{21} \gg N_e C_{21}^d$ and $N_e C_{23}^e$, $A_{31} \ll A_{21}$, and also $\Delta E_{12} \simeq \Delta E_{13}$ and $\Delta E_{23}/kT_e \ll 1$. If the plasma is isothermal, we have

$$\frac{F_{31}}{F_{21}} = \frac{A_{31}}{A_{32}[1 + (C_{12}^e/C_{13}^e)] + A_{31}(1C_{12}^e/C_{13}^e) + N_e C_{32}^d[1 + (C_{12}^e/C_{13}^e)] + N_e(C_{12}^e C_{31}^d/C_{13}^e)} \tag{3.28}$$

For $N_e \to 0$,

$$\frac{F_{31}}{F_{21}} \to \frac{C_{13}^e}{C_{12}^e} \tag{3.29}$$

with use of the additional condition $A_{32} \ll A_{31}$. For N_e large such that A_{31} and A_{32} are much smaller than the collisional processes considered, but not large enough such that $A_{21} \lesssim N_e C^d_{21}$ and $N_e C^e_{23}$, Eq. (3.28) becomes

$$\frac{F_{31}}{F_{21}} = \frac{1.16 \times 10^5 A_{31} \omega_3 T_e^{1/2}}{N_e\{\bar{\Omega}_{23}[1 + (\omega_2/\omega_3)] + \bar{\Omega}_{12}\}} \tag{3.30}$$

Thus, F_{31}/F_{21} depends only weakly on the electron temperature and is inversely proportional to the density.

The hypothetical four-level atom depicted in Fig. 23 has the density-sensitive ratio $(4 \to 2)/(3 \to 1)$. In this case, level 2 is the metastable level. At low densities, only the ground level 1 is significantly populated. Level 3 is easily excited from level 1, e.g., the transition might be ${}^2P_{1/2} \to {}^2D_{3/2}$. However, the excitation rate from level 1 to level 4 is much less than $1 \to 3$, e.g., the transition might be ${}^2P_{1/2} \to {}^2D_{5/2}$. At low densities, the intensity ratio $(4 \to 2)/(3 \to 1)$ is proportional to C^e_{14}/C^e_{13}, which is typically about 1/10. At somewhat higher densities, collisional mixing between levels 1 and 2 results in level 2 attaining a non-negligible population. The $1 \to 2$ excitation may be due to protons as well as electrons. The excitation $2 \to 4$, which might be ${}^2P_{3/2} \to {}^2D_{5/2}$, is strong and comparable to, or even greater than, $1 \to 3$. Therefore, the ratio $(4 \to 2)/(3 \to 1)$ increases. At very high densities, levels 1 and 2 attain a nearly statistical population distribution, and the $(4 \to 2)/(3 \to 1)$ ratio is no longer sensitive to density. Note that in all situations of interest, i.e., sensitivity to density, we are assuming that collisional processes are unimportant relative to radiative decay in depopulating levels 3 and 4. The region of useful density sensitivity is roughly where $N_e C^e_{12} \simeq A_{21}$. The decay $2 \to 1$ is typically a magnetic dipole transition, and some of the forbidden lines that result are strong lines in low-density solar and tokamak plasma spectra.[32,140–142]

The four-level case just discussed can also be made more quantitative. Applying Eq. (3.5) to the four levels gives

$$\frac{n_2}{n_1} = \frac{C^e_{12} + C^e_{13}B_{32} + C^e_{14}B_{42}}{A_{21}/N_e + (C^d_{21} + C^e_{23}B_{31} + C^e_{24}B_{41})} \tag{3.31}$$

where

$$B_{ji} = \frac{A_{ji}}{A_{j1} + A_{j2}} \tag{3.32}$$

The flux ratio F_{42}/F_{13} is given by

$$\frac{F_{42}}{F_{13}} = \frac{(N_e n_1 C^e_{14} + N_e n_2 C^e_{24})B_{42}}{(N_e n_1 C^e_{13} + N_e n_2 C^e_{23})B_{31}} \tag{3.33}$$

or

$$\frac{F_{42}}{F_{13}} = \frac{B_{42}[C^e_{14} + (n_2/n_1)C^e_{24}]}{B_{31}[C^e_{13} + (n_2/n_1)C^e_{23}]} \tag{3.34}$$

For $N_e \to 0$,

$$\frac{F_{42}}{F_{13}} \to \frac{B_{42}C^e_{14}}{B_{31}C^e_{13}} \tag{3.35}$$

Since $A_{42} \gg A_{41}$, $B_{42} \simeq 1$, and in many cases, $A_{31} \gg A_{32}$ and $B_{31} \simeq 1$, therefore for $N_e \to 0$, $F_{42}/F_{13} \to C^e_{14}/C^e_{13}$. For N_e large, F_{42}/F_{13} becomes independent of N_e, since A_{21}/N_e becomes very small compared to the other term in the denominator of Eq. (3.31).

Most of the density-sensitive line ratios fall into one or the other of the two rather general systems outlined in Fig. 23. Ideally, levels 2 and 3 are close in energy compared to their separation from level 1 in the three-level example, and levels 3 and 4 and levels 1 and 2 are close in energy compared to, say, the $2 \to 4$ energy separation in the four-level example. If these conditions are met, the ratios discussed are primarily sensitive to density, and not too sensitive to temperature. If these conditions are not met, e.g., if the $1 \to 2$ energy separation in the three-level example is much smaller than the $1 \to 3$ energy difference, then the ratio $(3 \to 1)/(2 \to 1)$ will be sensitive to temperature as well as density because of the Boltzmann factors. As it turns our purely from spectroscopic considerations, line ratios useful for the quiet corona and solar flares, i.e., $T_e > 10^6$ K, are in general not too sensitive to temperature. Line ratios useful for the transition region are sensitive to temperature as well as density for most cases similar to the three-level example. However, there are several useful transition region ratios of the three- and the four-level type, and they are not sensitive to temperature.

The applicability of the density diagnostics just outlined depends in part on accurate excitation rate coefficients for weak intercombination or forbidden transitions. As mentioned, resonances have a considerable effect on the excitation rates for such transitions. For cases more complicated than those just outlined Eq. (3.5) must be solved taking all processes into account.

3.3.2. The Solar Atmosphere. The general method for determining density discussed in Section 3.3.1 is not new and has in fact been used by astronomers working in the optical spectral region for some time.[143] The familiar diagnostic to astronomers is the O II ratio,

$$\frac{2s^2 2p^3(^4S_{3/2} - {}^2D_{5/2})}{2s^2 2p^3(^4S_{3/2} - {}^2D_{3/2})}$$

that is, 3730 Å/3727 Å. This example, from the N I isoelectronic sequence, is similar to the three-level case previously discussed. Both of the upper levels $^2D_{3/2}$ and $^2D_{5/2}$ results in intercombination transitions to $^4S_{3/2}$. However, the lifetime of the $^2D_{5/2}$ level is significantly longer than the $^2D_{3/2}$ lifetime, and

TABLE 5
The N I Isoelectronic Sequence

Ion	$({}^2D^0_{3/2} \rightarrow {}^4S^0_{3/2})$	$({}^2D^0_{5/2} \rightarrow {}^4S^0_{3/2})$
O II	3727.25[a]	3730.17
Ne IV	2421.84	2424.48
Mg VI	1805.97	1806.66
Si VIII	1445.76	1440.49
S X	1213.00	1196.26
Ar XII	1057.4*p*[b]	1021.2*p*
Ca XIV	951.2*p*	887.0*p*
Fe XX	824.1*p*	629.9*p*

[a] Wavelengths in angstroms.
[b] *p* = predicted wavelength.

therefore, within a limited range of densities, the ${}^2D_{3/2}$ level can be regarded as an allowed level relative to ${}^2D_{5/2}$. Density sensitivity for the O II ratio occurs around $10^3\,\text{cm}^{-3}$.[144] This density is typical of planetary nebulas and some H II regions. At solar densities, the O II diagnostic would be useless because the ${}^2D_{3/2}$ and ${}^2D_{5/2}$ levels are statistically mixed.

However, if more highly ionized ions in the N I sequence are considered, examples can be found that are useful at solar densities. As we proceed along the sequence, the wavelengths of the two lines we are considering become shorter, and the radiative decay rates increase rapidly in magnitude. Also, the ions are formed at higher temperatures. However, the excitation rate coefficients actually decrease, and the combination of these effects results in a rapid increase in the critical density for sensitivity, $N_e \approx A_{31}/(C^d_{32} + C^d_{31})$ (see Fig. 23). Wavelengths for the N I sequence lines are given in Table 5, and the behavior of decay rate, excitation rate coefficient, temperature of line formation, and critical density are given in Fig. 25. In addition to lines from the ${}^2D$ levels, lines from the ${}^2P$ terms within the $2s^2 2p^3$ configuration can also be used.

Figure 26 shows ratios of intercombination lines in the N I sequence. For each ion, there is a range of critical densities over which both line ratios can be used as a density diagnostic. However, to be useful from the practical standpoint, there must exist a plasma at the critical densities and also at the temperatures where the ion abundance is large. For example, a plasma at a density of $10^9\,\text{cm}^{-3}$ is ideal for the S X lines (Fig. 26), but if the temperature were only 10^5 K, there would be no S X produced in the plasma, and, of course, the lines would not be present in the spectrum. As mentioned previously, the solar corona has a density near $10^9\,\text{cm}^{-3}$ and a temperature of about 10^6 K. The S X is quiet abundant at 10^6 K, and S X lines have been used to deduce densities in the corona.[145]

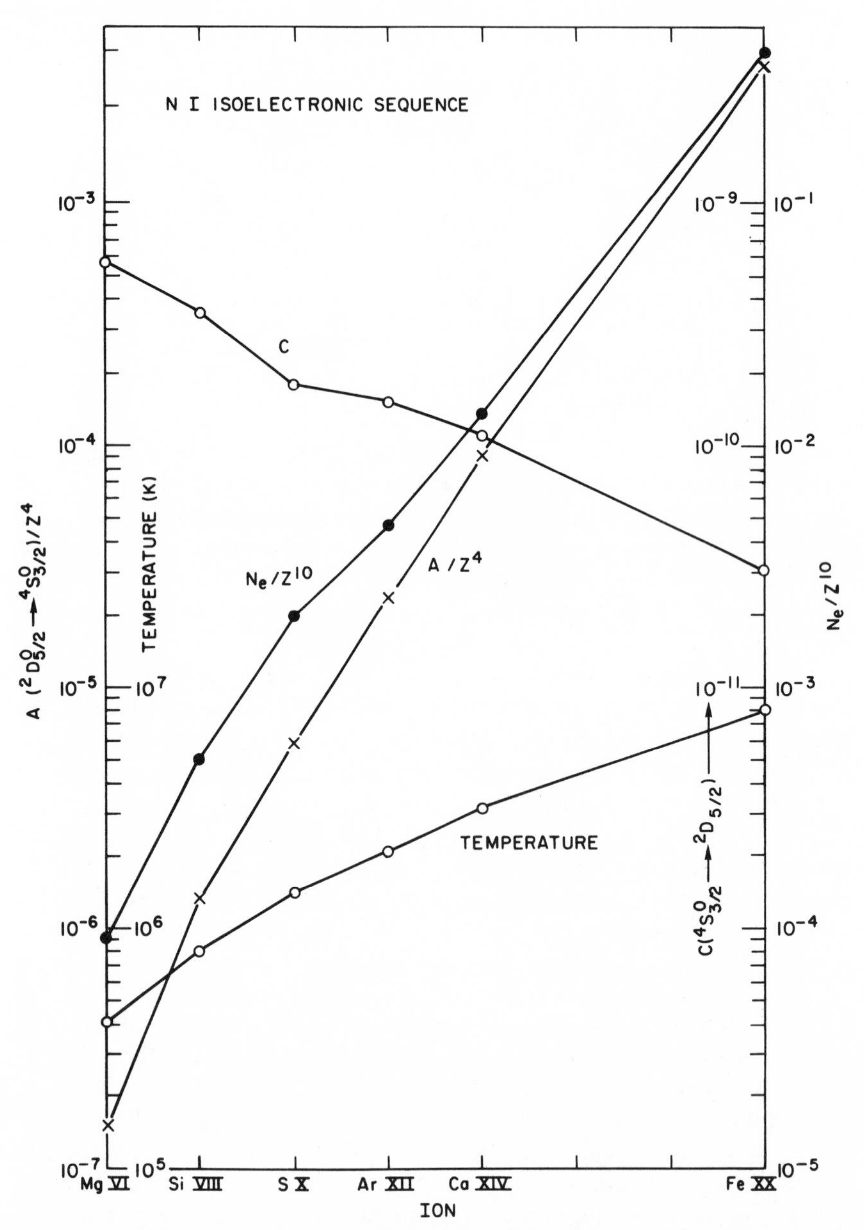

FIGURE 25. Behavior of indicated excitation-rate coefficient, radiative decay rate, density at which deexcitation of the $^2D_{5/2}$ level is affected by collisional processes, and temperature of maximum ion abundance, as a function of the ion along the N I isoelectronic sequence.

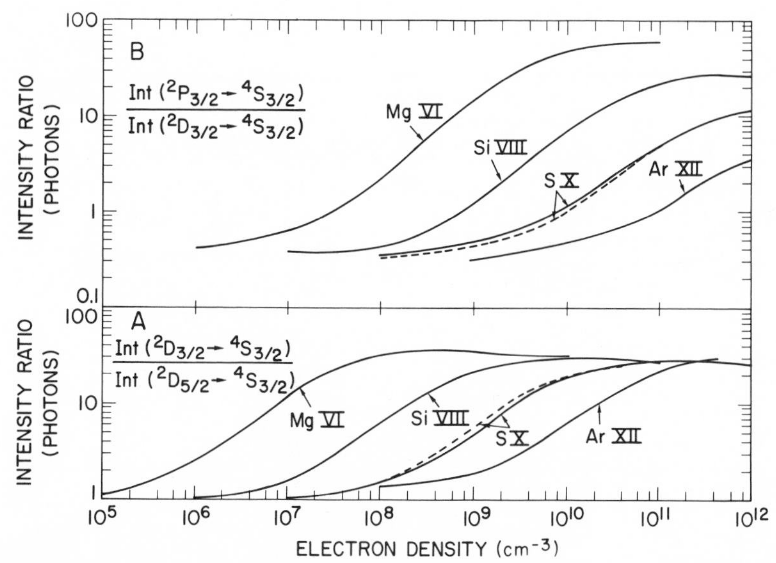

FIGURE 26. Density-sensitive line ratios of the N I isoelectronic sequence. The dashed lines for S X show the effect of including protons as well as electron impact excitation. (Courtesy *Astrophys. J.*)

Figure 27 shows the S X lines in a coronal spectrum taken rather far outside the solar limb.[140] Many of the lines in this peculiar spectrum are either forbidden or intercombination lines. They rival in strength strong allowed lines, such as the C IV doublet at 1550 Å, because there is very little plasma at 10^5 K far outside the solar limb, compared to plasma at 10^6 K. The C IV ion reaches maximum ion abundance at 10^5 K, while most of the forbidden and intercombination lines are emitted at temperatures near 10^6 K. Note in Fig. 26 that at a density of $10^9\,cm^{-3}$, the line ratios from neighboring ions, such as Si VIII or Ar XII, are in the high- and low-density limits, respectively. The high densities deduced from the S X line ratios are much greater than expected if the plasma were distributed uniformly along the line of sight. The obvious interpretation is that the plasma is clumped, or confined, to flux tubes. Such density measurements as these enable us to determine the pressure in coronal flux tubes, which is important for understanding the physical mechanisms that heat and maintain the plasma in the tubes.

We have considered the nitrogen isoelectronic sequence in some detail. However, a similar discussion could be presented for many other isoelectronic sequences as well. By now, many sequences have in fact been examined by a number of authors. The C III seems to be the first ion considered for application to the solar transition region by Jordan[146,147] and independently by Munro, Dupree, and Withbroe.[148] Reviews and discussions of possible diagnostics for the solar atmosphere are given by Feldman, Doschek, and

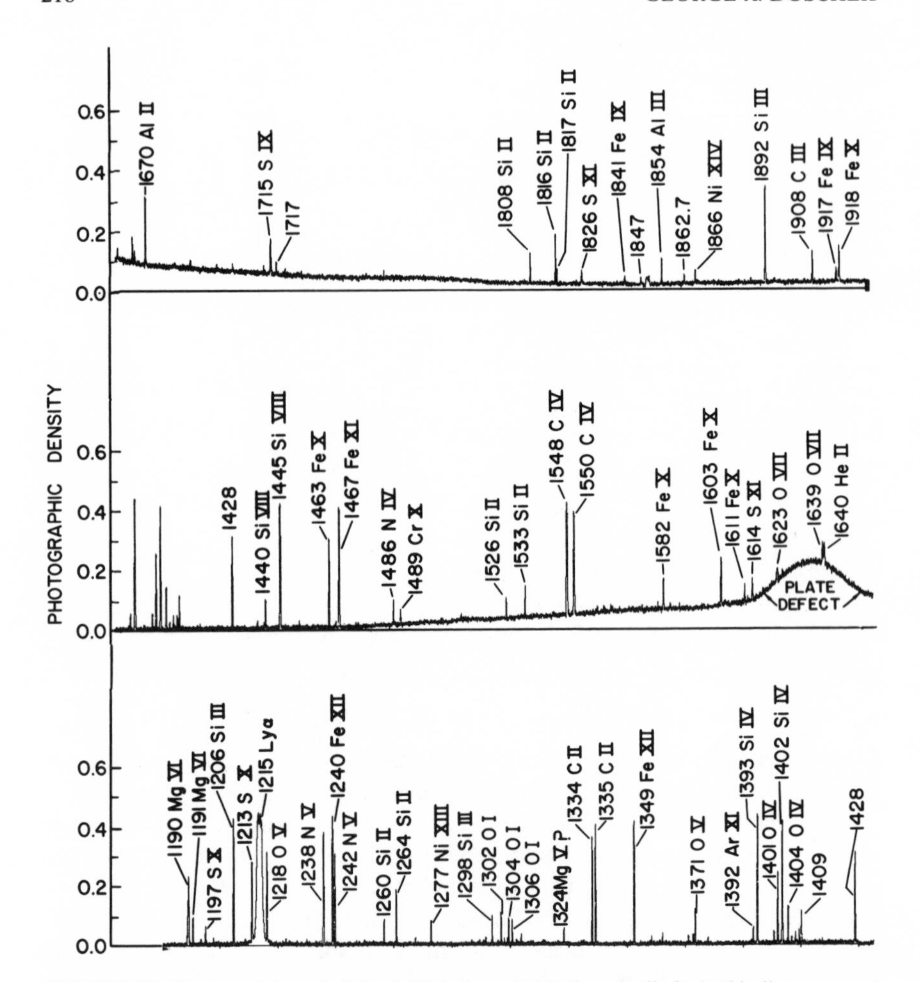

FIGURE 27. A spectrum recorded about 30 arc sec outside the solar limb. At this distance, most of the plasma is coronal at a temperature near 2×10^6 K. Many forbidden lines of highly ionized iron are present (see Feldman and Doschek for a line list and identifications).

Behring,[149] Doschek and Feldman,[150] Dere *et al.*,[27] Jordan,[151] Dere and Mason,[152] and Feldman.[153] There are many line ratios that fall throughout the wavelength region from about 80–2000 Å that provide useful density diagnostics for every combination of temperature and density that might be found in the solar atmosphere. Some of the ions useful over different ranges of temperature and density are shown in Fig. 28. A summary of electron pressures, i.e., values of $N_e T_e$ in different regions of the solar atmosphere derived using line ratio techniques, is given in Table 6. We discuss later the derivation of some of these results in more detail.

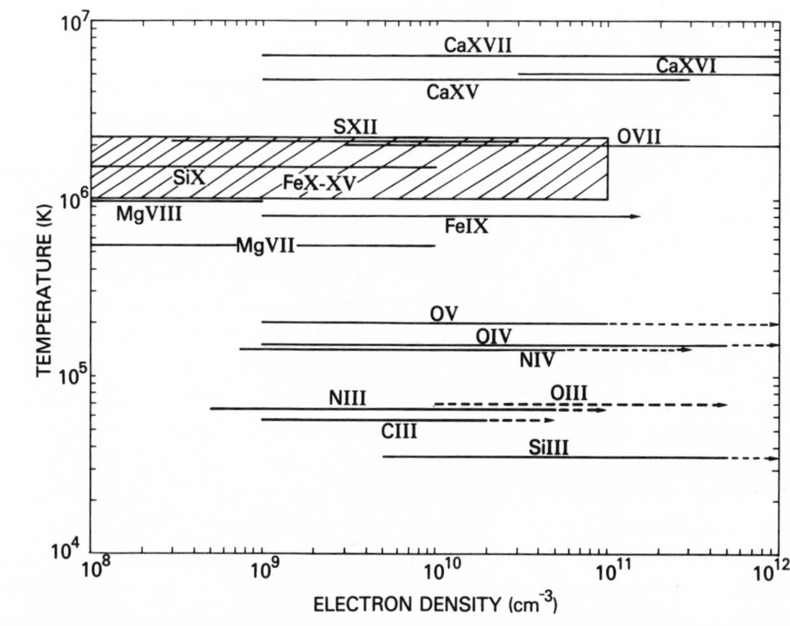

FIGURE 28. Range of density sensitivity of ions formed at different temperatures. Data are from Ref. (152).

TABLE 6
Electron Pressures

Region	Log $(N_e T_e)$
Quiet sun (chromosphere)	15.2
Quiet sun (transition region)	15.1
Quiet sun (corona)	15.0 ($h > 20$ arc-sec)
Coronal hole (transition region)	15.1 (14.8)
Active region (chromosphere)	15.2
Active region (transition region)	15.9
Active region (corona)	15.2–16.0
Prominence (transition region)	14.9
Sunspot (transition region)	15.0
Surges (flare-related activity) (transition region)	15.0–16.2
Flares (transition region)	16.0–18.0
Flares (coronas, 10^6 K $< T_e <$ 6×10^6 K)	16.8–18.7

Finding suitable density diagnostics for the lower transition region ($\sim 2 \times 10^4$–2×10^5 K) is difficult because of spectroscopic parameters. The energy differences between allowed lines and intercombination lines are comparable to kT_e, and therefore many line ratios are temperature as well as density sensitive. A further complication is that many of the allowed lines fall below 1000 Å, the short wavelength cutoff for such instruments as the NRL Skylab spectrograph. Nevertheless, there are four ions for which density-sensitive line ratios can be found that are nearly independent of temperature (for reasons described in Section 3.3.1). The ions are N III, O IV, Si III, and S IV. The transitions involved are the $2s^2 2p\ ^2P^0$–$2s2p^2\ ^4P^e$ lines near 1750 and 1400 Å for N III and O IV, respectively; the $3s3p\ ^3P^0$–$3p^2\ ^3P^e$ lines of Si III near 1300 Å; and the $3s^2 3p\ ^2P^0$–$3s3p^2\ ^4P^e$ lines of S IV, also near 1400 Å. Many of the lines are intercombination lines, and resonances are very important in calculating excitation rate coefficients for all four ions. The density sensitivity of the N III and O IV lines has been considered by Flower and Nussbaumer,[154] Nussbaumer and Storey,[155] and Feldman and Doschek.[156] More recently, O IV has been investigated by Nussbaumer and Storey[157] and Hayes.[158] The Si III lines in solar spectra have been studied by Nicolas *et al.*[159] and Dufton *et al.*[160], and the S IV lines have been examined by Bhatia, Doschek, and Feldman;[161] Bhadra and Henry;[162] Dufton and Kingston;[112] and Dufton *et al.*[163]

The difficulty in applying ratios of these lines to solar problems is that at quiet-sun densities, they are all in the low-density limit, i.e., there is no density sensitivity. Agreement with theory and observation in this case is good. At high densities ($> 5 \times 10^{11}\ \mathrm{cm}^{-3}$), the ratios are near the high-density limit where again there is no sensitivity. Unfortunately, the density in flares often appears to lie in the high-density limit for N III, O IV, and Si III. Another problem is that the overall density sensitivity is not high, e.g., one of the O IV ratios changes by a factor of only 7.7 between a density of 10^9 and $10^{12}\ \mathrm{cm}^{-3}$, a three-order of magnitude change in density. The densities obtained using the N III and O IV ratios in active regions and flare spectra are usually much less than densities obtained using other techniques. Part of this difficulty may lie in the fact that regions of different densities are involved in the emission, and line intensities are averaged over these regions.[164] Nevertheless, more accurate calculations for N III and O IV ions are desirable.

Another method for obtaining densities for the lower transition region has been developed by Feldman, Doschek, and Rosenberg.[165] The method involves calculating the ratio of an intercombination line, such as O IV ($^2P^0_{3/2}$–$^4P^e_{5/2}$), to an allowed line of another ion, formed at the same temperature as O IV. The simple calculation of this ratio requires the assumption of ionization equilibrium and a knowledge of the relevant element abundances if the two different ions are also different elements. Although this adds additional uncertainties to the theoretical line ratios, the advantage of this technique is that some of the ratios are much more sensitive to density than the ratios just discussed. Furthermore, some of the errors that may result from errors in the element abundances or values of the fractional ion abundances can be partly eliminated by calculating the value of the ratio in the low-density limit and then comparing this value to observational data obtained from a plasma known to have a very low density. The calculated ratio can be adjusted, or "normalized," to agree with the observed ratio at low densities.

At higher densities, variations in the ratio then depend on only the excitation rate coefficients and radiative decay rates, although ionization equilibrium is assumed. If ionization equilibrium is not assumed, the ratio is still a useful density diagnostic, but in this case, the ratio will also be model dependent and vary according to the particular calculations and physical parameters considered. Resonances are very important in all of these calculations, since they strongly affect the collisional excitation rate coefficients of the intercombination lines used in the method.

A summary of the density sensitivity of some excited levels useful for the 1100–2000 Å region and lower solar transition region is given in Fig. 29. Plotted is the quantity n_j/N_eN_z, where j refers to the excited level indicated in the figure and n_j is defined by Eq. (3.4). The sources of the atomic data are: C III, O V, Dufton *et al.*;[166] N IV, Dufton, Doyle, and Kingston;[167] O III, Bhatia *et al.*[168] and Baluja *et al.*[110,169] S IV, Bhatia *et al.*[161] Dufton and Kingston,[112] and Dufton *et al.*;[163] N III and O IV, Nussbaumer;[170] and S V, Feldman *et al.*[171] and van Wyngaarden and Henry.[172]

To understand the behavior of n_j/N_eN_z shown in Fig. 29, consider a hypothetical two-level atom for which the excited level (level 2) is metastable. Equation (3.5) then reduces to

$$n_1 N_e C_{12}^e = n_2(A_{21} + N_e C_{21}^d) \tag{3.36}$$

Using Eqs. (3.6) and (3.7), i.e., $n_1 + n_2 = N_z$, Eq. (3.36) can be rewritten as

$$\frac{n_2}{N_e N_z} = \frac{C_{12}^e}{A_{21} + N_e C_{12}^e[1 + \omega_1 \exp(\Delta E_{12}/kT_e)/\omega_2]} \tag{3.37}$$

If $N_e \to 0$, $n_2/N_eN_z \to C_{12}^e/A_{21}$, a quantity that depends on only temperature. For very high densities, $n_2/N_eN_z \propto 1/N_e$, as shown in Fig. 29. For $N_e \to 0$, $n_2/N_eN_z \propto \exp(-\Delta E_{12}/kT_e)/\sqrt{T_e}$ [see Eq. (3.15)]. The values of n_j/N_eN_z plotted in Fig. 29 were calculated at the temperatures of maximum emission of the lines indicated, assuming ionization equilibrium.

The quantity n_j/N_eN_z is directly proportional to the flux ratio of the intercombination line, arising from level j, to an allowed line formed at the same temperature. To see this, refer to Eq. (3.4) for the intercombination line and use Eq. (3.10) for the allowed line and consider a case where the two lines

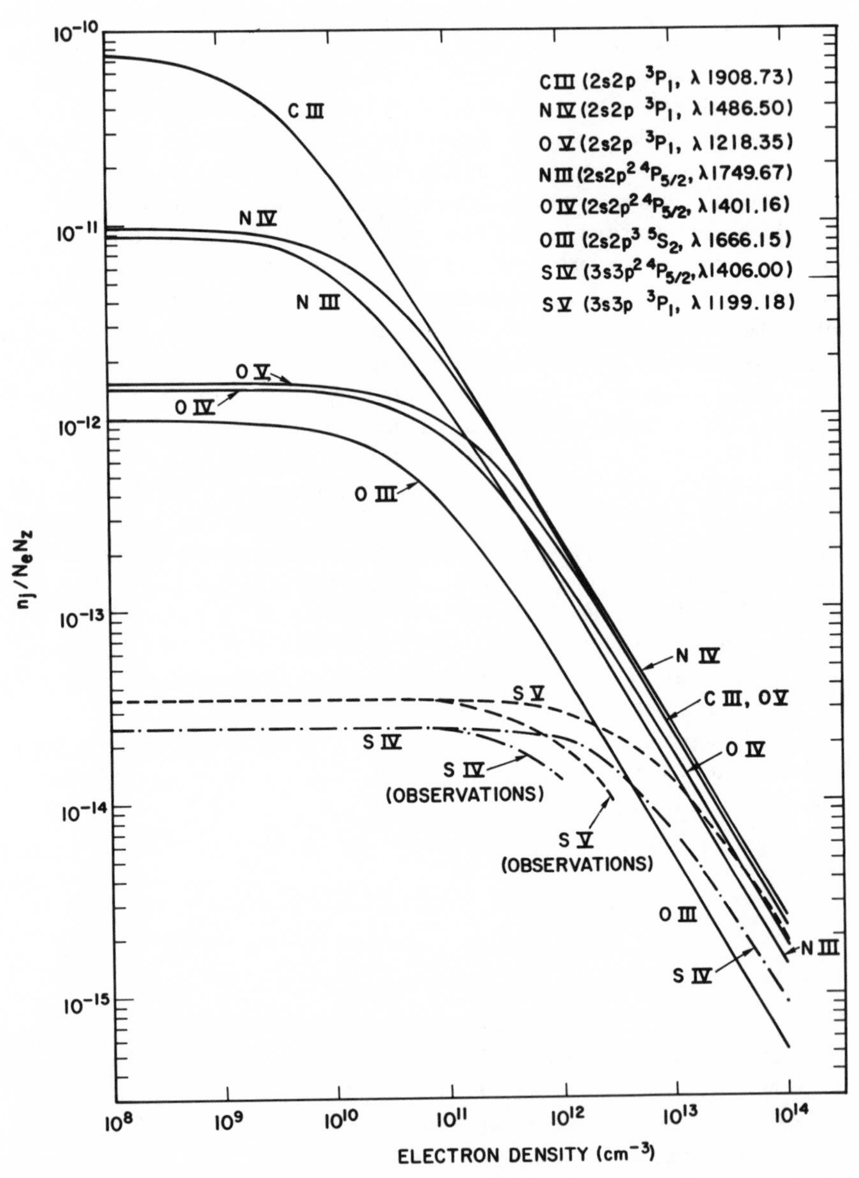

FIGURE 29. Populations of levels and ions indicated as a function of electron density. Wavelengths of key lines are also given (See text for discussion.)

arise in different ions. Assuming only two-level ions, the flux in the intercombination line is

$$F^I_{12} \propto n^I_2 \Delta V \tag{3.38}$$

and the flux in an allowed line formed at the same temperature is

$$F^A_{12} \propto N_e N^A_z \Delta V \tag{3.39}$$

[For simplicity, the subscript (12) for flux is defined as the same for both ions.] If ionization equilibrium is assumed, the ion abundance N^A_z can be related to the intercombination line's ion abundance N^I_z by a factor β (that will contain the relative element abundance if the two lines arise in ions of different elements), i.e., $N^A_z = \beta N^I_z$. Then, combining Eqs. (3.38) and (3.39) gives

$$\frac{F^I_{12}}{F^A_{12}} \propto \frac{1}{\beta} \frac{n^I_2}{N_e N^I_z} \tag{3.40}$$

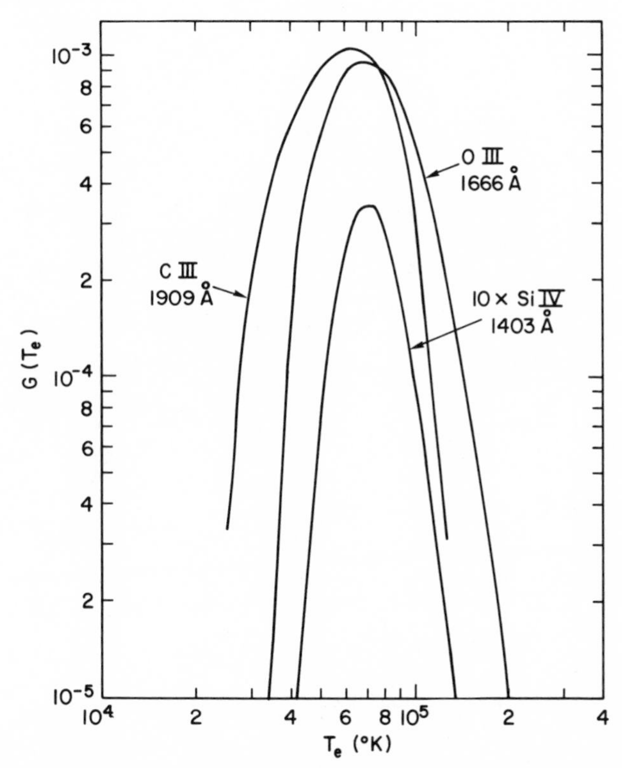

FIGURE 30. Contribution functions [Eq. (3.42)] for the three lines indicated. All three lines are formed at about the same temperature, and therefore ratios of the 1909 and 1666 Å lines to the 1403 Å line can be used to deduce electron density under certain conditions. (Courtesy *Astrophys. J.*)

Figure 29 shows that the only good density diagnostic below $10^{10}\,cm^{-3}$ is the C III line at 1909 Å. The other lines become useful above 10^{10} and $10^{11}\,cm^{-3}$. Below $10^{10}\,cm^{-3}$, both the O III 1666 Å line and the Si IV lines at 1394 and 1403 Å can be used as allowed lines with the C III line. (The O III line behaves as an allowed line at these densities.) Figure 30 shows the contribution function $G(T_e)$ for the C III, O III, and Si IV lines. The contribution function is the product of the temperature-dependent terms in C_{ij}^e times the fractional ion abundance $f(T_e)$, i.e.,

$$G(T_e) = f(T_e)\frac{\exp(-\Delta E_{ij}/kT_e)}{T_e^{1/2}} \tag{3.41}$$

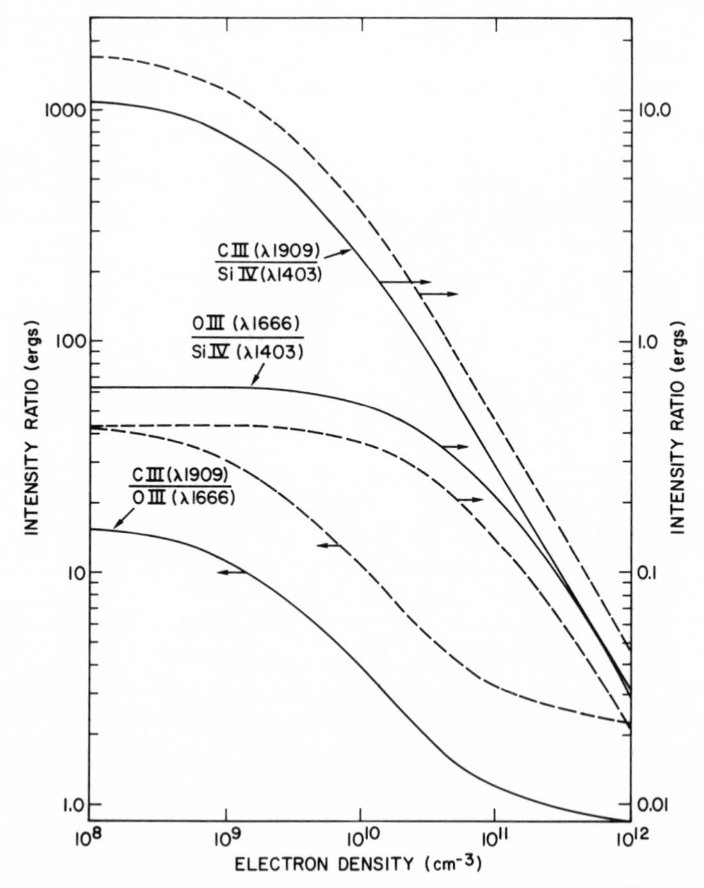

FIGURE 31. Density-sensitive line ratios useful for transition region plasmas. The contribution functions of the lines are shown in Fig. 30. The solid curves are theoretical calculations, the dashed curves are the "normalized" curves (see text). (Courtesy *Astrophys. J.*)

where $f(T_e)$ is defined by Eq. (3.21). It thus represents the radiating efficiency of a line as a function of temperature. It can be seen from Fig. 30 that these three lines are formed at essentially the same temperature. The line ratios shown in Fig. 31 have been used[173] to obtain densities of about $1.8 \times 10^{10}\,\text{cm}^{-3}$ in quiet-sun regions and coronal holes at the temperature of formation of C III ($\simeq 6.0 \times 10^4\,\text{K}$).

At densities above $\simeq 5 \times 10^{10}\,\text{cm}^{-3}$, several intercombination lines become density sensitive. These lines have been used to obtain densities in active regions,[174–177] flares,[165,178,179] and a surge.[180] In the 15 June 1973 flare observed from Skylab,[165] a density $> 10^{13}\,\text{cm}^{-3}$ was derived from an apparent surge that accompanied the flare. More typically, however, densities in active regions and flares fall between 10^{11} and $10^{12}\,\text{cm}^{-3}$. We remind the reader that these densities apply to a specific region of the solar atmosphere, i.e., the lower transition zone at temperatures $> 4 \times 10^4\,\text{K}$ and $< 2 \times 10^5\,\text{K}$. Sometimes the densities derived using different line pairs formed at similar temperatures are different by factors of two or three. Part of these differences may reflect the atmospheric structure, i.e., the lines are not formed in regions of the same density.[164] However, some of the error probably also lies in uncertainties in relative element abundances, atomic data, and instrumental calibration and data reduction.

As mentioned, the only intercombination line useful at densities below $10^{10}\,\text{cm}^{-3}$ is the 1909 Å C III line. There are also some other useful C III lines. The group of lines labeled g_i in Fig. 24 fall near 1175 Å. The ratio of these lines to the 1909 Å line (line a in Fig. 24) is both density and temperature sensitive, but all the lines can be observed by a single instrument and have been observed in solar and stellar spectra. Similarly, line e in Fig. 24 falls near 1247 Å and has been observed in solar spectra. The ratio of this line to 1909 Å is also both temperature and density sensitive. Close-coupling cross sections for all of these C III lines have been calculated, and the results from Dufton *et al.*[166] are shown in Fig. 32. If the 1175 Å multiplet and the 1247 and 1909 Å lines can be observed simultaneously, then Fig. 32 shows that both the temperature and the density can be determined. It is difficult to accomplish this using the solar spectra obtained with the NRL Skylab instrument, because the instrumental efficiency is decreasing rapidly around 1175 Å and is poorly known. Finally, we note that at densities $< 10^9\,\text{cm}^{-3}$, the ratio of line f in Fig. 24 to any of the other lines, say, g_5, becomes density sensitive. This happens because the 1909 Å line behaves like an allowed line for $N_e < 10^9\,\text{cm}^{-3}$, and consequently the $2s2p\ ^3P_1$ population decreases rapidly relative to the $2s2p\ ^3P_0$ and $2s2p\ ^3P_2$ populations as the density decreases. In solar spectra, however, the ratios f/g_i are always in the high-density limit.

The 1247 Å/1909 Å ratio has been observed to vary by as much as an order of magnitude in different solar regions (see Fig. 33). If a temperature of

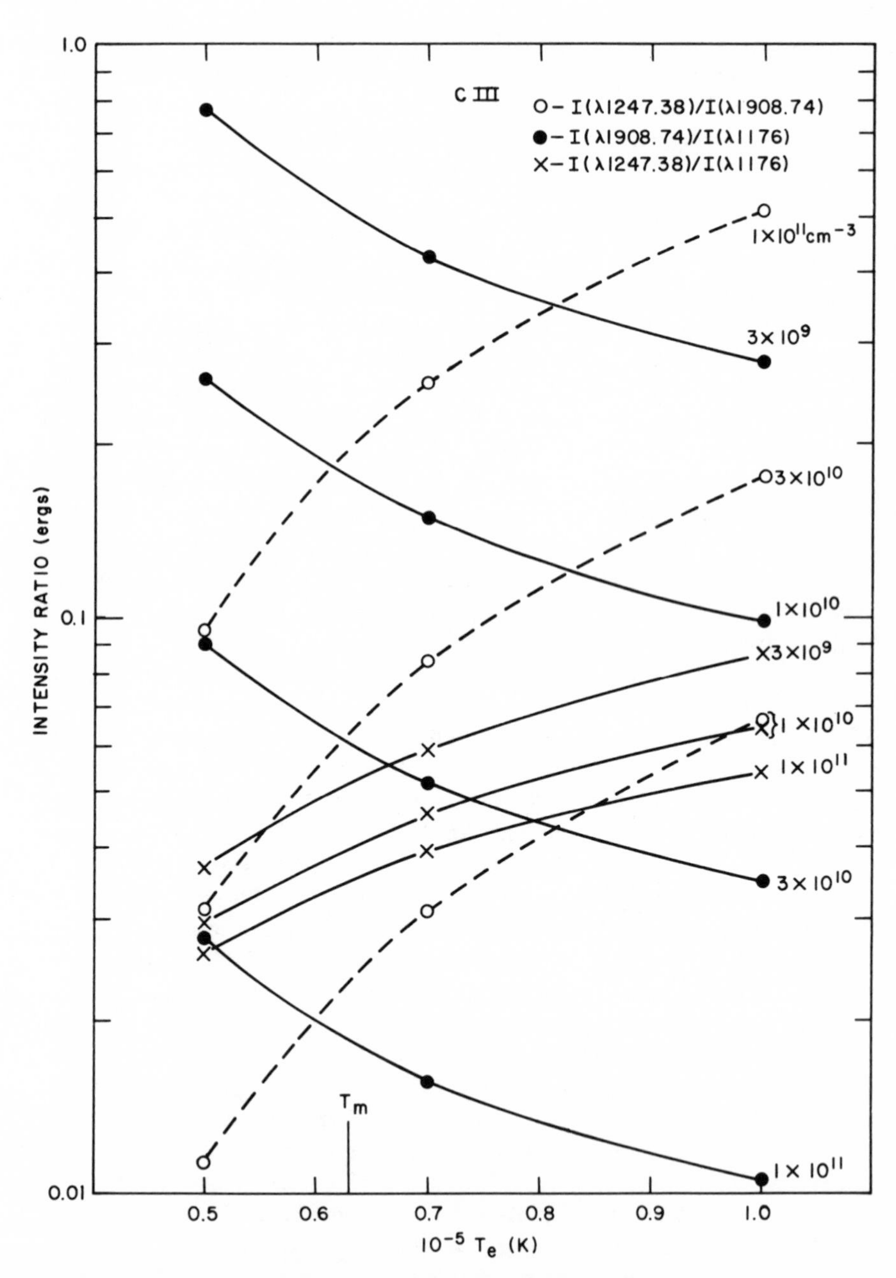

FIGURE 32. Density- and temperature-sensitive C III line ratios. The T_M is the temperature of maximum ion abundance in ionization equilibrium. The temperature variation of the line ratios is shown evaluated at different densities. Intensity ratios are from Ref. (166.)

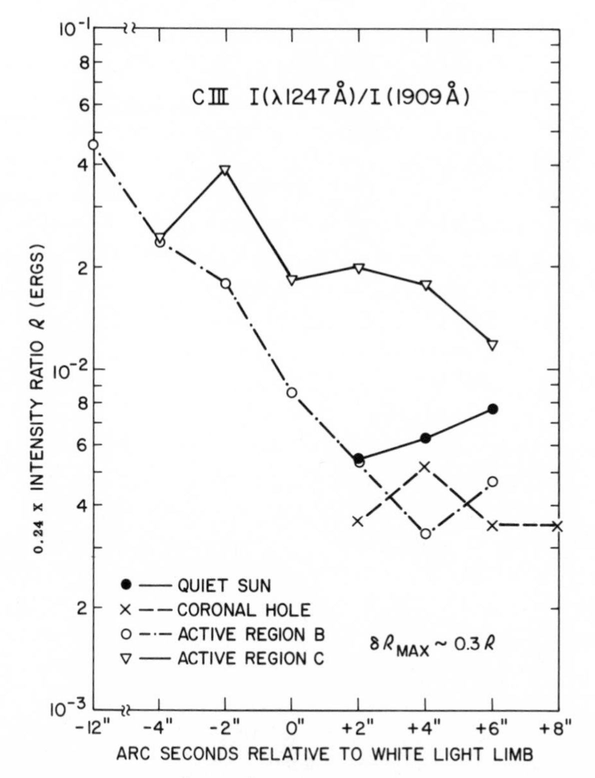

FIGURE 33. Intensity ratio 1247 Å/1909 Å as a function of position relative to the solar white-light limb for the solar regions indicated (0 arc sec is at the white-light limb; positive arc sec values are outside the limb; negative values are inside the limb.) The large values of the ratio for the active regions indicate high densities ($\simeq 10^{11}$ cm^{-3}).

formation is assumed for C III, this ratio can be used to obtain densities, and the results agree with densities obtained using other ratios. In Fig. 33, measurements refer to observations made just inside and outside the solar limb. These measurements, when combined with Fig. 32, show that the quiet-sun and coronal hole regions have densities of $\simeq 10^{10}\,\text{cm}^{-3}$ and that the highest density obtained (-12 arc sec, active region *B*) has a density of $10^{11}\,\text{cm}^{-3}$ (assuming $T_e = T_M$ in Fig. 32).

Once a density has been derived using an intercombination line, the intensity of an allowed line formed at about the same temperature can be used with Eq. (3.10) to derive the volume of emitting plasma within the spectrograph field of view. Volumes of transition region plasma derived in this manner are extremely small,[177] with characteristic lengths on the order of, and less than, 700 km, depending on the atmospheric region considered. For the 15 June flare, previously mentioned, regions as small as 60 km in length were detected.[181] That small regions on the order of 700 km in size do exist in the atmosphere at temperatures of about 10^5 K was confirmed by Brueckner and colleagues[41] from high spatial resolution rocket observations.

Let us now turn to density measurements for coronal temperatures $\simeq 10^6$ K in solar flares. Figure 28 shows that there are many available ions for coronal temperatures. These ions emit density-sensitive emission lines over a wide wavelength range, from the x-ray region to $\simeq 600$ Å. The region between 170 and 600 Å is particularly rich in density-sensitive lines; Table 1 shows that this region was covered by the NRL slitless spectrograph on Skylab. A number of excellent solar flare spectra were obtained, and a large number of diagnostic possibilities were investigated by Dere and Mason;[152] Dere *et al.*,[27] and Feldman, Doschek, and Behring.[149] One of the most sensitive ratios for $T_e \simeq 10^6$ K involves lines of Fe IX near 240 Å; this ratio is described in detail elsewhere.[182] Density diagnostics using coronal forbidden lines ($\lambda > 1100$ Å) were developed by Gabriel and Jordan,[34] who found poor agreement between theoretical-level populations in the ground configuration, calculated using the best available atomic data, and populations derived from observation for Si VIII, Si IX, Fe XI, and Fe XII. The authors suggested that configuration mixing and resonances may explain at least part of the discrepancy. The particular transitions involved are difficult to calculate accurately.

Unfortunately, there are no useful ratios for $T_e \simeq 20 \times 10^6$ K because there are no sufficiently metastable levels.[150] The highest temperature ions that are useful are Ca XVI, Ca XVII ($T_e \simeq 6 \times 10^6$ K), and Fe XXI ($T_e \simeq 10 \times 10^6$ K). Figure 34 shows a flare spectrum between 200 and 225 Å; note the two Ca XVI lines at 208.6 and 224.5 Å. The Ca XVI is a member of the B I sequence, and density sensitivity occurs as for the four-level case in Fig. 23. The theoretical line ratio as a function of density calculated by Dere *et al.*[27] is shown in Fig. 35, and densities of about 3×10^{11} cm^{-3} are obtained for different flares. (For similar types of calculations for other coronal ions, see also Mason and Bhatia[183] and Bhatia and Mason.[184]) The Ca XVII is a member of the Be I sequence, and theoretical line ratios have been calculated by Doschek *et al.*[185] and more recently by Bhatia and Mason[186] and Dufton *et al.*[187] The densities derived from Ca XVII lines are not in satisfactory agreement with densities deduced from other ions, such as Ca XVI. Part of the difficulty may be due to uncertainties in the observational data.

Density-sensitive Fe XXI lines in the C I sequence do not fall within the range of the NRL instrument. The best region in which to observe them is the region around 100 Å, where the $2s^2 2p^2$–$2s2p^3$ transitions fall. However, only a few relatively low-resolution flare spectra in the 100 Å region exist.[19] There are also density-sensitive Fe XXI lines near 12 Å, but it is not clear how useful

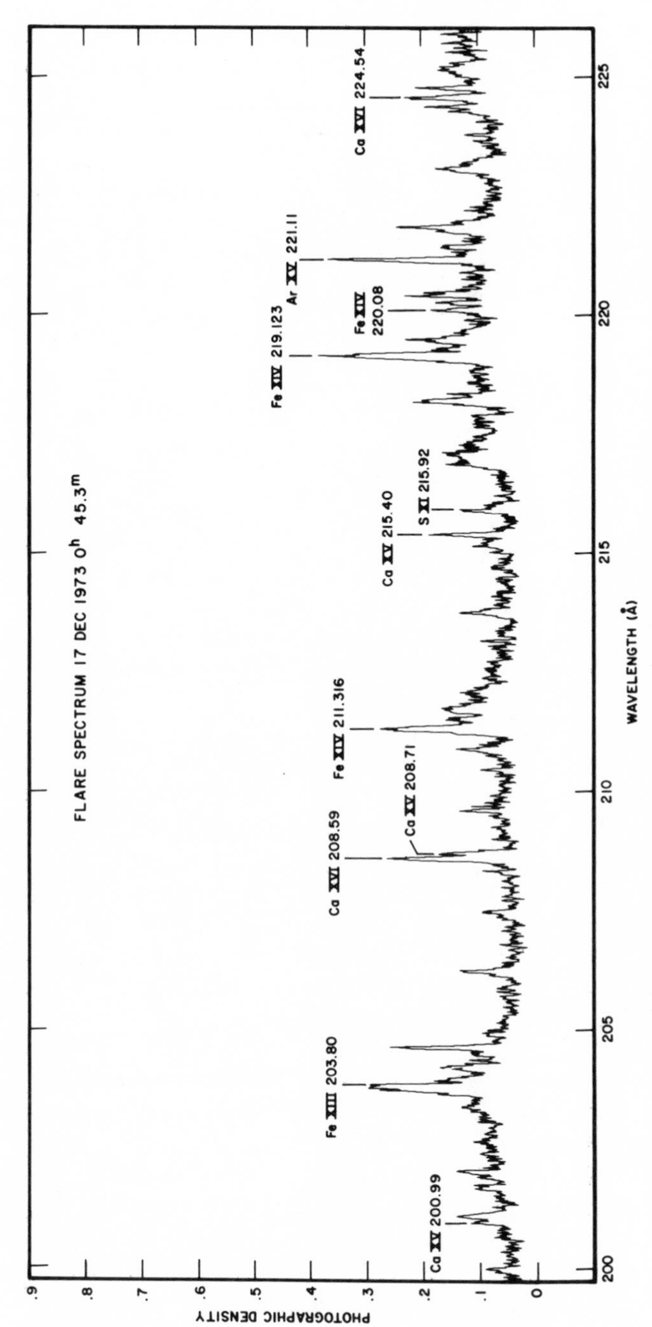

FIGURE 34. Solar flare spectrum recorded by the NRL spectroheliograph on Skylab. Density-sensitive line ratios for different temperature regions can be constructed for many lines in the 170–600 Å region. Spectrum from Ref. (27). (Courtesy *Astrophys. J.*)

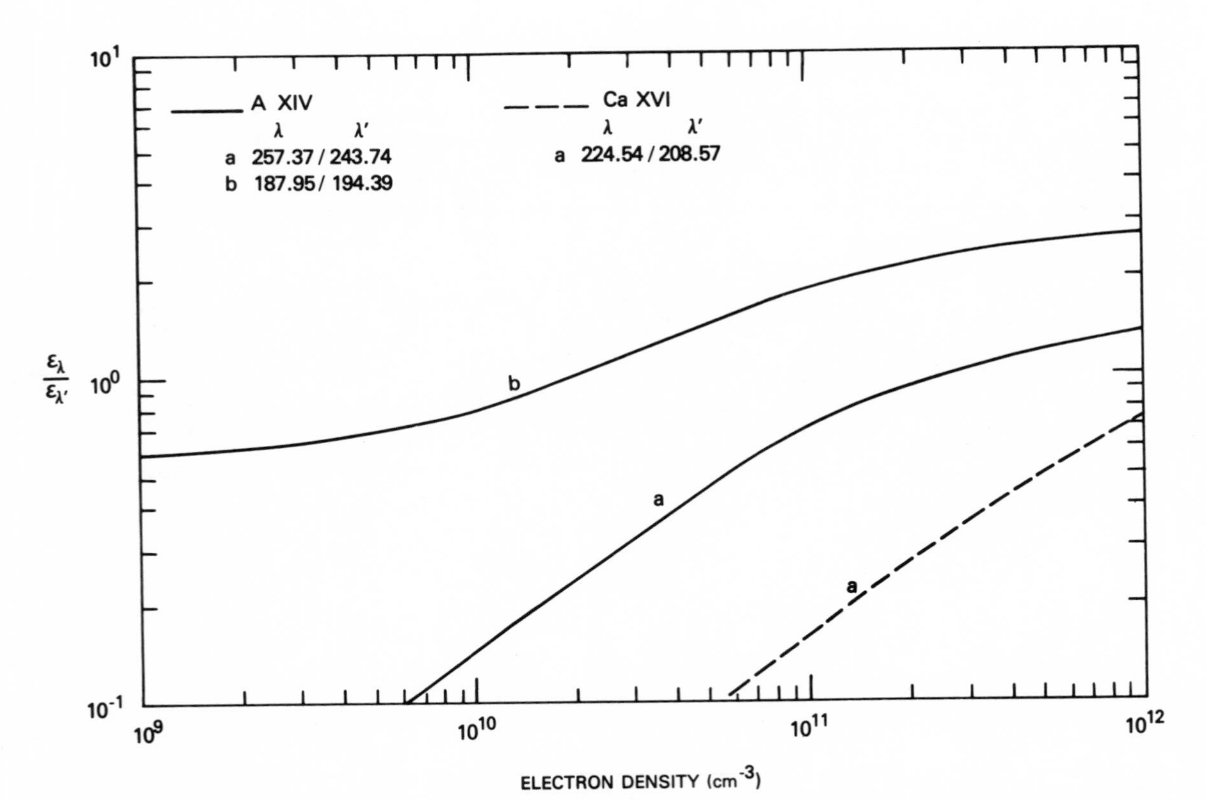

FIGURE 35. Emission ratios as a function of density for the lines and ions indicated. The Ca XVI is formed at a temperature of about 6×10^6 K, and it is useful as a density diagnostic in flare plasmas. Calculations are from Ref. (27). (Courtesy *Astrophys. J.*)

they will ultimately be because of severe blending. The Fe XXI lines are discussed in detail by Mason *et al.*,[188] who give collision strengths, radiative transition probabilities, and comparisons to previous observations.

At x-ray wavelengths ($\lesssim 25$ Å), some of the He-like ions are useful for diagnostic purposes. The density sensitivity of the ratio of the forbidden line of the He-like ion, $1s^2\,{}^1S_0 - 1s2s\,{}^3S_1$, to the intercombination line, $1s^2\,{}^1S_0 - 1s2p\,{}^3P_1$ was first recognized by Gabriel and Jordan.[189] In an earlier paper,[190] they identified the forbidden lines of He-like ions in solar x-ray spectra and also pointed out the importance of the satellite lines discussed in Section 3.2. In general, Gabriel and Jordan can be credited with pioneering the exploration of the x-ray region for spectroscopic plasma diagnostics useful for solar applications.

A term diagram for the relevant transitions in He-like ions is shown in Fig. 36. Wavelengths of the lines for O VII, Ca XIX, and Fe XXV are also shown. For low z ions, such as O VII, the quadrupole line (x) is negligibly weak compared to the other lines. At low densities, the upper levels of w, x, y, and z are populated by electron impact excitation from $1s^2\,{}^1S_0$. Although the ratio of collision rates $C^e(1\,{}^1S_0 \rightarrow 2\,{}^3S_1)/C^e(1\,{}^1S_0 \rightarrow 2\,{}^3P)$ is ~ 0.2,[191] line z is stronger than lines x and y because the $1s2p\,{}^3P$ levels can decay to $2\,{}^3S_1$

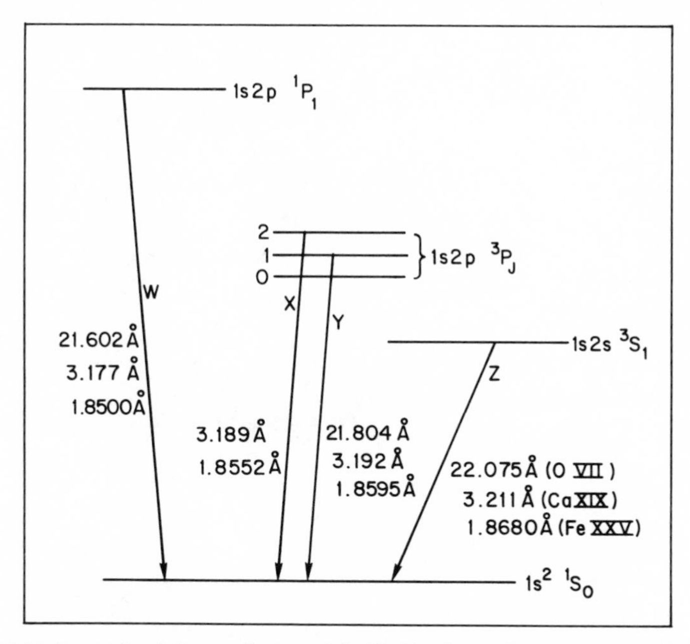

FIGURE 36. Energy-level diagram for ions of the He I isoelectronic sequence. The $2\,{}^3P_2 \rightarrow 1\,{}^1S_0$ line (×) is negligibly weak for O VII.

as well as to $1\,^1S_0$. The most metastable level is $2\,^3S_1$, which decays to $1\,^1S_0$ by relativistically induced magnetic dipole radiation.[189,192] When the density is increased, eventually the excitation rate $N_eC^e(2\,^3S_1 \rightarrow 2\,^3P)$ can compete with $A(2\,^3S_1 \rightarrow 1\,^1S_0)$ in depopulating $2\,^3S_1$, and the ratio $R = z/(x+y)$ decreases. Figure 37 shows a graph of R versus N_e for three He-like ions. Values of R were obtained from atomic data and the expressions given by

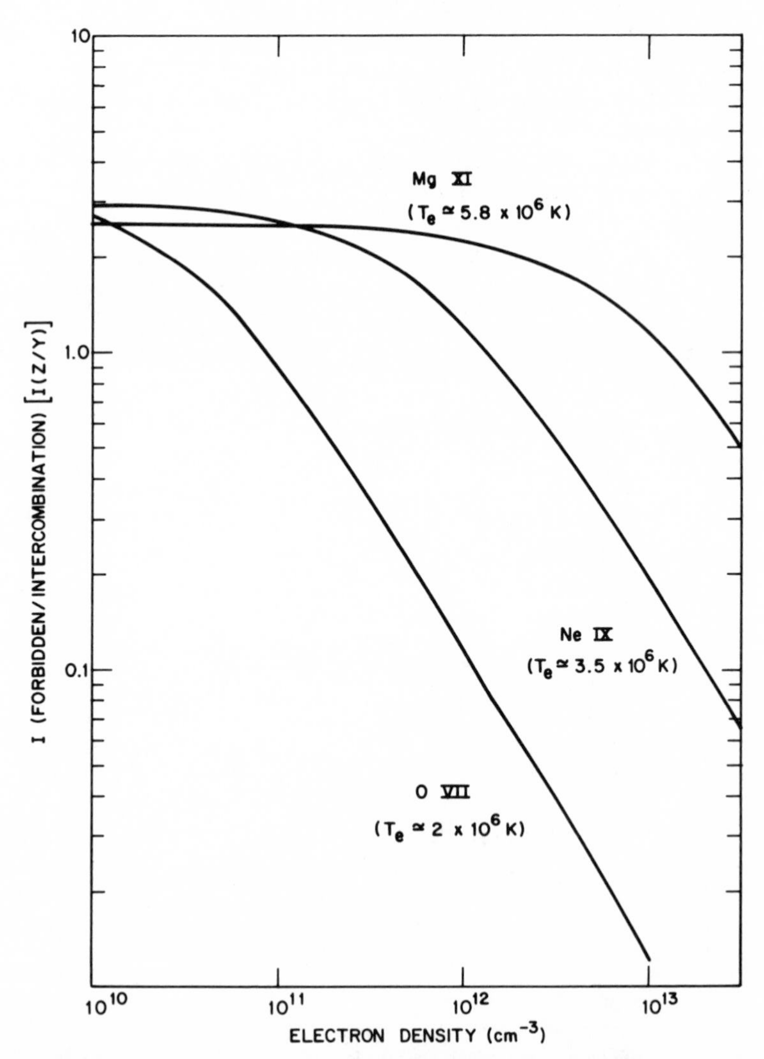

FIGURE 37. Ratio of the forbidden to the intercombination line for three ions of the He I isoelectronic sequence. Density sensitivity commences at higher densities for the heavier ions.

Gabriel and Jordan[61]

$$R = \frac{A(2\,^3S \rightarrow 1\,^1S)}{N_e C^e(2\,^3S \rightarrow 2\,^3P)(1+F) + A(2\,^3S \rightarrow 1\,^1S)} \left(\frac{1+F}{B} - 1 \right) \tag{3.42}$$

where

$$F = \frac{C^e(1\,^1S \rightarrow 2\,^3S)}{C^e(1\,^1S \rightarrow 2\,^3P)} \tag{3.43}$$

and

$$B = \frac{1}{3} \frac{A(2\,^3P_1 \rightarrow 1\,^1S)}{A(2\,^3P_1 \rightarrow 1\,^1S) + A(2\,^3P \rightarrow 2\,^3S)} + \frac{5}{9} \frac{A(2\,^3P_2 \rightarrow 1\,^1S)}{A(2\,^3P_2 \rightarrow 1\,^1S) + A(2\,^3P \rightarrow 2\,^3S)} \tag{3.44}$$

The values of R obtained from these sources for the low-density limit have been adjusted in order to fit solar observations. This was accomplished by modifying the theoretical value of F. The effective value of F used in the expression for R derived by Gabriel and Jordan[61,189] should, in fact, be somewhat different from the simple ratio of the excitation rates, because of the effects of recombination and cascades from higher levels, which are neglected in the derivation of R given by Gabriel and Jordan.[191–194] However, the semiempirical approach given in that paper is still reasonably valid. The best presently available atomic data for the lighter He-like ions appears to be the data published by Pradhan, Norcross, and Hummer,[191,195] although other calculations that may improve these data are in progress.[196] The Pradhan, Norcross, and Hummer results show that resonances are also important for the He-like ions.

For about 10 years after the density diagnostic value of He-like ions was proposed, no solar observations were obtained that conclusively gave for any He-like ion a ratio R substantially less than the low-density limit, except possibly for one Mg XI flare measurement by Neupert.[197] Most solar atmospheric regions are at densities lower than that required for sensitivity, at the temperatures of formation of the He-like ions. Figure 37 shows that the density at which sensitivity begins increases rapidly with increasing atomic number. The He-like ions of sulfur, argon, calcium, and iron are not useful for solar density diagnostics. However, the Neupert[197] measurement was obtained from a flaring region, where the density is known from other diagnostics to reach values $> 10^{11}\,\mathrm{cm}^{-3}$.

The first observations of a large variation in the ratio R were obtained by McKenzie *et al.*[198] from P78-1 solar flare and active-region observations of the O VII lines recorded by Aerospace Corp. spectrometers. A number of such observations have been made,[7] and an example is shown in Fig. 38. In the active-region spectrum, the ratio R is about 2.8, implying from Fig. 37 a density of $1.1 \times 10^{10}\,\mathrm{cm}^{-3}$. In the two flare spectra, the ratio is clearly less,

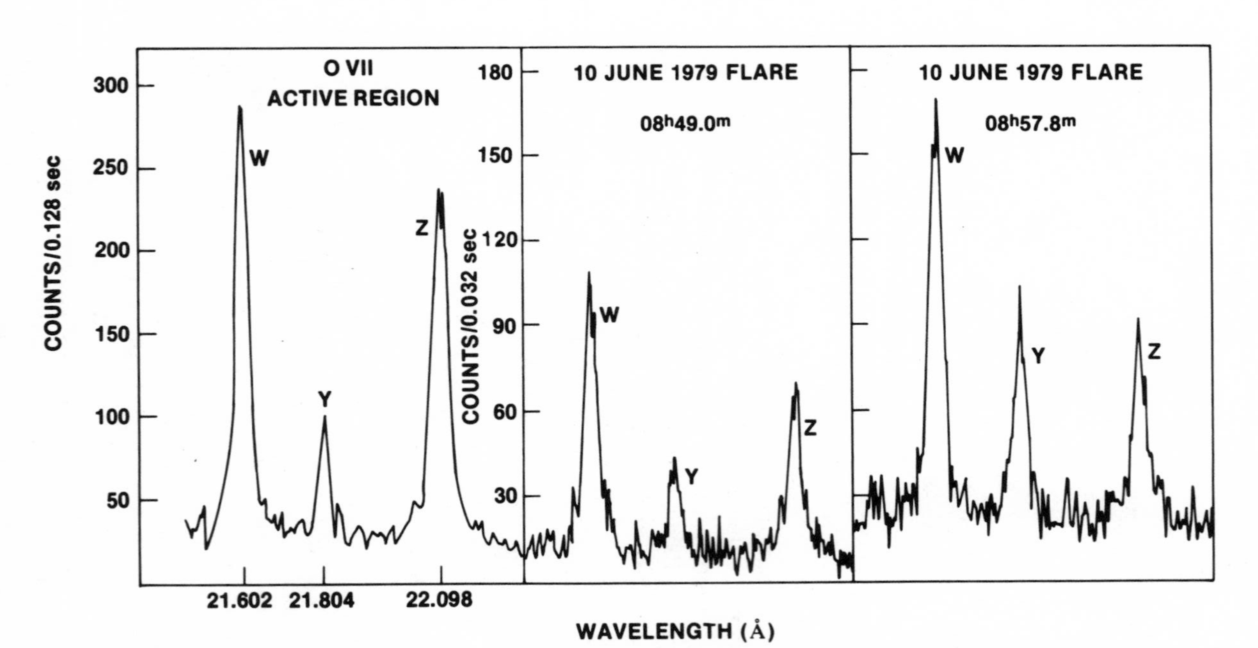

FIGURE 38. The O VII spectra of an active region and a solar flare. The ratio of line z to y is density sensitive. For light ions, such as O VII, the decay rate from $1s2p\ ^3P_2$ to $1s2s\ ^3S_1$ is much stronger than to $1s^2s\ ^1S_0$. Consequently, line x is negligibly weak for O VII. Data are from Ref. (198).

implying densities of 3.7×10^{10} and $1.2 \times 10^{11}\,\mathrm{cm}^{-3}$. The O VII observations[7] of two flares with high time resolution have shown that there is a strong, short-lived density enhancement near the time of peak hard x-ray flux, where densities reached $\simeq 10^{12}\,\mathrm{cm}^{-3}$ at the temperature of formation of O VII ($\simeq 2 \times 10^6\,\mathrm{K}$). More recently, densities have been obtained from the Ne IX lines recorded by the SMM spectrometers.[199]

In summary, there are many useful density diagnostics that can be applied to most regions of the solar atmosphere. Some of them have been discussed in detail. It should be emphasized that resonances play a very important role for these diagnostics. As an example of their importance, consider the intercombination lines of the ions S IV (1406 Å) and S V (1199 Å). Distorted wave calculations are shown for these ions in Fig. 29. Note the curves associated with the calculations marked S IV and S V observations. Bhatia *et al.*[161] and Feldman *et al.*[171] have found that the S IV and S V theoretical distorted wave data give densities that are about four to five times greater than the densities determined from other line ratios. They suggested that the neglect of resonances in distorted wave calculations may account for part of this discrepancy. This in fact has been confirmed for S IV and Si III.[112,160] The curves marked observations are the curves that would result if the S IV and S V densities were forced to agree with densities derived by other diagnostics and at least for S IV, show the effect of resonances on the computed population of the $3s3p^2\,{}^4P_{5/2}$ level.

3.3.3. Application of Density Diagnostics to Stellar Atmospheres and Other Sources. Stellar spectra in the UV have been obtained from the IUE. The region between about 1100 and 2000 Å can be observed using both low- and high-spectral resolution. The high-resolution mode gives a spectral resolution of about 0.1 Å, but the fluxes are considerably lower in this mode; low resolution is about 7 Å, not really acceptable for transition region density diagnostics.

At low resolution, it is difficult to apply many of the diagnostics discussed in Section 3.3.2 because some of the lines are weak and lie close in wavelength to other lines. For example, the N III lines are blended with Fe II and other cold lines; the Si IV, S IV, and O IV complex near 1403 Å is blended; and the O III lines near 1666 Å are blended with Al II and perhaps C I lines. The unblended lines are C III 1175 Å, N V 1242 Å, Si IV 1394 Å, N IV 1486 Å, C IV 1549 Å, Si III 1892 Å, and C III 1909 Å. The N V lines may be blended in some cases with an Fe XII forbidden line. The N IV line may also contain S I blends in some cases. Fortunately, the C III lines appear to be unblended, and this is most important, since the C III lines are the only good density diagnostics between 1100 and 2000 Å for densities $< 10^9\,\mathrm{cm}^{-3}$. Many giant and supergiant stars have atmospheres at considerably lower densities than the solar

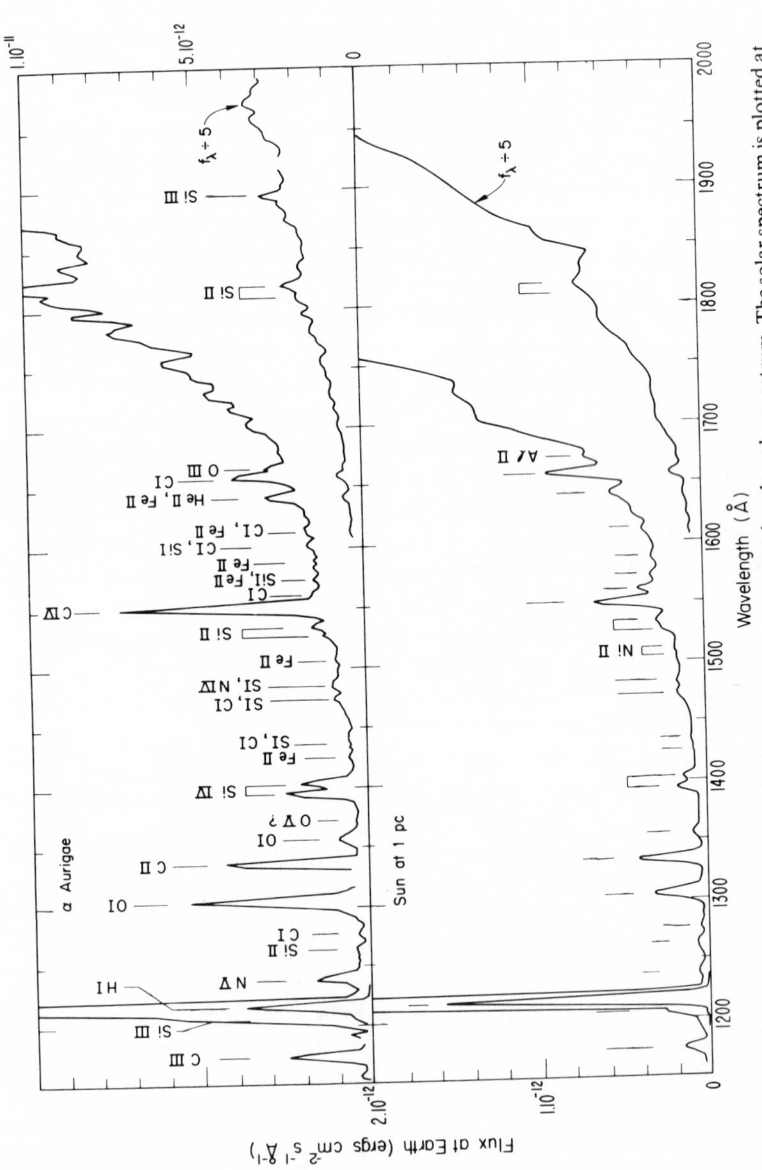

FIGURE 39. IUE spectrum of the star α Aurigae (Capella) compared to the solar spectrum. The solar spectrum is plotted at the same spectral resolution as the IUE spectrum, and the flux scale for the solar spectrum is defined assuming that the sun is at a distance of 1 parsec. Data are from Ref. (89). (Courtesy *Astrophys. J.*)

atmosphere. The other diagnostic lines have application in certain cases where densities might become quite high, e.g., stellar flares.

It is important to keep in mind that the stellar observations represent the flux from the entire disk of the star. If the star has quiet and active regions, such as the solar atmosphere, the observed fluxes in the lines represent some sort of average over the different atmospheric regions. If the sun represents any guide at all, this will certainly be the case for most stars. Figure 39 shows the spectrum of Capella (α Aur) obtained by Ayres and Linsky,[89] compared to an averaged solar spectrum as it would appear at a distance of 1 parsec. Capella is a binary system, containing G6 and F9 giants (luminosity class III). The transition region emission lines come from the atmosphere of the secondary (F9 III), which rotates more rapidly than the primary.[89] Analyses have shown that the degree of chromospheric and transition region activity for late-type stars is correlated with stellar rotation.[200]

The only type of transition region density diagnostic for $\lambda < 2000$ Å that can be applied to the low-resolution IUE spectra is the type where an intercombination line, such as C III 1909 Å, is compared to another allowed line formed at a similar temperature, such as Si IV 1394 Å. In applying such a diagnostic to stellar spectra, three potential difficulties should be considered: the relative element abundances in the stellar atmosphere may be somewhat different than in the sun's atmosphere; the distribution of plasma with temperature in the stellar transition region,i.e., the amount of plasma at a given temperature, may be different than in the solar transition region; and the possibility of nonionization equilibrium in the stellar atmosphere should be considered. As far as the first difficulty is concerned, there is not much to be said except that if the stellar abundances have been determined from previous observations in the visible, they should be used in calculating line ratios in the UV. In the following paragraphs, we consider the potential problem of plasma distribution in more detail; nonequilibrium is considered in Section 3.4.

In analyzing solar spectra, the distribution of plasma with temperature is often expressed through a quantity defined as the differential emission measure E i.e.,

$$E = N_e^2 \frac{dV}{dT_e} \tag{3.45}$$

where V is the plasma volume. We assume constant pressure is valid. (The solar transition region extends over such a small height interval above the limb that this assumption is most likely valid.) Then, with the scaled electron pressure p defined as $p = N_e T_e$, we have

$$E = \frac{p^2}{T_e^2} \frac{dV}{dT_e} \tag{3.46}$$

The reason the function E is considered is that it arises in the expression for the flux in an optically thin allowed line. Combining Eqs. (3.10), (3.20), and (3.21) leads to

$$F_{12}=\frac{\Delta E_{12}}{4\pi R^2}A_H\langle f(T_e)C_{12}^e\rangle p^2\int\frac{1}{T_e^2}\frac{dV}{dT_e}dT_e \tag{3.47}$$

where the average is over the narrow temperature range of formation of the transition region lines. If ionization equilibrium is assumed, the function E, and hence dV/dT_e, can be derived from the intensities of allowed transition region lines that span a wide temperature range, from a set of equations like Eq. (3.47). The function dV/dT_e, combined with $p=N_eT_e$, specifies the structure of the atmosphere, particularly if $V=ah$, where a is a constant area and height h is above the solar or stellar limb or along a flux tube. Model solar atmospheres have been calculated by a number of authors, e.g., Dupree,[201] Jordan,[202] and Gabriel.[203]

A stellar differential emission measure can be compared to the quiet-sun solar differential emission measure in a straightforward manner by constructing a graph that compares the relative intensities of solar and stellar transition region lines that are formed at different temperatures. Since only relative intensities are important for this comparison, we may define the sum of the intensities of the two strong C IV lines at 1548 and 1551 Å as unity. Using this scale, the solar intensities of the other important transition region lines are

TABLE 7

Quiet-Sun Emission Lines at +2 and +4 arc sec Outside the Solar Limb

Ion	Wavelength λ(Å)[a]	Transition	Intensity relative to C IV (1549 Å)
O V	1218.35	$2s^2\,{}^1S_0–2s2p\,{}^3P_1$	0.10
N V	1241	$2s\,{}^2S_{1/2}–2p\,{}^2P$	0.18
O V	1371.29	$2s2p\,{}^1P_1–2p^2\,{}^1D_2$	0.011
O IV	1401	$2s^22p\,{}^2P–2s2p^2\,{}^4P$	0.085
Si IV	1402.77	$3s\,{}^2S_{1/2}–3p\,{}^2P_{1/2}$	0.17
S IV	1406.00	$3s^23p\,{}^2P_{3/2}–3s3p^2\,{}^4P_{5/2}$	0.0078
N IV	1486.50	$2s^2\,{}^1S_0–2s2p\,{}^3P_1$	0.018
C IV	1549	$2s\,{}^2S_{1/2}–2p\,{}^2P$	1.0
O III	1666.15	$2s^22p^2\,{}^3P_2–2s2p^3\,{}^5S_2$	0.051
N III	1750	$2s^22p\,{}^2P–2s2p^2\,{}^4P$	0.054[b]
Al III	1854.72	$3s\,{}^2S_{1/2}–3p\,{}^2P_{3/2}$	0.099
Si III	1892.03	$3s^2\,{}^1S_0–3s3p\,{}^3P_1$	0.82
C III	1908.73	$2s^2\,{}^1S_0–2s2p\,{}^3P_1$	0.19

[a] Integer wavelengths refer to the entire multiplet.
[b] Blended with Ni II and Fe II.

thereby defined relative to C IV. These relative solar intensities are given in Table 7.

A particular stellar spectrum can now be compared to the solar spectrum in the following way. A similar scale is defined for the stellar spectrum such that the intensity of the two C IV lines is unity. Then the ratio $r = F_{ij}^{\text{stellar}}/F_{ij}^{\text{solar}}$ is defined for each line as the intensity ratio of the stellar line to the same solar line. Thus, if in our system of units F_{ij}^{stellar} for Si IV 1394 Å = 0.6 and $F_{ij}^{\text{solar}} = 0.3$, then $r = 2$ for the Si IV 1394 Å line.

Using this definition, a number of stellar spectra are compared in Fig. 40 to the solar spectrum defined in Table 7. The sources of the stellar spectra are Ayres and Linsky,[89] Bopp and Stencel,[204] Linsky *et al.*,[205] Simon and Linsky,[206] Simon, Linsky, and Schiffer,[207] Basri, Linsky, and Eriksson,[208] and Hartmann, Dupree, and Raymond.[209] The allowed lines are indicated by crosses; the intercombination lines by dots; the solid lines connect the allowed-line crosses and define the stellar differential emission measure relative to the solar one. The solid line is also connected to the 1892 Å Si III dot because at the densities we are mostly concerned with, this line behaves as an allowed line. Densities derived from the stellar spectra can be obtained by measuring the increment of distance (departure) along the ordinate that separates the 1909 Å C III dot from the solid line and then measuring up or down by the same increment along the ordinate of Fig. 29 from the point that defines $n_i/N_e N_z$ for C III at 10^{10} cm^{-3}, i.e., the quiet-sun solar density. In this way, the densities given in Fig. 40 have been derived. (We comment further on the results for β Dra.) The densities derived as described depend on the following conditions and assumptions: the intensity calibration as a function of wavelength is correct for both the IUE and NRL Skylab instruments; the relative element abundances in the stellar atmospheres are the same as in the sun; ionization equilibrium is valid; and the quiet-sun transition region electron pressure is $N_e T_e = 6 \times 10^{14}$ cm^{-3} K.

It is relatively straightforward from Figs. 29 and 40 and Table 7 to ascertain the effect that violating any of these assumptions would have on the results, with the exception of the ionization equilibrium assumption. For example, consider the star HD 199178. The results in Fig. 40 have been derived using a Skylab calibration given by Nicolas *et al.*[210] A somewhat different calibration was derived by Doschek *et al.*[37] Use of the Doschek *et al.* calibration would give a density of 3.5×10^{10} cm^{-3} for HD 199178 instead of the density shown in Fig. 40. Similarly, if the quiet sun's electron pressure were twice the chosen value of 6×10^{14} cm^{-3} K, then the density for HD 199178 would be 4.5×10^{10} cm^{-3}, derived using the Nicolas *et al.*[210] calibration, and 6.3×10^{10} cm^{-3} using the Doschek *et al.*[37] calibration. The errors are larger for stars with densities below 10^{10} cm^{-3} because the C III 1909 Å density sensitivity begins to decrease below this value. It is clear that a good instrumental calibration is necessary in order to apply successfully the density diagnostics to low-density stellar atmospheres.

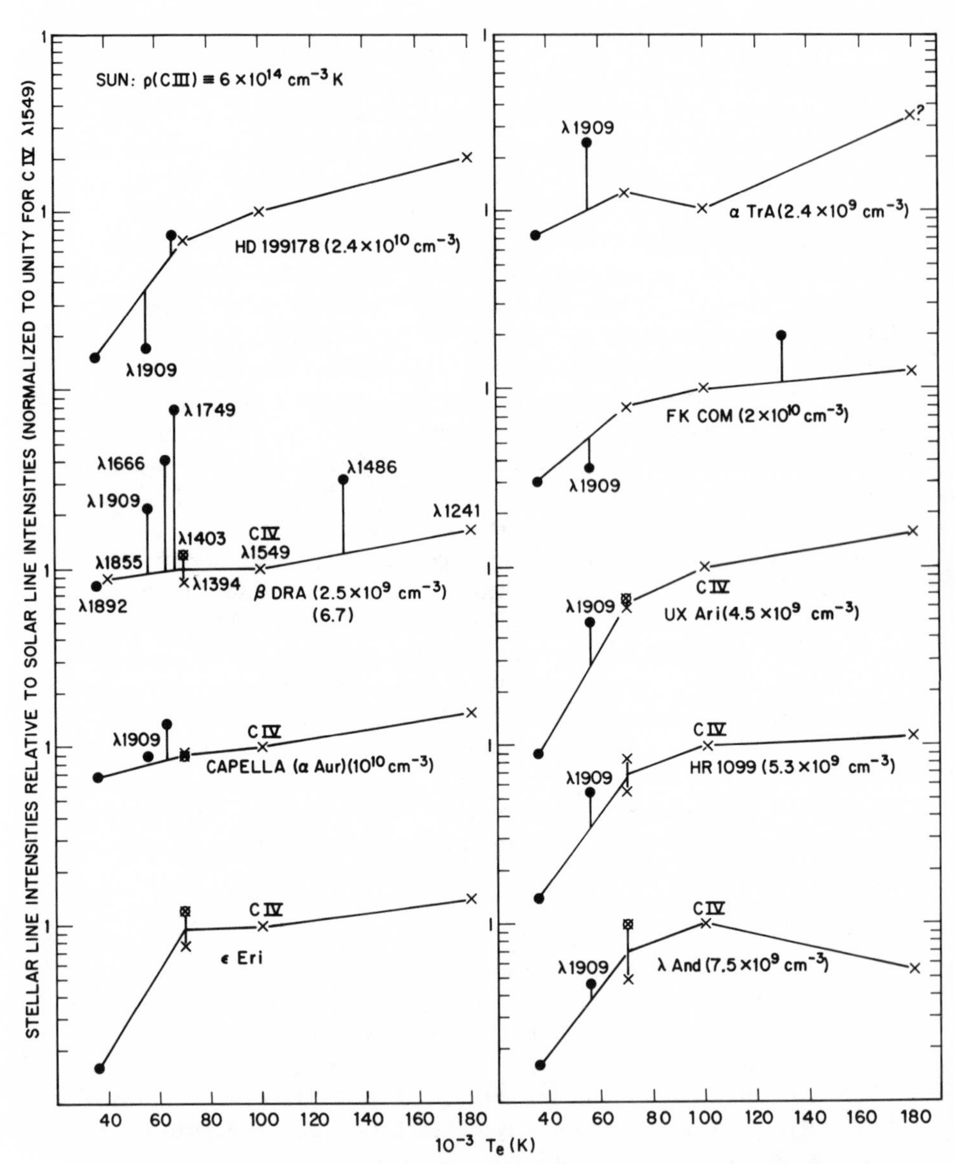

FIGURE 40. IUE stellar spectra compared to the quiet-sun solar spectrum (see text for discussion and sources of stellar spectra). The densities determined for each star pertain to the temperature region near 6×10^4 K.

We note in Fig. 40 that the shapes of the stellar differential emission measures can be quite different from the quiet-sun solar case, particularly at the low temperature of formation of Si III. The qualitative differences between the stellar and solar atmospheres can be estimated at a glance from such diagrams as those in Fig. 40.

Similar diagrams for different solar regions are shown in Fig. 41. Spectra of a surge, active region and a prominence are compared to the quiet-sun spectrum. We note differences in the differential emission measures as well as much enhanced densities in the surge and active regions relative to the quiet sun. The densities are so high that nearly all the intercombination lines are affected and not just the 1909 Å line. Densities can be derived from the vertical departures of each of the intercombination lines by using Fig. 29. The prominence density is about the same as the quiet-sun density ($\simeq 10^{10}\,\mathrm{cm}^{-3}$), but the surge and active-region densities are $\simeq 10^{11}\,\mathrm{cm}^{-3}$ or higher. Densities obtained using different lines agree rather well.

Note in Fig. 41 that the 1909 Å C III line is always the line most affected, i.e., its density is reduced by more than the other lines from the expected value at $10^{10}\,\mathrm{cm}^{-3}$, i.e., it has the largest departure. From Fig. 29, it is clear that the 1909° C III line will be the most affected, because it becomes density sensitive at a lower density than any of the other lines, as we have mentioned previously. With this in mind, we can see immediately that the departures for the 1666 Å O III, 1486 Å N IV, and 1752 Å N III lines are too large to be due to a density effect in the β Dra spectrum in Fig. 40. If these departures were due to a reduced density, then the departure for 1909 Å would have to be much larger. But even worse, if the solar quiet-sun pressure is really $6 \times 10^{14}\,\mathrm{cm}^{-3}\,\mathrm{K}$, then the populations of the upper levels of the O III, N IV, and N III lines are nearly in the low-density limit, and departures as large as those observed are impossible under the assumptions we have just stated. It is more likely, as indicated by Basri *et al.*,[(208)] that the C IV and Si IV lines are optically thick and photons have been lost from them; therefore, these allowed lines have spuriously low intensities. In fact, very recently, high-resolution, IUE spectra of β Dra have been analyzed by Brown *et al.*[(211)] The high-resolution spectra show that the Si IV and C IV lines are optically thick. They have derived densities using the C III 1909 Å/O III 1666 Å ratio and the C III 1909 Å/Si III 1892 Å ratio, obtaining $N_e \lesssim 1 \times 10^{10}\,\mathrm{cm}^{-3}$ and $N_e < 2 \times 10^{9}\,\mathrm{cm}^{-3}$, respectively.

In summary, for the spectral region between 1175 and 2000 Å, the 1909 Å C III line is the only reasonable density diagnostic for densities less than $10^{10}\,\mathrm{cm}^{-3}$. If sufficient spectral resolution and sensitivity are available, then the 1175 Å C III multiplet may also be useful. For outbursts on flare stars, some of the other intercombination lines may be useful if high enough densities are produced.

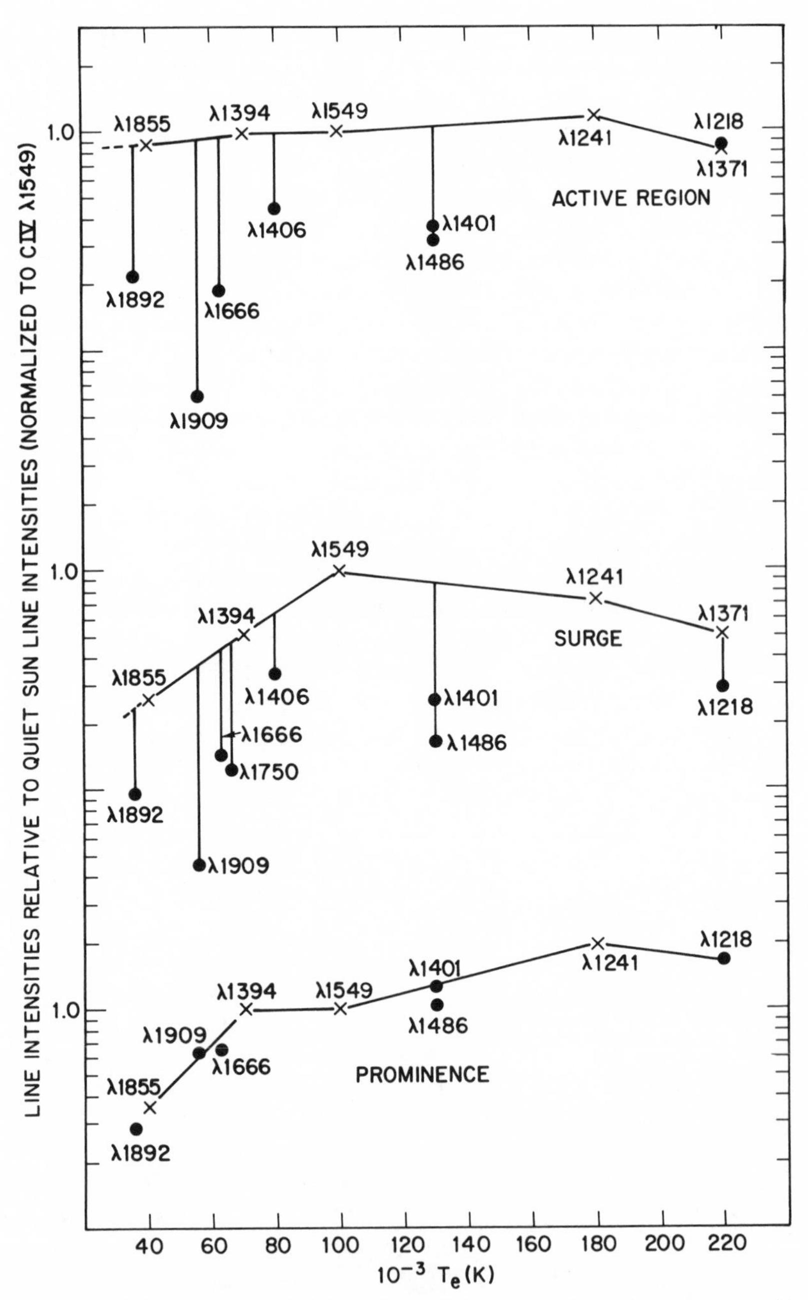

FIGURE 41. Skylab solar spectra of an active region, surge, and prominence compared to a typical quiet solar spectrum (see text for discussion).

However, there is a good stellar diagnostic for low densities and wavelengths greater than 2000 Å. This is the group of C II lines near 2325 Å. These lines are due to $^4P^0 \rightarrow {}^2P^e$ transitions in the B I isoelectronic sequence, and density sensitivity occurs as for the N III lines near 1750 Å and the O IV lines near 1400 Å discussed previously. These lines have been observed in the spectra of cool stars by Stencel *et al.*[212] The range of density sensitivity according to these authors is 10^7–10^9 cm^{-3}; thus, they are good diagnostics for low-density stellar atmospheres, such as the atmospheres of supergiants. For example, densities obtained from the measurements of Stencel *et al.*[212] are $\simeq 2.5 \times 10^8$ cm^{-3} at $T_e \simeq 10^4$ K for α Boo and α Tau.

Some quite interesting density diagnostics for low-density nonstellar astrophysical objects have also become available from IUE observations. We are referring to such objects as planetary nebulas, H II regions, and possible circumstellar shells around such systems as eruptive symbiotics. Densities measured from optical (visible spectral region) density diagnostics are typically 10^3 cm^{-3} for planetary nebulas; in some shells, densities on the order of 10^5–10^6 cm^{-3} may occur.

Wavelengths of two lines of the N I isoelectronic sequence have been given previously in Table 5 for several ions of the sequence. As we discussed earlier, the S X lines are a good density diagnostic for the solar corona. The

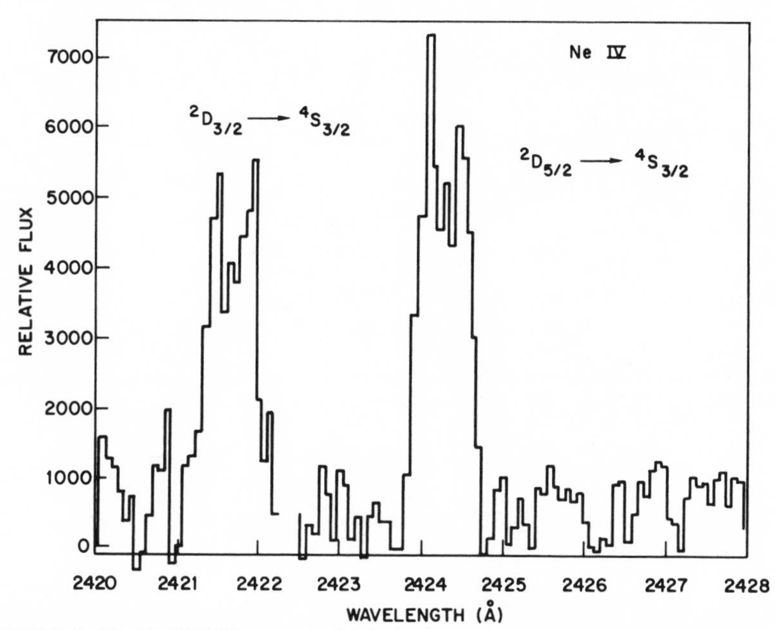

FIGURE 42. The Ne IV IUE spectrum of the planetary nebula NGC 3242, obtained in Ref. (214). The lines are split due to internal expansion of the nebula.

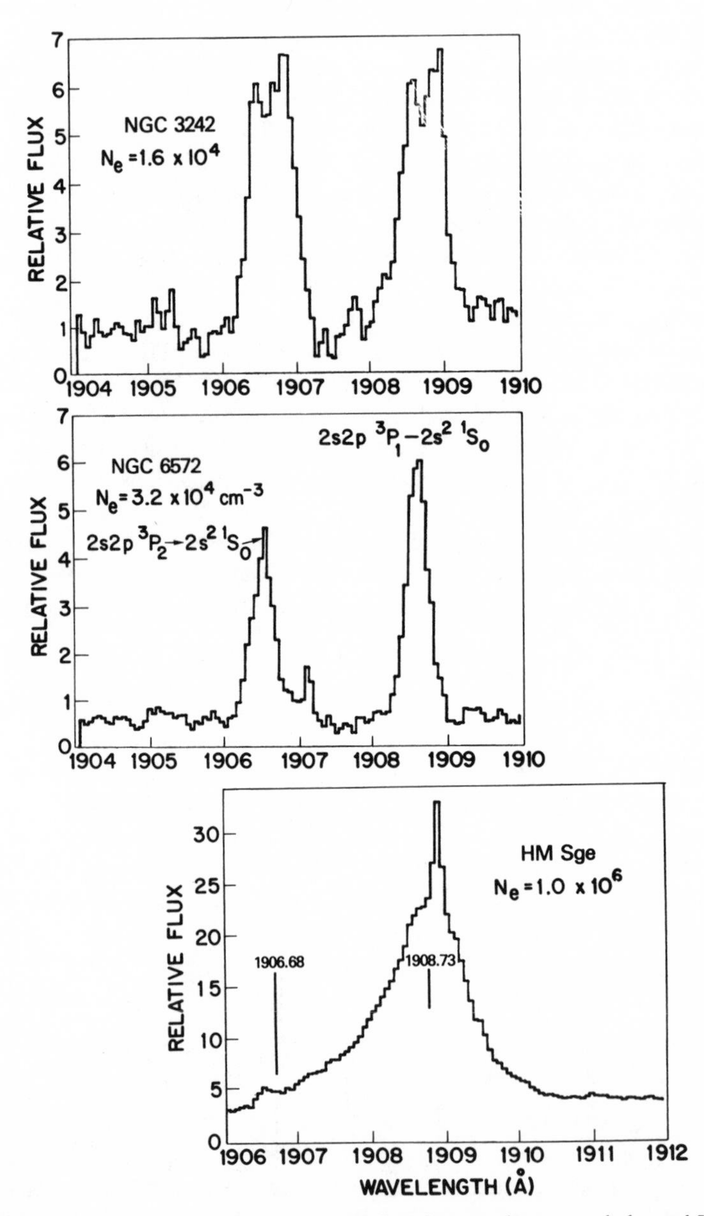

FIGURE 43. IUE spectra of the indicated C III lines for two planetary nebulas and HM Sge. Nebular motions determine the shapes of the two optically thin spectral lines. Spectra are from Refs. (218 and 219.)

calculations show that the Ne IV lines might be useful for densities between 10^4 and 10^5. This is indeed the case, and the Ne IV lines were identified in the IUE spectrum of the planetary nebulas NGC 7662 by Lutz and Seaton.[213] They derived a density of $1.1 \times 10^4\ \mathrm{cm}^{-3}$ at the temperature of formation of the Ne IV lines (1.3×10^4 K). Feibelman[214] has recently obtained spectra of these lines in several planetary nebulas, including NGC 3242, 6818, and 7009. The spectrum of NGC 3242 is shown in Fig. 42. Notice that the lines are split due to the Doppler effect produced by internal expansion of the nebula. For this nebula, the ratio appears to be close to the low-density limit.

One of the most interesting diagnostics for such objects as planetary nebulas again involves C III. Consider line b in Fig. 24. This is a magnetic quadrupole transition with a very small transition probability.[215] However, at low densities, such as $10^3\ \mathrm{cm}^{-3}$, line b behaves like an allowed line because the collision rate is much less than the decay rate. The wavelength of line b falls quite close to the wavelength of line a, i.e., 1907 Å. The ratio b/a is similar to the three-level example illustrated in Fig. 23. In the low-density limit, the ratio is about 5/3; at high densities the, ratio falls as shown on the left-hand side of Fig. 23. The ratio is negligible at solar densities, and the quadrupole line is not observed in solar spectra. However, it is a strong line in the IUE spectra of some planetary nebulas. The idea of using the C III quadrupole line as a density diagnostic was actually suggested some time ago by Osterbrock.[216] The calculation of the line ratio as a function of density has been carried out by Nussbaumer and Schild,[217] who also obtained IUE observations of these lines for HM Sge. Also, Feibelman *et al.*[218,219] have obtained IUE spectra of 13 planetary nebulas and HM Sge and have used the 1907 Å/1909 Å ratio with the Nussbaumer and Schild[217] calculations to derive electron densities. Figure 43 shows some spectra and derived densities from their work. Note that in addition to density, the line profiles give considerable information about the dynamics within the nebulas. Note also the high density for the unusual object HM Sge, which is suspected of being a protoplanetary in its early stages of development.

3.4. *Ionization Equilibrium*

In the preceding sections, ionization equilibrium (IE) [see Eq. (3.3)] has been assumed in the application of spectroscopic diagnostics except for the dielectronic recombination temperature diagnostic (Section 3.2.3) and a few density diagnostics where the transition energies are small compared to kT_e. For these cases, actual departures from IE will not affect the results significantly. However, there are other cases where a departure from IE may have considerable influence on a derived temperature and density, and we consider some of the effects of a departure in this section.

Before proceeding, we note that in the case of the solar atmosphere, there is no concrete spectroscopic evidence for a departure from IE in any region of the atmosphere accessible to recent observations. That is, existing spectroscopic data can be interpreted in terms of IE, and no inconsistencies result that can be interpreted in *only* terms of a departure from IE. There are observations of Doppler shifts of transition region lines that can be interpreted as anisotropic motions or flows of transition region plasma. If the Doppler shifts do in fact reflect net flows of plasma and if the flows occur across a steep temperature gradient, then departure from IE may occur. However, the solar Doppler shifts need not be interpreted in terms of a *net* flow of plasma, so these shifts do not provide unambiguous evidence for even the possibility of a departure from IE.

Consider a solar or stellar flux tube filled with plasma that reaches a maximum temperature T_M. We assume that the footpoints of this tube are anchored in the chromosphere and that heat conduction occurs along the flux tube toward the chromosphere. Energy is supplied to the loop by an unknown mechanism and conducted into the chromosphere where it is radiated away. A temperature gradient will exist along the flux tube, and thus plasma will exist at all temperatures between T_M and the temperature of the chromosphere ($\simeq 10^4$ K). We assume that the loop plasma is in static equilibrium, i.e., the temperature and density at any point along the flux tube are constant in time. The physics governing such flux tubes has in recent years been subject to intensive investigations.[220–223]

There are two distinct but related ways in which the plasma in this flux tube can depart from IE. Firstly, the energy source can be either increased or decreased substantially, which will result in either *in situ* plasma heating or cooling. Plasma motions in the tube may also occur as a consequence of energy source variations. If the temperature variations in the tube occur on a time scale faster than the ionization and recombination time scales, the plasma will depart from IE, and the distribution of ions will depend on initial and boundary conditions and will be solutions of Eq. (3.1). If heating or cooling occurs in a transient and random manner, the plasma at each point in the tube will be continually subjected to variations in temperature and density and may never reach ionization equilibrium. On the other hand, the plasma may be regarded as in a quasi-dynamic equilibrium from the hydrodynamical point of view over time scales larger than the time scale for variations in energy input. That is, no net flow of plasma in or out of the tube need occur, and over a sufficiently large time scale, the time-averaged temperature and density at each point in the tube would remain constant. A self-consistent hydrodynamical treatment of this problem has recently been carried out by Mariska *et al.*[224] As a starting point in the calculations, these authors chose a solar flux tube with an electron pressure of 6×10^{14} K cm^{-3} and a maximum temperature T_M of 6×10^5 K.

They considered both impulsive heating and cooling of this tube. The result is that the fractional ion abundances can depart from equilibrium values by as much as an order of magnitude. However, if the heating is followed by cooling, or vice versa, in quasi equilibrium, this departure may be considerably reduced, i.e., the departure may be reduced to a factor of two to four.

When large departures from ionization balance occur, i.e., an order of magnitude, it appears from the calculation that temperature-sensitive line ratios would indicate a departure from IE. That is, line ratios would give temperatures uncharacteristic of IE values. However, when the departures of ion abundances are less, i.e., factors of two to four, then the calculations show that many observationally accessible temperature-sensitive line ratios are rather close to IE values, and departure from IE may be difficult to detect observationally. Estimates of emission measures from allowed lines emitted from such a loop could be in error by factors of two to four, and the differential emission measure would also be expected to reflect the departure from IE. More work on the spectroscopic consequences of transient heating and cooling within flux tubes is needed to clarify these issues.

The consequences of transiently cooling plasma on the spectral output of lower transition region lines, such as C IV 1550 Å, have been investigated by Dere *et al.*[225] These authors find that cooling plasma, resulting from a postulated thermal instability in coronal plasma, can explain the time- and space-varying character of these emission lines in the high-resolution HRTS spectra (see Table 1).

Another method of producing a departure from IE in our idealized flux tube is to consider the case where the energy input is constant in time but heat is injected into the loop asymmetrically, i.e., most of the energy is deposited into one side of the loop. Boris and Mariska[226] and Mariska and Boris[227] have treated this problem in a hydrodynamically self-consistent manner and found that a net plasma flow is set up in the tube. The plasma flows into the tube from one footpoint and out through the other, and total mass is conserved. The flow velocity, temperature, and density are constant in time at each point in the tube.

From the spectroscopic standpoint, these flows occur across the very steep temperature gradient transition region of these tubes and result in departures from IE. This particular spectroscopic problem has been studied by a number of authors, e.g., Raymond and Dupree;[228] Dupree, Moore, and Shapiro;[229] Francis;[230] and Roussel-Dupre and Beerman.[231] The result in all these calculations is that significant variations in temperature-sensitive line ratios can occur. Results from Raymond and Dupree[228] are shown in Fig. 44. In Fig. 44, velocities that result in plasma flowing into the chromosphere, or equivalently into lower-temperature regions, are positive (downflows). Velocities that result in plasma flowing toward higher temperatures are negative

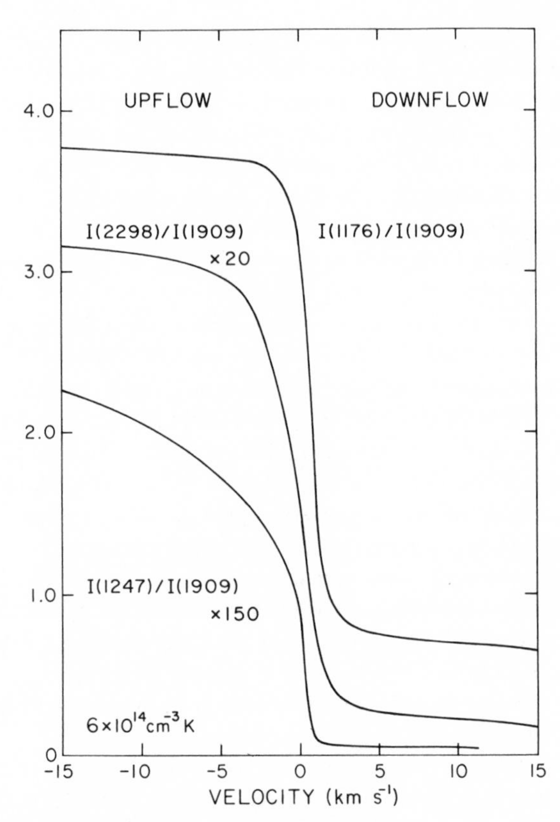

FIGURE 44. The effect of mass flows through a thin transition region on III line ratios. The ordinate is intensity ratio. Calculations are from Ref. (228) (see text for discussion). (Courtesy *Astrophys. J.*)

(upflows). The point is that if the transition region is a very thin region, i.e., on the order of a few hundred kilometers thick, ions may find themselves in a lower- or higher-temperature region than would occur in IE, because ionization and recombination times are longer than the times required to flow through a substantial part of the transition region. Equation (3.23) then shows that line ratios may be significantly altered because of the Boltzmann factors in the excitation rate coefficients. Figure 44 shows that large variations in line ratios are possible if the transition region is thin.

However, if the *observed* transition region is much thicker than that chosen for these flow calculations, i.e., contained in magnetic structures isolated from the corona, departures in transition region line ratios would not be so large as those derived by Raymond and Dupree[228] even for substantial flow velocities. In principle, if the 1175, 1247, and 1909 Å C III lines or similar lines could be observed simultaneously, the effect of a departure from IE should be detectable. The point is that the three C III lines allow both a temperature and density determination. If the temperature departs significantly from the IE value, then a departure from IE may be indicated.

So far, we have considered primarily departures from IE in transition region plasmas. Departures in coronal plasmas have not been well investigated. An exception is solar flares, where there are difficulties in interpreting the pertinent x-ray line ratios, but where IE appears nevertheless to hold, at least up to the He-like ionization stages (see Section 3.2.3).

Two problems not discussed so far are the accuracy of the IE calculations themselves and determining relative element abundances using UV spectroscopic techniques. The latter problem is well covered in the literature [e.g., Pottasch,[232] Jordan and Pottasch,[233] and Dupree[201]] and will not be discussed further here. The state of IE calculations has been discussed by Dere and Mason.[152] We simply make the following remark here: one of the best sets of currently available IE calculations for low-density plasmas appears to be that by Jacobs *et al.*,[234] although more recent calculations for calcium and iron have been published by Doyle and Raymond,[235] and a set of calculations for astrophysically abundant elements has been recently published by Shull and Van Steenberg.[236] The IE calculations by Summers[237] for the high-Z ions, such as iron, appear to give temperatures of maximum ion abundance that are too high by about a factor of two. Autoionization processes are important in calculating both dielectronic recombination and ionization rate coefficients. The effect on ionization rate coefficients of autoionization following excitation to metastable levels may be important in many cases and needs further investigation. For example, Cowan and Mann[238] have shown that this process can be about 1.5 times greater than the direct ionization rate for Fe XVI for $kT_e = 300$–500 eV. Finally, charge exchange may significantly alter IE calculations in some cases.[239]

ACKNOWLEDGEMENTS

This work was supported by NRL/ONR and a NASA grant from the Office of Solar and Heliospheric Physics. The author would like to thank Dot Waters for typing the final version of the manuscript. He thanks Uri Feldman for critical comments on the manuscript.

REFERENCES

1. J. H. Parkinson, *Nature* **236**, 68 (1972).
2. R. J. Hutcheon, J. P. Pye, and K. D. Evans, *Astron. Astrophys.* **51**, 451 (1976).

3. Yu. I. Grineva. V. I. Karev, V. V. Korneev, V. V. Krutov, S. L. Mandelstam, L. A. Vainstein, B. N. Vasilyev, and I. A. Zhitnik, *Solar Phys.* **29**, 441 (1973).
4. E. V. Aglizki, V. A. Boiko, A. Ya. Faenov, V. V. Korneev, V. V. Krutov, S. L. Mandelstam, S. A. Pikuz, U. I. Safronova, J. A. Sylwester, A. M. Urnov, L. A. Vainstein, and I. A. Zhitnik, *Solar Phys.* **56**, 375 (1978).
5. V. V. Korneev, V. V. Krutov, S. L. Mandelstam, B. Sylwester, I. P. Tindo, A. M. Urnov, B. Valnicek, and I. A. Zhitnik, *Solar Phys.* **68**, 381 (1980).
6. V. V. Korneev, I. A. Zhitnik, S. L. Mandelstam, and A. M. Urnov, *Solar Phys.* **68**, 391 (1980).
7. G. A. Doschek, U. Feldman, P. B. Landecker, and D. L. McKenzie, *Astrophys. J.* **249**, 372 (1981).
8. U. Feldman, G. A. Doschek, R. W. Kreplin, and J. T. Mariska, *Astrophys. J.* **241**, 1175 (1980).
9. D. L. McKenzie and P. B. Landecker, *Astrophys. J.* **248**, 1117 (1981).
10. D. L. McKenzie, P. B. Landecker, R. M. Broussard, H. R. Rugge, R. M. Young, U. Feldman, and G. A. Doschek, *Astrophys. J.* **241**, 409 (1980).
11. D. L. McKenzie and P. B. Landecker, *Astrophys. J.* **254**, 309 (1982).
12. J. L. Culhane, A. H. Gabriel, L. W. Acton, C. G. Rapley, K. J. H. Phillips, C. J. Wolfson, E. Antonucci, R. D. Bentley, R. C. Catura, C. Jordan, M. A. Kayat, B. J. Kent, J. W. Leibacher, A. N. Parmar, J. C. Sherman, L. A. Springer, K. T. Strong, and N. J. Veck, *Astrophys. J.* (*Lett.*) **244**, L141 (1981).
13. L. W. Acton, J. L. Culhane, A. H. Gabriel, C. J. Wolfson, C. G. Rapley, K. J. H. Phillips, E. Antonucci, R. D. Bentley, R. W. Hayes, E. G. Joki, C. Jordan, M. A. Kayat, B. Kent, J. W. Leibacher, R. A. Nobles, A. N. Parmar, K. T. Strong, and N. J. Veck, *Astrophys. J.* (*Lett.*) **244**, L137 (1981).
14. K. J. H. Phillips, J. W. Leibacher, C. J. Wolfson, J. H. Parkinson, B.C. Fawcett, B. J. Kent, H. E. Mason, L. W. Acton, J. L. Culhane, and A. H. Gabriel, *Astrophys. J.* **256**, 774 (1982).
15. K. Tanaka, T. Watanabe, K. Nishi, and K. Akita, *Astrophys. J.* (*Lett.*) **254**, L59 (1982).
16. W. E. Behring, L. Cohen, and U. Feldman, *Astrophys. J.* **175**, 493 (1972).
17. W. E. Behring, L. Cohen, U. Feldman, and G. A. Doschek, *Astrophys. J.* **203**, 521 (1976).
18. M. Malinovsky and L. Heroux, *Astrophys. J.* **181**, 1009 (1973).
19. S. O. Kastner, W. M. Neupert, and M. Swartz, *Astrophys. J.* **191**, 261 (1974).
20. W. M. Burton and A. Ridgeley, *Solar Phys.* **14**, 3 (1970).
21. F. F. Freeman and B. B. Jones, *Solar Phys.* **15**, 288 (1970).
22. W. M. Burton, C. Jordan, A. Ridgeley, and R. Wilson, *Astron. Astrophys.* **27**, 101 (1973).
23. B. C. Boland, E. P. Dyer, J. G. Firth, A. H. Gabriel, B. B. Jones, C. Jordan, R. W. P. McWhirter, P. Monk, and R. F. Turner, *Mon. Not. R. Astron. Soc.* **171**, 697 (1975).
24. E. Chipman and E. C. Bruner, Jr. *Astrophys. J.* **200**, 765 (1975).
25. G. W. Cushman, L. Farwell, G. Godden, and W. A. Rense, *J. Geophys. Res.* **80**, 482 (1975).
26. G. J. Rottman, F. Q. Orrall, and J. A. Klimchuk, *Astrophys. J.* (*Lett.*) **247** L135 (1981).
27. K. P. Dere, H. E. Mason, K. G. Widing, and A. K. Bhatia, *Astrophys. J.* (*Suppl.*) **40**, 341 (1979).
28. K. G. Widing, in *Solar Gamma, X, and EUV Radiation* (S. R. Kane, ed.), Symposium no. 68 of the International Astronomical Union, p. 153 (1975).
29. J. E. Vernazza and E. M. Reeves, *Astrophys. J.* (*Suppl.*) **37**, 485 (1978).
30. W. H. Parkinson and E. M. Reeves, *Solar Phys.* **10**, 342 (1969).
31. J. L. Kohl, *Astrophys. J.* **211**, 958 (1977).
32. C. Jordan, *Solar Phys.* **21**, 381 (1971).
33. A. H. Gabriel, W. R. S. Garton, L. Goldberg, T. J. L. Jones, C. Jordan, F. J. Morgan, R. W. Nicholls, W. J. Parkinson, H. J. B. Paxton, E. M. Reeves, C. B. Shenton, R. J. Speer, and R. Wilson, *Astrophys. J.* **169**, 595 (1971).

34. A. H. Gabriel and C. Jordan, *Mon. Not. R. Astron. Soc.* **173**, 397 (1975).
35. G. E. Brueckner, J.-D. F. Bartoe, K. R. Nicolas, and R. Tousey, *Nature* **226**, 1132 (1970).
36. G. E. Brueckner and K. R. Nicolas, *Solar Phys.* **29**, 301 (1973).
37. G. A. Doschek, U. Feldman, M. E. VanHoosier, and J.-D. F. Bartoe, *Astrophys. J.* (*Suppl.*) **31**, 417 (1976).
38. G. A. Doschek, U. Feldman, and Leonard Cohen, *Astrophys. J. (Suppl)* **33**, 101 (1977).
39. G. E. Brueckner and O. Kjeldseth Moe, *Space Res.* **12**, 1595 (1973).
40. G. E. Brueckner and J.-D. F. Bartoe, *Solar Phys.* **38**, 133 (1974).
41. G. E. Brueckner, J.-D. F. Bartoe, and M. E. VanHoosier, in *Proceedings of the Nov. 7–10, 1977 OSO-8 Workshop*, p. 380, University of Colorado Press (1977).
42. G. E. Brueckner, in *Highlights of Astronomy* (P. A. Wayman, ed.), vol 5, p. 557, International Astronomical Union (1980).
43. H. C. McAllister, *Solar Phys.* **21**, 27 (1971).
44. L. R. Doherty and H. C. McAllister, *Astrophys. J.* **222**, 716 (1978).
45. E. C. Bruner, Jr., E. G. Chipman, B. W. Lites, G. J. Rottman, R. A. Shine, R. G. Athay, and O. R. White, *Astrophys. J.* (*Lett.*) **210**, L97 (1976).
46. R. Roussel-Dupre, M. H. Francis, and D. E. Billings, *Mon. Not. R. Astron. Soc.* **187**, 9 (1979).
47. G. E. Artzner, R. M. Bonnet, P. Lemaire, J. C. Vial, A. Jouchoux, J. Leibacher, A. Vidal-Madjar, and M. Vite, *Space Sci. Instrum.* **3**, 131 (1977).
48. R. M. Bonnet, P. Lemaire, J. C. Vial, G. Artzner, P. Gouttebroze, A. Jouchoux, J. W. Leibacher, A. Skumanich, and A. Vidal-Madjar, *Astrophys. J.* **221**, 1032 (1978).
49. E. Tandberg-Hanssen, R. G. Athay, J. M. Beckers, J. C. Brandt, E. C. Bruner, R. D. Chapman, C.-C. Cheng, J. B. Gurman, W. Henze, C. L. Hyder, A. G. Michalitisianos, R. A. Shine, S. A. Schoolman, and B. E. Woodgate, *Astrophys. J.* (*Lett.*) **244**, L127 (1981).
50. C.-C. Cheng, E. Tandberg-Hanssen, E. C. Bruner, L. Orwig, K. J. Frost, P. J. Kenny, B. E.Woodgate, and R. A. Shine, *Astrophys. J.* (*Lett.*) **248**, L39 (1981).
51. J. B. Rogerson, L. Spitzer, J. F. Drake, K. Dressler, E. B. Jenkins, D. C. Morton, and D. G. York, *Astrophys. J.* (*Lett.*) **181**, L97 (1973).
52. T. P. Snow, Jr., *Earth Extraterrestrial Sci.* **3**, 1 (1976).
53. T. P. Snow, Jr. and E. B. Jenkins, *Astrophys. J.* (*Suppl.*) **33**, 269 (1977).
54. A. Boggess, F. A. Carr, D. C. Evans, D. Fischel, H. R. Freeman, C. F. Fuechsel, D. A. Klinglesmith, V. L. Krueger, G. W. Longanecker, J. V. Moore, E. J. Pyle, F. Rebar, K. O. Sizemore, W. Sparks, A. B. Underhill, H. D. Vitagliano, D. K. West, F. Macchetto, B. Fitton, P. J. Barker, E. Dunford, P. M. Gondhalekar, J. E. Hall, V. A. W. Harrison, M. B. Oliver, M. C. W. Sandford, P. A. Vaughan, A. K. Ward, B. E. Anderson, A. Boksenberg, C. I. Coleman, M. A. J. Snijders, and R. Wilson, *Nature* **275**, 2 (1978).
55. C. R. Canizares, G. W. Clark, D. Bardas, and T. Markert, *SPIE J.* **106**,154 (1977).
56. C. R. Canizares and P. F. Winkler, *Astrophys. J.* (*Lett.*) **246**, L33 (1981).
57. A. B. C. Walker, Jr., *Space Sci. Rev.* **13**, 672 (1972).
58. G. A. Doschek, *Space Sci. Rev.* **13**, 765 (1972).
59. A. Burgess, *Astrophys. J.* (*Lett.*) **141**, 1588 (1965).
60. M. J. Seaton and P. J. Storey, in *Atomic Processes and Applications* (P. G. Burke and B. L. Moiseiwitsch, eds.), Chap. 6, p. 135, North-Holland, Amsterdam (1976).
61. A. H. Gabriel and C. Jordan, in *Case Studies in Atomic Collision Physics* (E. W. McDaniel and M. C. McDowell, eds.), vol. 2, Chap. 4, North-Holland, Amsterdam (1972).
62. P. J. Storey, *Mon. Not. R. Astron. Soc.* **195**, 27P (1981).
63. W. Grotrian, *Naturwissenschaften* **27**, 214 (1939).
64. B. Edlén, *Z. Astrophys.* **22**, 30 (1942).
65. G. A. Doschek and U. Feldman, *Astrophys. J.* (*Lett.*) **212**, L143 (1977).

66. R. H. Munro and B. V. Jackson, *Astrophys. J.* **213**, 874 (1977).
67. A. J. Hundhausen, in *Coronal Holes and High-Speed Wind Streams* NASA Skylab Solar Workshop 1 (Jack B. Zirker, ed.), Chap. 7, University Press, Colorado (1977).
68. K. P. Dere, *The Extreme Ultraviolet Structure of Solar Active Regions,* Ph.D. Dissertation, Catholic University of America, Washington, D.C. (1980).
69. Z. Svestka, *Solar Flares*, Chap. 4., D. Reidel , Dordrecht, Holland (1976).
70. E. Tandberg-Hanssen, *Solar Prominences*, Chap. 1, D. Reidel, Dordrecht, Holland (1974).
71. N. R. Sheeley, Jr., D. J. Michels, R. A. Howard, and M. J. Koomen, *Trans. Am. Geophys. Union* **62**, 153 (1981).
72. E. G. Gibson, *The Quiet Sun*, Chap. 5, NASA SP-303, U.S. Government Printing Office, Washington, D.C (1973).
73. U. Feldman, G. A. Doschek, and J. T. Mariska, *Astrophys. J.* **229**, 369 (1979).
74. U. Feldman, *Astrophys. J.* **275**, 367 (1983).
75. J. V. Hollweg, S. Jackson, D. Galloway, *Solar Phys.* **75**, 35 (1982).
76. R. G. Athay and O. R. White, *Astrophys. J.* **229**, 1147 (1979).
77. R. G. Athay and T. E. Holzer, *Astrophys. J. 255*, 743 (1982).
78. J. Ionson, *Astrophys. J.* **254**, 318 (1982).
79. R.C. Canfield, J. C. Brown, G. E. Brueckner, J. W. Cook, I. J. D. Craig, G. A. Doschek, A. G. Emslie, J. C. Henoux, B. W. Lites, M. E. Machado, and J. H. Underwood, in *Solar Flares* (P. A. Sturrock, ed.), Chap. 6, Colorado Associated University Press, CO (1980).
80. K. G. Widing and C.-C Cheng, *Astrophys. J.* (*Lett.*) **194**, L111 (1974).
81. C.-C. Cheng and K. G. Widing, *Astrophys. J.* **201**, 735 (1975).
82. P. B. Landecker and D. L. McKenzie, *Astrophys. J.* (*Lett.*) **241**, L175 (1980).
83. C.-C. Cheng, *Solar Phys.* **55**, 413 (1977).
84. K. G. Widing and K. P. Dere, *Solar Phys.* **55**, 431 (1977).
85. G. A. Doschek, U. Feldman, R. W. Kreplin, and L. Cohen, *Astrophys. J.* **239**, 725 (1980).
86. S. R. Kane, C. J. Crannell, D. Datlowe, U. Feldman, A. H. Gabriel, H. S. Hudson, M. R. Kundu, C. Matzler, D. Neidig, V. Petrosian, and N. R. Sheeley, Jr., in *Solar Flares* (P. A. Sturrock, ed.), Chap. 5, Colorado Associated University Press, CO (1980).
87. A. L. Kiplinger, B. R.Dennis, A. G. Emslie, K. J. Frost and L. E. Orwig, *Astrophys. J.* (*Lett.*) **265**, L99 (1983).
88. U. Feldman, C.-C. Cheng, and G. A. Doschek, *Astrophys. J.* **255**, 320 (1982).
89. T. R. Ayres and J. L. Linsky, *Astrophys. J.* **241**, 279 (1980).
90. P. F. Winkler, C. R. Canizares, G. W. Clark, T. H. Markert, K. Kalata, and H. W. Schnopper, *Astrophys. J.* (*Lett.*) **246**, L27 (1981).
91. T. Ayres, in *The Universe at Ultraviolet Wavelengths,* NASA Conference Pub. 2171 (R. D. Chapman, ed.), p. 237 (1981).
92. A. K. Dupree, in *Highlights of Astronomy*, vol. 5 (P. A. Wayman, ed.), p. 263, D. Reidel, Dordrecht, Holland (1980).
93. A. Boggess, W. A. Feibelman, and C. W. McCracken, in *The Universe at Ultraviolet Wavelengths*, NASA Conference Pub. 2171 (R. D. Chapman, ed.), p. 663 (1981).
94. J. P. Harrington, J. H. Lutz, and M. J. Seaton, *Mon. Not. R. Astron. Soc.* **195**, 21p (1981).
95. J. C. Raymond, in *The Universe at Ultraviolet Wavelengths*, NASA Conference Pub. 2171 (R. D. Chapman, ed.), p. 595 (1981).
96. C. Jordan, *Mon. Not. R. Astron. Soc.* **142**, 501 (1969).
97. V. L. Jacobs, J. Davis, J. E Rogerson, and M. Blaha, *J. Quant. Spectrosc. Radiat. Transfer* **19**, 591 (1978).
98. C. Jordan, *Mon. Not. R. Astron. Soc.* **148**, 17 (1970).
99. V. L. Jacobs, J. Davis, P. C. Kepple, and M. Blaha, *Astrophys. J.* **211**, 605 (1977).
100. A. H. Gabriel, *Mon. Not. R. Astron. Soc.* **160**, 99 (1972).

101. C. P. Bhalla, A. H. Gabriel, and L. P. Presnyakov, *Mon. Not. R. Astron. Soc.* **172**, 359 (1975).
102. L. A. Vainstein and U. I. Safronova, *At. Data Nucl. Data Tables* **21**, 49 (1978).
103. G. A. Doschek, J. T. Mariska, and U. Feldman, *Mon. Not. R. Astron. Soc.* **195**, 107 (1981).
104. U. Feldman and G. A. Doschek, *Astrophys. J.* (*Lett.*) **212**, L147 (1977).
105. M. Blaha, *Astron. Astrophys.* **16**, 437 (1972).
106. D. R. Flower and H. Nussbaumer, *Astron. Astrophys.* **42**, 265 (1975).
107. U. Feldman and G. A. Doschek, *Astron. Astrophys.* **61**, 295 (1977).
108. G. A. Doschek and U. Feldman, *Astrophys. J.* **254**, 371 (1982).
109. K. A. Berrington, P. G. Burke, P. L. Dufton, and A. E. Kingston, *J. Phys. B* **10**, 1465 (1977).
110. K. L. Baluja, P. G. Burke, and A. E. Kingston, *J. Phys. B* **13**, 829 (1980).
111. R. D. Cowan, *J. Phys. B* **13**, 1471 (1980).
112. P. L. Dufton and A. E. Kingston, *J. Phys. B* **13**, 4277 (1980).
113. M. Bitter, K. W. Hill, N. R. Sauthoff, P. C. Efthimion, E. Meservey, W. Roney, S. von Goeler, R. Horton, M. Goldman, and W. Stodiek, *Phys. Rev. Lett.* **43**, 129 (1979).
114. TFR Group, J. Dubau, and M. Loulergue, *J. Phys. B* **15**, 1007 (1981).
115. E. Kallne, J. Kallne, and J. E. Rice, *Phys. Rev. Lett.* **49**, 330 (1982).
116. F. Bely-Dubau, P. Faucher, L. Steenman-Clark, M. Bitter, S. von Goeler, K. W. Hill, C. Camhy-Val, and J. Dubau, *Phys. Rev. A* **26**, 3459 (1982).
117. G. A. Doschek, U. Feldman, and R. D. Cowan, *Astrophys. J.* **245**, 315 (1981).
118. K. J. H. Phillips, J. R. Lemen, R. D. Cowan, G. A. Doschek, and J. W. Leibacher, *Astrophys. J.* **265**, 1120 (1983).
119 F. Bely-Dubau, A. H. Gabriel, and S. Volonté, *Mon. Not. R. Astron. Soc.* **186**, 405 (1979).
120 F. Bely-Dubau, A. H. Gabriel, S. Volonté, *Mon. Not. R. Astron. Soc.* **189**, 801 (1979).
121 F. Bely-Dubau, J. Dubau, P. Faucher, and A. H. Gabriel, *Mon. Not. R. Astron. Soc.* **198**, 239 (1982).
122. F. Bely-Dubau, P. Faucher, L. Steenman-Clark, J. Dubau, M. Loulergue, A. H. Gabriel, E. Antonucci, S. Volonté, and C. G. Rapley, *Mon. Not. R. Astron. Soc.* **201**, 1155 (1982).
123. L. A. Vainstein and U. I. Safronova, *At. Data Nucl. Data Tables* **25**, 311 (1980).
124. U. Feldman, G. A. Doschek, and R. W. Kreplin, *Astrophys. J.* **238**, 365 (1980).
125 L. Steenman-Clark, F. Bely-Dubau, and P. Faucher, *Mon. Not. R. Astron. Soc.* **191**, 951 (1980).
126. M. Bitter, S. Von Goeler, K. W. Hill, R. Horton, D. Johnson, W. Roney, N. Sauthoff, E. Silver, and W. Stodiek, *Phys. Rev. Lett.* **47**, 921 (1981).
127. U. I. Safronova and T. G. Lisina, *At. Data Nucl. Data Tables* **24**, 49 (1979).
128. A. L. Merts, R. D. Cowan, and N. H. Magee, Jr., Los Alamos Scientific Laboratory Report LA-6220-MS (1976).
129. R. D. Cowan, *The Theory of Atomic Structure and Spectra,* University of California Press, Berkeley (1981).
130. R. D. Cowan, private communication (1980).
131. A. Temkin and A. K. Bhatia, Chap. 1, this volume.
132. G. A. Doschek, R. W. Kreplin, and U. Feldman, *Astrophys. J.* (*Lett.*) **233**, L157 (1979).
133. J. Dubau, M. Loulergue, and L. Steenman-Clark, *Mon. Not. R. Astron. Soc.* **190**, 125 (1980).
134. J. Dubau, A. H. Gabriel, M. Loulergue, L. Steenman-Clark, and S. Volonte, *Mon. Not. R. Astron. Soc.* **195**, 705 (1981).
135. A. N. Parmar, J. L. Culhane, C. G. Rapley, E. Antonucci, A. H. Gabriel, and M. Loulergue, *Mon. Not. R. Astron. Soc.* **197**, 29P (1981).
136. A. L. Merts, J. B. Mann, W. D. Robb, and N. H. Magee, Jr., Los Alamos Scientific Laboratory Report No. LA-8267-MS (1980).
137. G. A. Doschek and U. Feldman, *Astrophys. J.* **251**, 792 (1981).
138. J. Dubau and S. Volonté, *Rep. Prog. Phys.* **43**, 199 (1980).

139. J. P. Harrington. J. H. Lutz, and M. J. Seaton, *Mon. Not. R. Astron. Soc.* **195**, 21P (1981).
140. U. Feldman and G. A. Doschek, *J. Opt. Soc. Am.* **67**, 726 (1977).
141. G. D. Sandlin, G. E. Brueckner, and R. Tousey, *Astrophys. J.* **214**, 898 (1977).
142. S. Suckewer and E. Hinnov, *Phys. Rev. A* **20**, 578 (1979).
143. L. H. Aller, C. W. Ufford, and J. H. Van Vleck, *Astrophys. J.* **109**, 42 (1949).
144. D. E. Osterbrock, *Astrophysics of Gaseous Nebulae*, Chap. 5, p. 110, W. H. Freeman, San Francisco (1974).
145. U. Feldman, G. A. Doschek, J. T. Mariska, A. K. Bhatia, and H. E. Mason, *Astrophys. J.* **226**, 674 (1978).
146. C. Jordan, in *Highlights in Astronomy* (C. de Jager, ed.), p. 519, D. Reidel, Dordrecht, Holland (1971).
147. C. Jordan, *Astron. Astrophys.* **34**, 69 (1974).
148. R. H. Munro, A. K. Dupree, and G. L. Withbroe, *Solar Phys.* **19**, 347 (1971).
149. U. Feldman, G. A. Doschek, and W. E. Behring, *Space Sci. Rev.* **22**, 191 (1978).
150. G. A. Doschek and U. Feldman, NRL Report 8307 (1979).
151. C. Jordan, in *Progress in Atomic Spectroscopy, Part B*, Chap. 33 (W. Hanle and H. Kleinpoppen, eds.), Plenum, New York (1979).
152. K. P. Dere and H. E. Mason, in *Solar Active Regions* (F. Q. Orrall, ed.), Chap. 6, Colorado Associated University Press, CO (1981).
153. U. Feldman, *Physica Scripta* **24**, 681 (1981).
154. D. R. Flower and H. Nussbaumer, *Astron. Astrophys.* **45**, 145 (1975).
155. H. Nussbaumer and P. J. Storey, *Astron. Astrophys.* **71**, L5 (1979).
156. U. Feldman and G. A. Doschek, *Astron. Astrophys.* **79**, 357 (1979).
157. H. Nussbaumer and P. J. Storey, *Astron. Astrophys.* **115**, 205 (1982).
158. M. Hayes, *J. Phys. B* **16**, 285 (1983).
159. K. R. Nicolas, J.-D. F. Bartoe, G. E. Brueckner, and M. E. VanHoosier, *Astrophys. J.* **233**, 741 (1979).
160. P. L. Dufton, A. Hibbert, A. E. Kingston, and G. A. Doschek, *Astrophys. J.* **274**, 420 (1983).
161. A. K. Bhatia, G. A. Doschek, and U. Feldman, *Astron. Astrophys.* **86**, 32 (1980).
162. K. Bhadra and R. J. W. Henry, *Astrophys. J.* **240**, 368 (1980).
163. P. L. Dufton, A. Hibbert, A. E. Kington, and G. A. Doschek, *Astrophys. J.* **257**, 338 (1982).
164. G. A. Doschek, *Astrophys. J.*, **279**, 446 (1984).
165. U. Feldman, G. A. Doschek, and F. D. Rosenberg, *Astrophys. J.* **215**, 652 (1977).
166. P. L. Dufton, K. A. Berrington, P. G. Burke, and A. E. Kingston, *Astron. Astrophys.* **62**, 111 (1978).
167. P. L. Dufton, J. G. Doyle, and A. E. Kingston, *Astron. Astrophys.* **78**, 318 (1979).
168. A. K. Bhatia, G. A. Doschek, and U. Feldman, *Astron. Astrophys.* **76**, 359 (1979).
169. K. L. Baluja, P. G. Burke, and A. E. Kingston, *J. Phys. B* **14**, 119 (1981).
170. H. Nussbaumer, private communication (1978).
171. U. Feldman, G. A. Doschek, and A. K. Bhatia, *Astrophys. J.* **250**, 799 (1981).
172. W. L. Van Wyngaarden and R. J. W. Henry, *Astrophys. J.* **246**, 1040 (1981).
173. G. A. Doschek, U. Feldman, A. K. Bhatia, and H. E. Mason, *Astrophys. J.* **226**, 1129 (1978).
174. U. Feldman and G. A. Doschek, *Astrophys. J.* (*Suppl.*) **37**, 443 (1978).
175. J. W. Cook and K. R. Nicolas, *Astrophys. J.* **229**, 1163 (1979).
176. A. K. Dupree, P.V. Foukal, and C. Jordan, *Astrophys. J.* **209**, 621 (1976).
177. K. P. Dere, *Solar Phys.* **75**, 189 (1982).
178. U. Feldman and G. A. Doschek, *Astron. Astrophys.* **65**, 215 (1978).
179. K. P. Dere and J. W. Cook, *Astrophys. J.* **229**, 772 (1979).
180. G. A. Doschek, U. Feldman, and H. E. Mason, *Astron. Astrophys.* **78**, 342 (1979).
181. G. A. Doschek, U. Feldman, and F. D. Rosenberg, *Astrophys. J.* **215**, 329 (1977).

182. U. Feldman, G. A. Doschek, and K. G. Widing, *Astrophys. J.* **219**, 304 (1978).
183. H. E. Mason and A. K. Bhatia, *Mon. Not. R. Astron. Soc.* **184**, 423 (1978).
184. A. K. Bhatia and H. E. Mason, *Mon. Not. R. Astron. Soc.* **190**, 925 (1980).
185. G. A. Doschek, U. Feldman, and K. P. Dere, *Astron. Astrophys.* **60**, L11 (1977).
186. A. K. Bhatia and H. E. Mason, *Astron. Astrophys. Suppl. Ser.* **52**, 115 (1983).
187. P. L. Dufton. A. E. Kingston, J. G. Doyle, and K. G. Widing, *Mon. Not. R. Astron. Soc.* **205**, 81 (1983).
188. H. E. Mason, G. A. Doschek, U. Feldman, and A. K. Bhatia, *Astron. Astrophys.* **73**, 74 (1979).
189. A. H. Gabriel and C. Jordan, *Mon. Not. R. Astron. Soc.* **145**, 241 (1969).
190. A. H. Gabriel and C. Jordan, *Nature* **221**, 947 (1969).
191. A. K. Pradhan, D. W. Norcross, and D. G. Hummer, *Astrophys. J.* **246**, 1031 (1981).
192. R. Marrus and R. W. Schmieder, *Phys. Rev. A* **5**, 1160 (1972).
193. R. Mewe and J. Schrijver, *Astron. Astrophys.* **65**, 99 (1978).
194. N. J. Peacock and H. P. Summers, *J. Phys. B* **11**, 3757 (1978).
195. C. Jordan and N. J. Veck, *Solar Phys.* **78**, 125 (1982).
196. A. E. Kingston, private communication (1983).
197. W. M. Neupert, *Solar Phys.* **18**, 474 (1971).
198. D. L. McKenzie, R. M. Broussard, P. B. Landecker, H. R. Rugge, R. M. Young, G. A. Doschek, and U. Feldman, *Astrophys. J.* (*Lett.*) **238**, L43 (1980).
199. C. J. Wolfson, J. G. Doyle, J. W. Leibacher, and K. J. H. Phillips, *Astrophys. J.* **269**, 319 (1983).
200. R. Pallavicini, L. Golub, R. Rosner, G. S. Vaiana, T. Ayres, and J. L. Linsky, *Astrophys. J.* **248**, 279 (1981).
201. A. K. Dupree, *Astrophys. J.* **178**, 527 (1972).
202. C. Jordan, *Phil. Trans. R. Soc. London A* **281**, 391 (1976).
203. A. H. Gabriel, *Phil. Trans. R. Soc. London A* **281**, 339 (1976).
204. B. W. Bopp and R. E. Stencel, *Astrophys. J.* (*Lett.*) **247**, L131 (1981).
205. J. L. Linsky, T. R. Ayres, G. S. Basri, N. D. Morrison, A. Boggess, F. H. Schiffer III, A. Holm, A. Cassatella, A. Heck, F. Macchetto, D. Stickland, R. Wilson, C. Blanco, A. K. Dupree, C. Jordan, and R. F. Wing, *Nature* **275**, 19 (1978).
206. T. Simon and J. L. Linsky, *Astrophys. J.* **241**, 759 (1980).
207. T. Simon, J. L. Linsky, and F. H. Schiffer III, *Astrophys. J.* **239**, 911 (1980).
208. G. S. Basri, J. L. Linsky, and K. Eriksson, *Astrophys. J.* **251**, 162 (1981); J. L. Linsky, private communication (1982).
209. L. Hartmann, A. K. Dupree, and J. C. Raymond, *Astrophys. J.* **246**, 193 (1981).
210. K. R. Nicolas, G. E. Brueckner, R. Tousey, D. A. Tripp, O. R. White, and R. G. Athay, *Solar Phys.* **55**, 305 (1977).
211. A. Brown, C. Jordan, R. E. Stencel, J. L. Linsky, and T. R. Ayres, *Astrophys. J.* **283**, 731 (1984).
212. R. E. Stencel, J. L. Linsky, A. Brown, C. Jordan, K. G. Carpenter. R. F. Wing, and C. Czyzak. *Mon. Not. R. Astron. Soc.* **196**. 47p. (1981).
213. J. H. Lutz and M. J. Seaton, *Mon. Not R. Astron. Soc.* **187**, 1p (1979).
214. W. A. Feibelman, private communication (1981).
215. H. Nussbaumer and P. J. Storey, *Astron. Astrophys.* **74**, 244 (1979).
216. D. E. Osterbrock, *Astrophys. J.* **160**, 25 (1970).
217. H. Nussbaumer and H. Schild, *Astron. Astrophys.* **75**, L17 (1979).
218. W. A. Feibelman, A. Boggess, R. W. Hobbs, and C. W. McCracken, *Astrophys. J.* **241**, 725 (1980).
219. W. A. Feibelman, A. Boggess, C. W. McCracken, and R. W. Hobbs, *Astrophys. J.* **246**, 807 (1981).
220. R. Rosner, W. H. Tucker, and G. S. Vaiana, *Astrophys. J.* **220**, 643 (1978).

221. S. K. Antiochos, *Astrophys. J.* (*Lett.*) **232**, L125 (1979).
222. C. Jordan, *Astron. Astrophys.* **86**, 355 (1980).
223. A. W. Hood and E. R. Priest, *Astron. Astrophys.* **87**, 126 (1980).
224. J. T. Mariska, J. P. Boris, E. S. Oran, T. R. Young, Jr., and G. A. Doschek, *Astrophys. J.* **255**, 783 (1982).
225. K. P. Dere, J.-D. F. Bartoe, G. E. Brueckner, M. D. Dykton, and M. E. VanHoosier, *Astrophys. J.* **249**, 333 (1981).
226. J. P. Boris and J. T. Mariska, *Astrophys. J.* (*Lett.*) **258**, L49 (1982).
227. J. T. Mariska and J. P. Boris, *Astrophys. J.* **267**, 404 (1983).
228. J. C. Raymond and A. K. Dupree, *Astrophys. J.* **222**, 379 (1978).
229. A. K. Dupree, R. T. Moore, and P. R. Shapiro, *Astrophys. J.* (*Lett.*) **229**, L101 (1979).
230. M. H. Francis, *Solar Phys.* **69**, 239 (1981).
231. R. Roussel-Dupre and C. Beerman, *Astrophys. J.* **250**, 408 (1981).
232. S. R. Pottasch, *Space Sci. Rev.* **3**, 816 (1964).
233. C. Jordan and S. R. Pottasch, *Solar Phys.* **4**, 104 (1968).
234. V. L. Jacobs, J. Davis, J. E. Rogerson, M. Blaha, J. Cain, and M. Davis, *Astrophys. J.* **239**, 1119 (1980), and references therein.
235. J. G. Doyle and J. C. Raymond, *Mon. Not. R. Astron. Soc.* **196**, 907 (1981).
236. J. M. Shull and M. Van Steenberg, *Astrophys. J. Suppl.* **48**, 95 (1982).
237. H. P. Summers, *Appleton Laboratory Internal Memo* **367** (1974).
238. R. D. Cowan and J. B. Mann, *Astrophys. J.* **232**, 940 (1979).
239. S. L. Baliunas and S. E. Butler, *Astrophys. J.* (*Lett.*) **235**, L45 (1980).

INDEX